# Chaos and Forecasting

**NONLINEAR TIME SERIES AND CHAOS**

Editor: Howell Tong

---

Vol. 1: Dimension Estimation and Models
*ed. H. Tong*

Nonlinear Time Series and Chaos Vol. 2

# CHAOS AND FORECASTING

## Proceedings of the Royal Society Discussion Meeting

London 2–3 March 1994

Editor

**Howell Tong**

Institute of Mathematics and Statistics
University of Kent
UK

World Scientific
Singapore • New Jersey • London • Hong

*Published by*

World Scientific Publishing Co. Pte. Ltd.

P O Box 128, Farrer Road, Singapore 9128

*USA office:* Suite 1B, 1060 Main Street, River Edge, NJ 07661

*UK office:* 57 Shelton Street, Covent Garden, London WC2H 9HE

**CHAOS AND FORECASTING**



ISBN 981-02-2126-6



Printed in Singapore by Uto-Print

## From The Editor

The new field of deterministic chaos has been hailed as a revolution of thoughts and attracting ever increasing attention of many scientists and technologists from diverse disciplines including biology, computation, engineering, economics, mathematics, meteorology, physics, statistics and many others. It has become a truly multi-disciplinary area of research and even captured the imagination of the general public.

The Royal Society of Great Britain held a two-day discussion meeting on 2nd and 3rd March 1994, at which 15 papers were read, covering a vast range of subjects. It attracted well over one hundred participants from all walks of life; even some politicians and businessmen attended the meeting, perhaps out of curiosity. The meeting was organised by Professor Robert M. May, F.R.S., of Oxford University, Dr. Bryan Grenfell of Cambridge University and myself, with funding from the Hooke's Committee of the Royal Society, for which we are most grateful. The meeting was roughly divided into four areas: theoretical foundation (5 papers), economics/finance (2 papers), physical science (4 papers) and biological science (4 papers). Each topic was covered by a leading expert in the field. The meeting dealt with different basic approaches to the problem of chaos and forecasting, and with applications to both artificially-generated time series and real data from context in the above-mentioned diverse fields.

In view of the currency of the subject of chaos and its vast practical potential, especially in the science of forecasting, the Royal Society has come into an agreement with the World Scientific Publishing Company so that a single dedicated volume could be made available at a modest price. In most cases, the authors are now able to develop their respective topics at a much more leisurely pace than at the meeting; it is hoped that this should go some way towards making the topics accessible to a wider audience. It is my view that this book represents a rather special and rare occasion in which prominent scientists from different areas converge on the same theme and a welcome addition to the series "Nonlinear Time Series and Chaos". I am grateful to the Royal Society and the World Scientific Publishing Company for making this volume a reality. I owe my special thanks to Bob May and Bryan Grenfell, who persuaded me to initiate this academia/commerce joint venture. Last but not least, many thanks are due to Ms Jia-ling Leng, who assisted me with all stages of the editorial work.

# Contents

# Chaos and Forecasting

CHAPTER 1

# ORTHOGONAL PROJECTION, EMBEDDING DIMENSION AND SAMPLE SIZE IN CHAOTIC TIME SERIES FROM A STATISTICAL PERSPECTIVE

Bing Cheng and Howell Tong
Institute of Mathematics and Statistics
University of Kent at Canterbury
Canterbury, Kent CT2 7NF, U.K.

**Abstract**

By studying systematically the orthogonal projections, in a particular sense associated with a (random) time series admitting a possibly chaotic skeleton and in a sequence of suitably defined $\mathcal{L}_2$-spaces, we describe a geometric characterisation of the notion of embedding dimension within a statistical framework. The question of sample size requirement in the statistical estimation of the said dimension is addressed heuristically, ending with a pleasant surprise: the curse of dimensionality may be lifted except in the excessively stringent cases.

## 1. Introduction

In the study of chaos, substantial efforts have been expended on the detection of exotic attractors from data, which may be laboratory-based, field-based or simply computer generated. (See, e.g., the RSS Chaos Day papers ed. by Tong and Smith, 1992 and Warwick Workshop papers ed. by Drazin and King, 1992.) In most of these studies, the focus is first and foremost on the estimation of such summarising parameters as the correlation dimension, the Lyapunov exponents, the information dimension and so on. However, as argued in Cheng and Tong (1992) (and we shall argue even more strongly later), the most natural and logical order is to estimate the dimension of the Euclidean space in which the attractor resides, that is the embedding space, *before* putting the spot-light on the attractor itself. In fact, it is increasingly recognised that the embedding dimension has an impact on the estimation of the correlation dimension and the Lyapunov exponents of the attractor. (See, e.g., the RSS Chaos Day papers cited above.) Another powerful reason for not estimating first the correlation dimension of the attractor and its like is the *curse of dimensionality*: roughly $k^p$ data points are needed to yield a useful estimate of the correlation dimension, $p$, of an attractor, $k$ being a real positive constant. Indeed in one context, Smith (1988) has advised that a sample size as astronomical as $42^p$ is necessary. Just note that $42^4 = 3,111,696$!

In the same vein, Ruelle in his Claude Bernard Lecture to the Royal Society of London (1990) has advanced the rule that if the slope in the Grassberger-Procassia algorithm is measured over at least one decade one finds necessarily

$$\text{correlation dimension } \leq 2\log_{10} N \ , \tag{1}$$

$N$ being the sample size. [Eckmann and Ruelle (1992) gives a similar formula for the estimation of the Lyapunov exponent.]

However, the above *exponential* sample size requirement for the estimation of the correlation dimension seems to have led to some confusion in the literature. For example, Jensen (1992) has questioned the wisdom of adopting 4 as the 'embedding dimension' for the annual sunspot numbers with $N = 289$ on the grounds that '*it is not possible to estimate non-parametrically a function* $T_m : \mathbf{R^m} \to \mathbf{R}$ *unless one has of the order* $10^m$ *data points*'. He then concludes that '*for the sunspot data one has to confine oneself to* $m = 1, 2$ *and perhaps* $m = 3$.' By and large, we agree with his first statement because we have ourselves realized the exponential sample size requirement for the estimation of $T_m$. (Cheng and Tong, 1993b). That is, the sample size requirements for the estimation of $T_m$ and the correlation dimensions of the attractors defined by $T_m$ are comparable. However, as far as the estimation of the embedding dimension is concerned, confining $m$ to 1 or 2 is unduly pessimistic. For the moment, let us simply announce that under general conditions, the sample size requirement for the estimation of $m$ is *not* exponential but only (roughly) quadratic in $m$.

The above announcement might sound too good to be true, especially for those brought up with the curse of dimensionality. However, a moment's reflection should convince us that estimating $m$ is a much less demanding task than estimating either $T_m$ or the correlation dimension of the attractors of $T_m$. For, we can afford to stand back from the 'microscopic details' of $T_m$ and only concentrate on discerning the 'macroscopic clustering' of the data around some cylinder set. Moreover, the fact that $m$ is integer-valued whilst the correlation dimension is real-valued reinforces the point.

The rest of the paper is organised as follows. In Section 2, we describe cylinder sets associated with appropriate orthogonal projection of a time series. In Section 3, we introduce a distance function over the Cartesian product $\mathbf{N} \times \mathbf{N}$ as a measure of the '*goodness of fit*' of NLAR models of different orders. This formalises the macroscopic clustering mentioned earlier. This distance function leads, in Section 4, to a natural loss function with which consistent estimation of the embedding dimension may be obtained. Section 5 addresses the issue of sample size requirement. Section 6 reports some simulation results. Section 7 gives some concluding discussions.

## 2. Projections and cylinder sets

Let $\{X_t\}$ be a discrete-time stationary time series with $EX_t^2 < \infty$. Let $E[X_t|X_{t-1},\dots,X_{t-d}]$ denote the conditional expectation of $X_t$ given $(X_{t-1},\dots,X_{t-d})$. Define the residual variance by

$$\sigma^2(d) = E[X_t - E[X_t|X_{t-1},\dots,X_{t-d}]]^2 . \tag{2}$$

Define the generalized partial autocorrelation function (PACF) by

$$\phi(d) \ = \ \{1 - \sigma^2(d+1)/\sigma^2(d)\}^{\frac{1}{2}} . \tag{3}$$

**Definition 2.1**: $\{X_t\}$ is a nonlinear autoregressive process of order $d_0$, in short NLAR ($d_0$), if $\exists$ a non-negative integer $d_0 < \infty$ such that $\phi(d_0 - 1) \neq 0$ and $\phi(d) = 0$ for all $d \geq d_0$. If no such finite $d_0$ exists, then $\{X_t\}$ is said to be a nonlinear autoregressive process of infinite order, or NLAR ($\infty$).

For the space $\mathbf{R}^\tau$, $\tau \geq 1$ , define the metric $\| \ \|$ by

$$\| X - Y \| = \{(x_1 - y_1)^2 + \cdots + (x_\tau - y_\tau)^2\}^{\frac{1}{2}} , \tag{4}$$

where $X = (x_1,\dots,x_\tau)^T$ and $Y = (y_1,\dots,y_\tau)^T$. For the space $\mathbf{R}^{\tau+2}$, we define two types of projection $\overline{P}$ and $\underline{P}$ by

$$\overline{P}X = (x_1,\dots,x_{\tau+1})^T \tag{5}$$

and

$$\underline{P}X = (x_2,\dots,x_{\tau+2})^T , \tag{6}$$

where

$$X = (x_1,\dots,x_{\tau+2})^T \in \mathbf{R}^{\tau+2} . \tag{7}$$

For two random vectors $X = (X_1,\dots,X_n)^T$ and $Y = (Y_1,\dots,Y_m)^T$, define the conditional expectation of $X$ given $Y$ by

$$E[X|Y] = (E[X_1|Y], E[X_2|Y],\dots,E[X_n|Y])^T . \tag{8}$$

Let $Y_t^{(d)} = (X_{t+1}, X_t, X_{t-1},\dots,X_{t-d})^T$. Then the conditional expectation

$$\begin{aligned} &E[\underline{P}Y_t^{(d)}|X_{t-1},\dots,X_{t-d}] \\ &\quad = (E[X_t|X_{t-1},\dots,X_{t-d}], X_{t-1},\dots,X_{t-d})^T \\ &\quad = (F_d(X_{t-1},\dots,X_{t-d}), X_{t-1},\dots,X_{t-d})^T , \end{aligned} \tag{9}$$

where $F_d(X_{t-1}, \ldots, X_{t-d})$ denotes the conditional expectation of $X_t$ given $X_{t-1}, \ldots, X_{t-d}$.

**Lemma 2.1 (Projection Lemma)**: $\{X_t\}$ is an NLAR $(d_0)$ if and only if $F_{d_0-1} \neq F_{d_0}$ a.s. and $F_d \equiv F_{d_0}$ a.s. for $d > d_0$.

**Proof**: Trivial. □

**Remark 2.1**: Let $\varepsilon_t^{(d)}$ denote the difference $X_t - F_d(X_{t-1}, \ldots, X_{t-d})$. Then we have

$$\underline{P}Y_t^{(d)} = (F_d(X_{t-1}, \ldots, X_{t-d}), X_{t-1}, \ldots, X_{t-d})^T + (\varepsilon_t^{(d)}, 0, \ldots, 0)^T , \tag{10}$$

which is a point in the phase space $\mathbf{R}^{d+1}$. The larger $d$ is, the more structure has the dynamic associated with the function $F_d$. However, using $\mathbf{R}^{d+1}$, $d > d_0$, yields no further information about the structure of the dynamic associated with $F_{d_0}$ than using $\mathbf{R}^{d_0+1}$.

**Remark 2.2**: We may think of (2.9) as a stochastic perturbation of the deterministic dynamical system

$$\underline{P}Y_t^{(d)} = (F_d(X_{t-1}, \ldots, X_{t-d}), X_{t-1}, \ldots, X_{t-d})^T , \tag{11}$$

the term $(\varepsilon_t^{(d)}, 0, 0, \ldots, 0)^T$ representing dynamic noise. Chan and Tong (1994) have argued that (2.9), rather than (2.10), is a more appropriate and realistic framework for chaos study.

**Example 2.1**: Consider the stochastic logistic map

$$X_t = \alpha X_{t-1}(1 - X_{t-1}) + \varepsilon_t, \quad (0 \leq \alpha \leq 4) . \tag{12}$$

Here,

$$Y_t^{(1)} = (X_{t+1}, X_t, X_{t-1})^T, \quad E[\underline{P}Y_t^{(1)} | X_{t-1}] = (F_1(X_{t-1}), X_{t-1})^T .$$

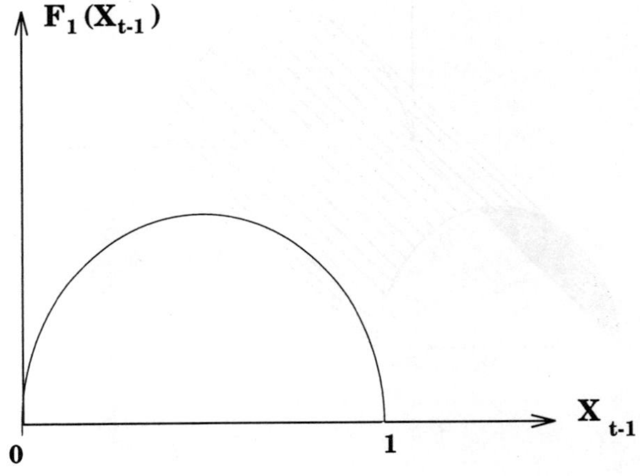


Fig. 2.1: The mapping $F_1 : X_{t-1} \to X_t$ in phase space $\mathbf{R}^2$.

Fig. 2.1 shows the mapping $F_1 : X_{t-1} \to X_t$ in phase space $\mathbf{R}^2$ with $\alpha = 3.95$. In phase space $\mathbf{R}^3$, we have

$$Y_t^{(2)} = (X_{t+1}, X_t, X_{t-1}, X_{t-2})^T \quad \text{and}$$

$$E[\underline{P}Y_t^{(2)} | X_{t-1}, X_{t-2}] = (F_2(X_{t-1}, X_{t-2}), X_{t-1}, X_{t-2})^T \, .$$

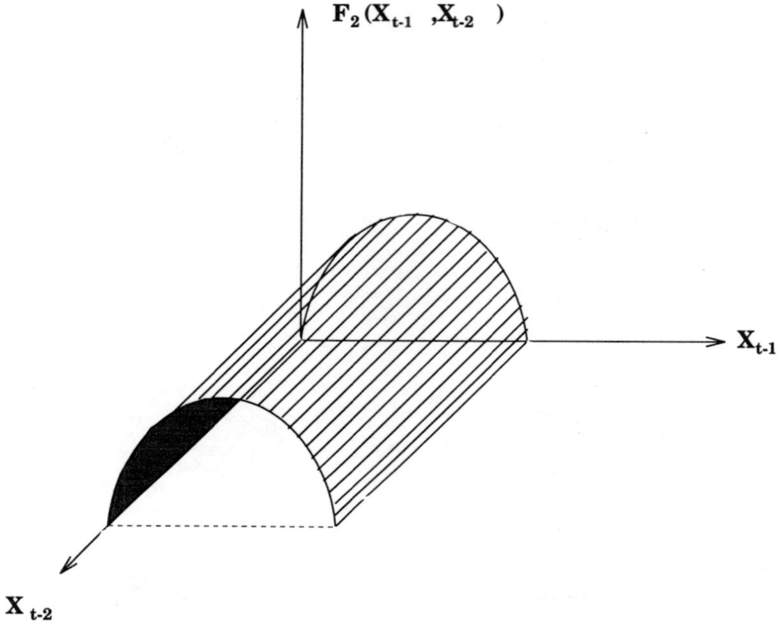


Fig. 2.2: The mapping $F_2 : (X_{t-1}, X_{t-2})^T \to X_t$ in phase space $\mathbf{R}^3$. Comparing the dynamic defined by $F_2$ in phase space $\mathbf{R}^3$ with that defined by $F_1$ in phase space $\mathbf{R}^2$, we can clearly see the cylinder set structure. We shall see that cylinder sets indicate redundant information. □

**Lemma 2.2 (Equivalent lemma)**: For $d \geq 1$, $E[\overline{P}Y_t^{(d)}|X_t, \ldots, X_{t-d+1}]$ and $E[\underline{P}Y_t^{(d)}|X_{t-1}, \ldots, X_{t-d}]$ have the same geometric structure, i.e. for each $\xi = (\xi_1, \ldots, \xi_d)^T \in \mathbf{R}^d$,

$$E[\overline{P}Y_t^{(d)}|X_t = \xi_1, X_{t-1} = \xi_2, \ldots, X_{t-d+1} = \xi_d]$$

$$= E[\underline{P}Y_t^{(d)}|X_{t-1} = \xi_1, X_{t-2} = \xi_2, \ldots, X_{t-d} = \xi_d]\ , \quad \text{a.s.}$$

**Proof**: Trivial. □

**Example 2.1 (continue)**: We may illustrate the above equivalence by reference to Figs. 2.3–2.5, which are self-explanatory. □

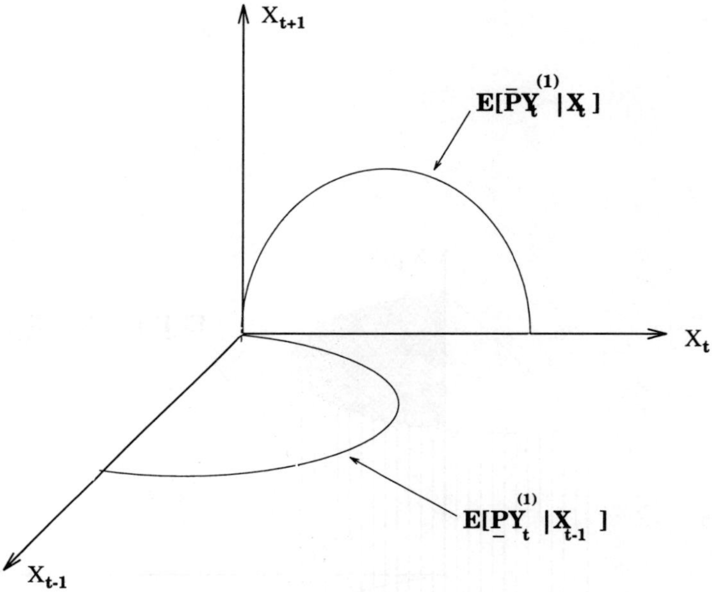


Fig. 2.3: $d = 1$.

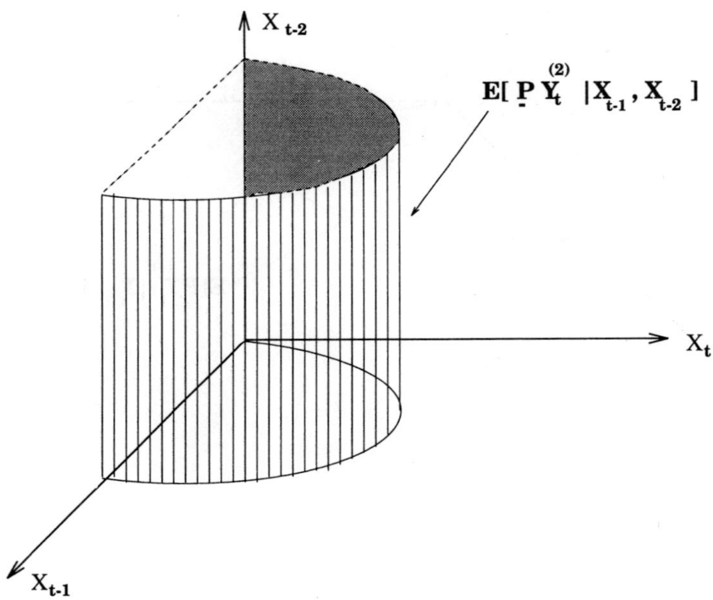


Fig. 2.4: $d = 2$.

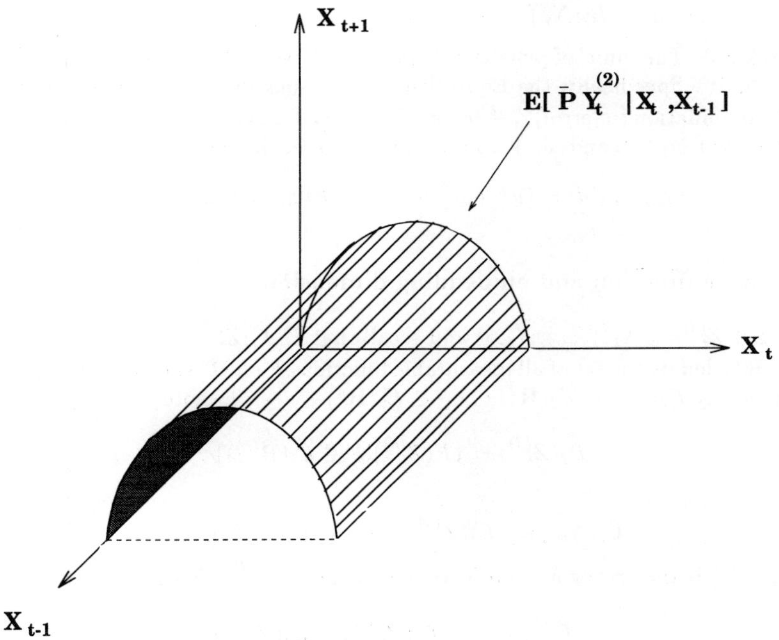


Fig. 2.5: $d = 2$.

**Definition 2.2**: In the phase space $\mathbf{R}^{d+1}$, we define the prediction error hyperplane of dimension $d$ by

$$\text{error}_t^{(d)} = (X_t - E[X_t|X_{t-1}, \ldots, X_{t-d}], X_{t-1}, \ldots, X_{t-d})^T .$$

Let

$$\delta(d) = E \parallel \text{error}_t^{(d)} \parallel^2 .$$

**Theorem 2.1 (Decomposition theorem)**:

$$\delta(d) = \sigma^2(d) + dE(X_t^2) . \tag{13}$$

**Proof**: $E \parallel \text{error}_t^{(d)} \parallel^2$

$$= E[X_t - E[X_t|X_{t-1}, \ldots, X_{t-d}]]^2 \; + \; E(X_{t-1}^2) + \cdots + E(X_{t-d}^2)$$

$$= \sigma^2(d) + dE(X_t^2) . \qquad \square$$

**Remark 2.3**: The 'unit' of penalty is fixed here. If we wish, we could consider a more flexible 'unit'. Specifically, the Equivalent lemma suggests that we may consider the convex combination $(w_1\text{error}_{t+1}^{(d)} + w_2\text{error}_t^{(d)}, w_1X_t + wX_{t-1}, \ldots, w_1X_{t-d+1} + w_2X_{t-d})$, where $w_1 > 0, w_2 > 0$ and $w_1 + w_2 = 1$. In this case, the mean square is

$$\delta_{(w_1,w_2)}(d) = (w_1^2 + w_2^2)\sigma^2(d) + dE[w_1X_t + w_2X_{t-1}]^2 . \tag{14}$$

## 3. Distance function and embedding dimension

Let $Z_t^{(d)} = (X_{t-1}, \ldots, X_{t-d})$ and recall that $E[X_t|Z_t^{(d)}]$ is denoted by $F_d(Z_t^{(d)})$. Let $\mathcal{L}_2(\mathbf{R}^d)$ denote the set of all measurable functions $F$ on $\mathbf{R}^d$ such that $E|F(Z_t^{(d)})|^2 < \infty$. Obviously $\mathcal{L}_2(\mathbf{R}^1) \subset \mathcal{L}_2(\mathbf{R}^2) \subset \cdots \subset \mathcal{L}_2(\mathbf{R}^d) \subset \cdots$. Denote

$$\mathcal{L}_2(Z_t^{(d)}) = \{F(Z_t^{(d)})|F \in \mathcal{L}_2(\mathbf{R}^{(d)})\} .$$

Then

$$\mathcal{L}_2(X_{t-1}) \subset \mathcal{L}_2(Z_t^{(2)}) \subset \cdots \subset \mathcal{L}_2(Z_t^{(d)}) \subset \cdots$$

and $F_d(Z_t^{(d)})$ is the orthogonal projection of $X_t$ in $\mathcal{L}_2(Z_t^{(d)})$. For integers $0 < d_1 \le d_2$, we have

$$F_{d_1}(Z_t^{(d_1)}) \in \mathcal{L}_2(Z_t^{(d_1)}) \subset \mathcal{L}_2(Z_t^{(d_2)})$$

and

$$F_{d_2}(Z_t^{(d_2)}) \in \mathcal{L}_2(Z_t^{(d_2)}) .$$

Therefore we may consider the closeness between $F_{d_1}(Z_t^{(d_1)})$ and $F_{d_2}(Z_t^{(d_2)})$ in $\mathcal{L}_2(Z_t^{(d_2)})$ by

$$\Delta(d_1, d_2) = E[F_{d_1}(Z_t^{(d_1)}) - F_{d_2}(Z_t^{(d_2)})]^2 . \tag{15}$$

Notice that $F_d$ is uniquely determined once $d$ is given. Thus $\Delta(.,.)$ is well defined and $\Delta(d_1, d_2)$ is the (squared) distance between the orthogonal projection of $X_t$ in $\mathcal{L}_2(Z_t^{(d_1)})$ and that in $\mathcal{L}_2(Z_t^{(d_2)})$.

**Definition 3.1**: The time series $\{X_t\}$ is said to have embedding dimension $d_0$ ($d_0 \ge 1$) if and only if

(i) $\Delta(d, d_0) \ne 0$ for all $d < d_0$,
and (ii) $\Delta(d, d_0) = 0$ for all $d \ge d_0$.

**Proposition 3.1**:

(i) $\Delta^{\frac{1}{2}}(.,.)$ is a properly defined distance function on $\mathbf{N} \times \mathbf{N}$, i.e.
$\Delta^{\frac{1}{2}}(d_1, d_2) = \Delta^{\frac{1}{2}}(d_2, d_1)$,
$\Delta^{\frac{1}{2}}(d, d) = 0$,
$\Delta^{\frac{1}{2}}(d_1, d_3) \le \Delta^{\frac{1}{2}}(d_1, d_2) + \Delta^{\frac{1}{2}}(d_2, d_3)$.

(ii) If for each $d \geq 1$, $F_d$ has bounded first partial derivatives on $\mathbf{R}^d$, then

$$\Delta(d_2, d_1) \leq c|d_2 - d_1| \, ,$$

where $c$ is a constant.

(iii) For $d_1 \leq d_2 \leq d_3$,

$$\Delta(d_2, d_3) \leq \Delta(d_1, d_3) \, .$$

That is for fixed $d_3$, $\Delta(d, d_3)$ is a decreasing function in $d$.

(iv) For any $d_1 \leq d_2$, we have

$$\Delta(d_1, d_2) = \sigma^2(d_1) - \sigma^2(d_2) \, . \tag{16}$$

(v) $\sum_{d=1}^{\infty} \Delta(d, d+1) < \infty$.

item[(vi)] There are infinitely many $d$ for which

$$\Delta(d, d+1) \ \leq \ \tilde{\kappa}/d,$$

where $\tilde{\kappa}$ is a constant.

(vi) $\forall \, d \leq D < \infty$, $\exists \, \kappa_D$, $0 < \kappa_D < \infty$, such that

$$\Delta(d, d+1) \ \leq \ \kappa_D/d.$$

**Proof**:

(i) The first two assertions are obvious. The third assertion follows using the Minkowski's inequality. Without loss of generality, let $d_1 \leq d_3$. If $d_2 \leq d_1$, then from (iii) we have $\Delta(d_1, d_3) \leq \Delta(d_2, d_3)$ and the assertion is true.

There are two cases left: $d_2 \geq d_3 \geq d_1$ and $d_3 \geq d_2 \geq d_1$. For the former case, the assertion follows from (iii). For the latter case, the assertion follows from (iv).

(ii) Suppose that $d_1 \leq d_2$. Then

$$\Delta(d_1, d_2) = E\{F_{d_2}(Z_t^{(d_2)}) - E[F(Z_t^{(d_2)})|Z_t^{(d_1)}]\}^2 \, .$$

Let $\zeta = F_{d_2}(Z_t^{(d_1)}, \underbrace{0, \ldots, 0}_{d_2 - d_1})$ . Obviously, $\zeta \in \mathcal{L}_2(Z_t^{(d_2)})$. Since $E[F_{d_2}(Z_t^{(d_2)})|Z_t^{(d_1)}]$ is the orthogonal projection of $F_{d_2}(Z_t^{(d_2)})$ in $\mathcal{L}_2(Z_t^{(d_1)})$, we have

$$\begin{aligned}
\Delta(d_1, d_2) &\leq E[F_{d_2}(Z_t^{(d_2)}) - \zeta]^2 \\
&= E[F_{d_2}(Z_t^{(d_1)}, X_{t-d-1}, \ldots, X_{t-d_2}) - F_{d_2}(Z_t^{(d_1)}, 0, \ldots, 0)]^2 \\
&= E[G(x_{d_1+1}^*)X_{t-d_1-1} + \cdots + G(x_{d_2}^*)X_{t-d_2}]^2 \, ,
\end{aligned}$$

where $G(x_j^*)$ denotes the partial derivative of $F_{d_2}$ with respect to the $j$th component and evaluated at $x_j^*$. Now,

$$\begin{aligned} E\left[\sum_{j=d_1+1}^{d_2} G(x_j^*)X_{t-j}\right]^2 &\le \text{constant}\times\, E\left[\sum_{j=d_1+1}^{d_2} G^2(x_j^*)X_{t-j}^2\right] \\ &\le \text{constant}\times \sum_{j=d_1+1}^{d_2} E(X_{t-j}^2) \\ &= \text{constant}\times\, (d_2-d_1)\,, \end{aligned}$$

and hence (ii) is proved.

(iii)
$$\begin{aligned} \Delta(d_2,d_3) &= E[F_{d_3}(Z_t^{(d_3)}) - F_{d_2}(Z_t^{(d_2)})]^2 \\ &= E[F_{d_3}(Z_t^{(d_3)}) - E[F_{d_3}(Z_t^{(d_3)})|Z_t^{(d_2)}]]^2\,. \end{aligned}$$

Since $F_{d_1}(Z_t^{(d_1)}) \in \mathcal{L}_2(Z_t^{(d_2)})$ and $E[F_{d_3}(Z_t^{(d_3)})|Z_t^{(d_2)}]$ is the orthogonal projection of $F_{d_3}(Z_t^{(d_3)})$ in $\mathcal{L}_2(Z_t^{(d_2)})$, therefore

$$\begin{aligned} \Delta(d_2,d_3) &\le E[F_{d_3}(Z_t^{(d_3)}) - F_{d_1}(Z_t^{(d_1)})]^2 \\ &= \Delta(d_1,d_3)\,. \end{aligned}$$

(iv)
$$\begin{aligned} \Delta(d_1,d_2) &= E[\{X_t - F_{d_1}(Z_t^{(d_1)})\} - \{X_t - F_{d_2}(Z_t^{(d_2)})\}]^2 \\ &= E[\varepsilon_t^{(d_1)} - \varepsilon_t^{(d_2)}]^2 \\ &= E[\varepsilon_t^{(d_1)}]^2 + E[\varepsilon_t^{(d_2)}]^2 - 2E[\varepsilon_t^{(d_1)}\varepsilon_t^{(d_2)}]\,. \end{aligned}$$

However,

$$\begin{aligned} E[\varepsilon_t^{(d_1)}\varepsilon_t^{(d_2)}] &= E[\varepsilon_t^{(d_2)}\{\varepsilon_t^{(d_2)} + F_{d_2}(Z_t^{(d_2)}) - F_{d_1}(Z_t^{(d_1)})\}] \\ &= E[\varepsilon_t^{(d_2)}]^2 + E[\varepsilon_t^{(d_2)}\{F_{d_2}(Z_t^{(d_2)}) - F_{d_1}(Z_t^{(d_1)})\}] \\ &= E[\varepsilon_t^{(d_2)}]^2 + E[\{F_{d_2}(Z_t^{(d_2)}) - F_{d_1}(Z_t^{(d_1)})\}E[\varepsilon_t^{(d_2)})|Z_t^{(d_2)}]] \\ &= E[\varepsilon_t^{(d_2)}]^2 \end{aligned}$$

and hence the result.

(v) Define

$$e_t^{(d)} = F_d(Z_t^{(d)}) - F_{d+1}(Z_t^{(d+1)})\,.$$

Clearly

$$E[e_t^{(d)}]^2 = \Delta(d, d+1) .$$

It is easy to check that for $d' \neq d$

$$E[e_t^{(d')} e_t^{(d)}] = 0.$$

We have therefore the following orthogonal decomposition

$$E[X_t | X_{t-1}] - E[X_t | X_{t-1}, X_{t-2}, \ldots] = \sum_{d=1}^{\infty} e_t^{(d)} .$$

It follows from this and the finiteness of $EX_t^2$ that

$$\sum_{d=1}^{\infty} E[e_t^{(d)}]^2 \ < \ \infty$$

This implies that

$$\sum_{d=1}^{\infty} \Delta(d, d+1) < \infty.$$

(vi) If there are only finitely many $d$ for which

$$\Delta(d, d+1) = 0(1/d) ,$$

then $\exists$ a $d^*$ and a constant $C > 0$ such that $\forall d \geq d^*$

$$\Delta(d, d+1) \geq C/d .$$

This implies that

$$\begin{aligned} \sum_{d=1}^{\infty} \Delta(d, d+1) \ &\geq \ \sum_{d \geq d^*} \Delta(d, d+1) \\ &\geq \ C \sum_{d \geq d^*} (1/d) \\ &= \ \infty . \end{aligned}$$

This is a contradiction and therefore (ii) is proved.

(vii) Trivial. □

Note that the bound in (vi) is almost sharp because, for example, $\sum_{d=2}^{\infty} \frac{1}{d((ln(d)))^2} < \infty$, and $\sum_{d=1}^{\infty} \frac{1}{d^{1-\epsilon}} = \infty$, where $\epsilon > 0$.

Note that for continuous parameters such as the bandwidth parameters in kernel smoothing, we may use the Euclidean norm as an appropriate distance function for parameter (e.g. the bandwidth) choice. However, for many discrete cases, the Euclidean norm is found to be unsuitable. For example, Akaike (1974) has used instead the Kullback-Leibler information to construct a suitable distance function for order determination. For our case, we have obtained an appropriate distance function, namely $\Delta^{\frac{1}{2}}(.,.)$ on $\mathbf{N} \times \mathbf{N}$, based on the projection of the skeleton from a low dimension to a high dimension as described in Section 2. Proposition 3.1 (ii) reveals the relation between $\Delta(.,.)$ and the Euclidean norm.

Suppose that henceforth $\{X_t\}$ is bounded. (C.f. Chan and Tong, 1994.) Let $d_2 \geq d_1$. We write

$$\Delta_N(d_1, d_2) = \sigma_N^2(d_1) - \sigma_N^2(d_2) , \tag{17}$$

where

$$\sigma_N^2(d) = (N - r + 1)^{-1} \Sigma(\varepsilon_t^{(d)})^2 , \tag{18}$$

the summation extending over $t \geq r$ and $r$ being a positive integer $\geq \max(d_1, d_2)$.

Trivially $E\Delta_N(d_1, d_2) = \Delta(d_1, d_2)$.

**Proposition 3.2**:

(a) The embedding dimension and the order of autoregressive models of $\{X_t\}$ coincide.

(b) Let $d_0(\geq 1)$ be the embedding dimension for $\{X_t\}$. Then the following hold.

   (i) If $d < d_0$ then $\Delta_N(d, d_0) \neq 0$ except for finitely many $N$;

   (ii) if $d \geq d_0$ then $\Delta_N(d_0, d) = 0$.

**Proof**:

(a) This is just a simple corollary of Proposition 3.1 (iv).

(b) For $d \geq d_0$,

$$F_d(Z_t^{(d)}) = F_{d_0}(Z_t^{(d_0)}) .$$

Therefore

$$\varepsilon_t^{(d)} = \varepsilon_t^{(d_0)} ,$$

and thus (ii) is proved.

For $d < d_0$, if for all but finitely many $N$

$$\Delta_N(d, d_0) = 0 ,$$

then, letting $N \to \infty$, we have, by any standard ergodic theorem,

$$0 = \lim_{N\to\infty} \Delta_N(d, d_0) = \sigma^2(d) - \sigma^2(d_0) = \Delta(d, d_0) .$$

This contradicts the definition of embedding dimension. Hence, (i) is proved. □

## 4. Loss functions for estimations of the embedding dimension

The squared distance function $\Delta(.,.)$ as defined by (3.1) naturally suggests the following loss function as a measure of goodness of fit

$$ASE(d) = N^{-1} \sum_{t=1}^{N} [F_{d,N}(Z_t^{(d)}) - F_{d_0}(Z_t^{(d_0)})]^2 , \tag{19}$$

where $d_0$ is the embedding dimension and $F_{d,N}(Z_t^{(d)})$ is the Nadaraya-Watson kernel estimate of $E[X_t|Z_t^{(d)}]$ based on the observations $(X_1, \ldots, X_N)$. Loss functions based on similar motivation have been used by Akaike (1974) and Shibata (1980) for the case where $F$ is linear. Manipulations of $U$-statistics as in Cheng and Tong (1992, 1993a) establish the following theorem. [Details of the proof are similar to the references just cited and are therefore omitted for the sake of brevity.]

**Theorem 4.1**: Under the same conditions as in Theorem 1 of Cheng and Tong (1992),

$$\begin{aligned} ASE(d) = \tilde{\Delta}_N(d, d_0) \; &- \; \{2\alpha(d)\tilde{\Delta}(d, d_0) - \beta(d)\sigma_N^2(d)\}(Nh_{d,N}^d)^{-1} \\ &+ o_p((Nh_{d,N}^d)^{-1}) , \end{aligned} \tag{20}$$

where

$$\begin{aligned} \tilde{\Delta}_N(d, d_0) &= N^{-1} \sum_{t=1}^{N} \{F_d(Z_t^{(d)}) - F_{d_0}(Z_t^{(d_0)})\}^2 , \\ \tilde{\Delta}(d, d_0) &= E[\{F_d(Z_t^{(d)}) - F_{d_0}(Z_t^{(d_0)})\}^2 / f(Z_t^{(d)})] , \\ \alpha(d) &= \{k(0)\}^d , \\ \beta(d) &= \{\int k^2(u)du\}^d, \end{aligned}$$

$f$ is the probability density function of $Z_t^{(d)}$, $k$ is the kernel (function) as defined in Cheng and Tong (1992) and $h_{d,N}$ is the bandwidth of the kernel (previously denoted as $B(N)$ in Cheng and Tong, 1992).

In a sense, the above result gives an affirmative answer to the question raised by one of us (B.C.) in his discussion of Hall and Johnstone (1992).

**Theorem 4.2**: Under the same conditions as in Theorem 4.1,

$$\lim_{N \to \infty} P\{\hat{d} = d_0\} = 1 ,$$

where $\hat{d}$ is the minimiser of $ASE(d)$.

**Proof**:

We imitate the proof of Theorem 2 of Cheng and Tong (1992).

For $d < d_0$, $ASE(d)$ is dominated by $\tilde{\Delta}_N(d, d_0)$, which is non-zero. For $d > d_0$, $\tilde{\Delta}_N(d, d_0) = 0$ and $\tilde{\Delta}(d, d_0) = 0$. Therefore

$$Nh_{d,N}^d\{ASE(d) - ASE(d_0)\} = c(d) + o_p(1) ,$$

where $c(d) > 0$. The conclusion of the theorem follows immediately. □

## 5. Sample size requirement for consistent estimation of embedding dimension

$ASE(d)$ is useful for theoretical discussion but it involves the unknown $d_0$. To make the notion of distance practically useful for the determination of embedding dimension, we need to remove $d_0$ and one way to do this is by a comparative approach. Returning to the basic squared distance function $\Delta(.,.)$, we may solve the problem systematically by noting that Part (iv) of Proposition 3.1 together with Proposition 3.2 suggests an obvious consistent estimate of $\Delta(d_1, d_2)$, $d_1 \le d_2$, namely

$$\hat{\Delta}(d_1, d_2) = RSS(d_1) - RSS(d_2) , \tag{21}$$

where, as in Cheng and Tong [1992, equation (2.5) but setting $W(x) \equiv 1$ and assuming boundedness for the time series],

$$RSS(d) = (N - r + 1)^{-1} \sum\{X_t - \hat{F}_{d,N}(Z_t^{(d)})\}^2 . \tag{22}$$

the sumation extending over $t \ge r$ and $\hat{F}_{d,N}$ is the kernel estimator of $F_d$.

We now appeal to some results in Cheng and Tong (1992, 1993a). Before doing so, we need to first of all simplify conditions (a)–(o) of Theorem 1 of their paper in 1992 for convenience. Specifically, we retain conditions (a)-(i), and replace the rest by the existence of bounded first partial derivatives of $F_d$ for each $d$ and

$(j')$ $\quad \beta_j = 0(\beta^j), \quad 0 < \beta < 1$ ;

$(k')$ $\quad h_{d,N} \in H_{d,N} = [aN^{-(1/(2d+1))-\xi}, \quad bN^{-(1/(2d+1))+\xi}]$ ,

where $a$ and $b$ are arbitrary real positive constants, $\xi$ is any real positive constant strictly less than $\{2(d+1)(2d+1)\}^{-1}$. The detailed justification is relegated to the appendix. We shall refer to the above simplified conditions collectively as Condition B.

From Theorem 3 of Cheng and Tong (1992), we have for each $d \ge 1$ and $h_{d,N} \in H_{d,N}$

$$RSS(d) = \sigma_N^2(d)\{1 - (2\alpha(d) - \beta(d))/Nh_{d,N}^d\} + o_p(1/Nh_{N,d}^d) . \tag{23}$$

Choose, for explicitness,

$$h_{d,N} = N^{-1/(2d+1)} ,$$

and write $Nh_{d,N}^d = N^{1/\eta(d)}$, where $\eta(d) = (2d+1)/(d+1)$. Note that (5.1) and (5.3) yield, for general $d_1$ and $d_2$, $\hat{\Delta}(d_1, d_2) = \Delta_N(d_1, d_2) + o_p(N^{-\frac{1}{2}})$. By standard arguments, we have $\Delta_N(d_1, d_2) = \Delta(d_1, d_2) + O_p(N^{-\frac{1}{2}})$.

Now, let $d_0$ denote the embedding dimension. Then we have from (5.3)

$$\begin{aligned} \hat{\Delta}(d_0, d_0+1) &= RSS(d_0) - RSS(d_0+1) \\ &= \sigma_N^2(d_0) - \sigma_N^2(d_0+1) - \sigma_N^2(d_0)\{2\alpha(d_0) - \beta(d_0)\}N^{-1/\eta(d_0)} \\ &+ \sigma_N^2(d_0+1)\{2\alpha(d_0+1) - \beta(d_0+1)\}N^{-1/\eta(d_0+1)} + o_p(N^{-1/\eta(d_0)}) \\ &+ o_p(N^{-1/\eta(d_0+1)}) . \end{aligned} \tag{24}$$

However, from Proposition 3.2,

$$\sigma_N^2(d_0) - \sigma_N^2(d_0+1) = \Delta_N(d_0, d_0+1) = 0 .$$

Now

$$N^{-1/\eta(d_0+1)} \div N^{-1/\eta(d_0)} \to \infty \quad \text{as} \quad N \to \infty ,$$

implying that

$$N^{-1/\eta(d_0)} = o(N^{-1/\eta(d_0+1)}) .$$

Hence, we have proved the following theorem.

**Theorem 5.1**: Under Condition B and boundedness of the time series,

$$\hat{\Delta}(d_0, d_0+1) = \sigma_N^2(d_0+1)\{2\alpha(d_0+1) - \beta(d_0+1)\}N^{-1/\eta(d_0+1)} + o_p(N^{-1/\eta(d_0+1)}) . \tag{25}$$

We now use Theorem 5.1 to throw some light on the sample size requirement for a consistent estimation of $d_0$. The derivation will be heuristic. The embedding dimension is characterised by $\Delta^{\frac{1}{2}}(d_0, d_0+1)$, which measures the distance between the projection of the skeleton $F_{d_0}$ in $R^{d_0+1}$ and the skeleton $F_{d_0+1}$ as illustrated in Fig. 2.2. This was summarised in Lemma 2.1 and Proposition 3.2. Since $\Delta^{\frac{1}{2}}$ is unknown, we use its consistent estimate $\hat{\Delta}^{\frac{1}{2}}$. Then Proposition 3.1 (v)-(vii) suggests that for each $d$ it is reasonable to set

$$\hat{\Delta}(d, d+1) = \Delta(d, d+1) + O_p(N^{-\frac{1}{2}}) \approx const/d + O_p(N^{-\frac{1}{2}}). \tag{26}$$

However, Theorem 5.1 yields

$$\hat{\Delta}(d_0, d_0+1) = const \times \rho^2(d_0)N^{-1/\eta(d_0+1)} + o_p(N^{-1/\eta(d_0+1)}), \tag{27}$$

where $\rho^2(d_0) = \sigma^2(d_0)/var(X_t)$, the normalised dynamic noise variance. Therefore, combining (5.6) and (5.7), we have up to $O_p(N^{\frac{1}{2}})$

$$N \approx (d_0\rho^2(d_0)/\kappa)^{\eta(d_0+1)} \leq \{d_0\rho^2(d_0)/\kappa\}^2 . \tag{28}$$

We will discuss the constant $\kappa$ and the term $O_p(N^{\frac{1}{2}})$ in the next section. Cheng and Tong (1992) have proved that by relying on a penalized form of $\tilde{\Delta}_N(d, d+1)$ and $\hat{\Delta}(d, d+1)$, namely the $CV$ criterion (-the definition will be recalled in (6.2)), a consistent estimate of $d_0$ may be obtained. Note that the difference between the $CV$ criterion (-or more precisely $\{CV(d) - CV(d+1)\}$) and $\hat{\Delta}(d, d+1)$ is $O_p(1/(Nh^d_{d,N}))$, which is only $o_p(N^{-1/2})$. Hence, as a practical guidance we may offer the advice that *for useful estimation of the embedding dimensions the sample size requirement is bounded by a constant multiple of* $\{(\textit{embedding dimension}) \times (\textit{normalised dynamic noise variance})\}^2$. First, note the presence of the dynamic noise variance. This is in contrast to the results for the estimation of correlation dimension obtained by Smith (1988) and Ruelle (1990), which deal with the noise-free case. Next, by far the more significant is the lifting of the curse of dimensionality! This also throws substantial light on the "better-than-originally-expected" simulation results reported in Tong (1994) and Yao and Tong (1993). Of course, the curse stays if we replace $const/d$ in (5.6) by $(const)^{-d}$. However, Proposition 3.1 (v)-(vii) suggests that the latter is excessively stringent. In fact, the same stringency will lead to the same curse even in the linear case, which is discussed in the next section.

By now, the case for determining the embedding dimension first before focusing on the attractors is clearly overwhelming not only on grounds of logic but also statistical soundness. Once the embedding dimension is determined, or even better once a parsimonious set of stochastic regressors is determined using, for example, the approach of Cheng and Tong (1992) and Yao and Tong (1994), we can set about searching for exotic attractors within this set. We can then face the horrendous task of map reconstruction and correlation dimension estimation etc. with perhaps a better chance, where the curse of dimensionality prevails. Of course, if we allow ourselves the facilities of parametric models such as the threshold models and others (See Tong, 1990), then the curse may disappear but the perennial problem of subjectivity returns.

We now summarise some of our simulation results with a view to suggesting some preliminary empirical guidance for sample size requirement in the estimation of embedding dimension.

## 6. Simulations

Notice that the requirement of sample size given by formula (5.8) is bounded by a constant multiple of $\{(\textit{embedding dimension}) \times (\textit{normalised dynamic noise variance})\}^2$, which is independent of the forms of the skeleton. As a typical illustration of our simulations with $d \geq 1$, we have used the following nonlinear models:

$$X_t = 0.1X_{t-1} + (-0.5 + 0.2exp[-0.1X^2_{t-d}]\,)X_{t-d} + \epsilon_t, \quad d = 1, 2, \cdots, \qquad (29)$$

where $\{\epsilon_t\}$ are independent random variables each with the Gaussian distribution $N(0, \sigma^2(d))$. In our simulations, we let $\sigma^2(d) = 0.1$ for each $d \geq 1$.

Following Cheng and Tong (1992), for each $d_0$, we estimate $d_0$ consistently by minimising the cross-validation criterion with respect to $d$:

$$CV(d) = (N-r+1)^{-1} \sum \{X_t - \hat{F}_{d,N,-t}(Z_t^{(d)})\}^2, \tag{30}$$

where $\hat{F}_{d,N,-t}(Z_t^{(d)})$ is the *leave-one-out* estimator of $F_d$ at $Z_t^{(d)}$. (See Cheng and Tong (1992) for details). In our simulations, $h \in H_{d,N}$ with $h = c \times N^{-1/(2d+1)}$. For different $d$ and $N$, the constant $c$ was adjusted between 1 and 10. In principle, a data-driven bandwidth may be preferred but the computations involved would be quite excessive relative to the computing power at our dispoal. Nevertheless our random checks suggest that the results are unlikely to be fundamentally different from those summarised in Tables 1-4. We set the true orders at $4, 8, 11$, and $13$. For each sample size $N$, $CV(d)$ was searched over d from 1 to 20 and the estimated order, $\hat{d}_{CV}$,was given by

$$\hat{d}_{CV} = Argmin_{1 \le d \le 20}\{CV(d)\}.$$

We used the NAG library (G05DDF and G05CBF) to generate independent samples of size $N$ from $N(0, \sigma^2(d))$ and 100 replications were generated for each $d$. All computations were run in a SUN SPARC 2 workstation. The results of the simulations are summarized in the following Tables.

Tables 1-4 are about here

Now, we define the frequency of 'success' for the true order $d$ using sample size $N$, $FREQ_N(d)$ say, by

$$FREQ_N(d) = \frac{\#(\hat{d}_{CV} = d) + \#(\hat{d}_{CV} = d+1)}{100}$$

Then, as a typical example, from Table 1, $FREQ_{100}(4) = (31+8)/100 = 39\% = 0.39$. Figure 6.1 depicts all the results.

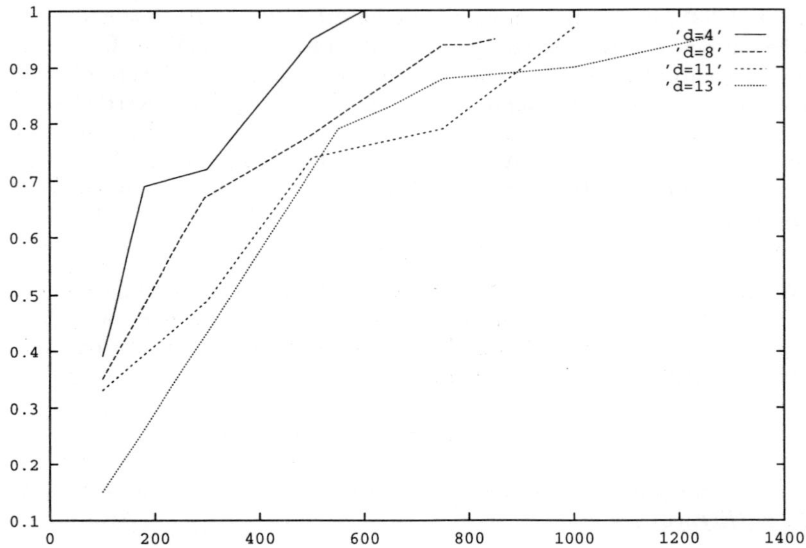


**Figure 6.1** $FREQ_N(d)$ $vs$ $N$

If we define the sample size requirement for order $d$, $N_{required}(d)$, by

$$FREQ_{N_{required}(d)}(d) \geq 95\%$$

i.e., there is at least 95% 'success' when sample size $N_{required}(d)$ is used, then from Tables 1-4, we know that

$$(*) \quad N_{required}(4) = 500; \qquad N_{required}(8) = 850;$$

$$(**) \quad N_{required}(11) = 1000; \qquad N_{required}(13) = 1250.$$

From Proposition 3.1 (i) we know that $\Delta^{\frac{1}{2}}$ is a distance function and if we bound $\hat{\Delta}^{\frac{1}{2}}(d, d+1)$ by $1 - 95\% = 5\% = 0.05$, i.e., bounded by the 'failure rate', then formula

(5.8) becomes

$$N_{required}(d) \;=\; N_0 \;+\; \{\frac{d\rho^2(d)}{\kappa}\}^{\eta}, \tag{31}$$

where we simply take $N_0$ and $\eta$ as constants and $\rho^2(4) = \sigma^2/var(X_t) = (0.1)^2/0.105 = 0.095 \approx 0.1$, $\rho^2(8) = (0.1)^2/0.1047 = 0.96 \approx 0.1$, $\rho^2(11) = (0.1)^2/0.1028 = 0.97 \approx 0.1$ , $\rho^2(13) = (0.1)^2/0.1021 = 0.98 \approx 0.1$ and $\kappa = (0.05)^2$. (Notice that in the derivation of (5.8), we have used $\hat{\Delta}$ not $\hat{\Delta}^{\frac{1}{2}}$ ) We may interpret $N_0$ as the 'base-line' sample size, which seems to be related to the term $O_p(N^{\frac{1}{2}})$ in (5.8). Since $\rho^2(d)/\kappa = 40$, we obtain

$$N_{required}(d) \;=\; N_0 \;+\; \{40d\}^{\eta}, \tag{32}$$

Since $N_{required}(4) \;=\; 500$ and $N_{required}(8) \;=\; 850$, it is easy to see that $N_0$ is between 120 and 240 (We allow $\pm 50$ oscillation here) and $\eta \;\approx\; 1.12$. This leads to the formula

$$N_{required}(d) \;=\; [120, 240] \;+\; \{40d\}^{1.12}, \tag{33}$$

where $[120, 240]$ means some integer between 120 and 240. Now we use this formula to predict the sample size requirements for $d = 11$ and $d = 13$. We get

$$N_{required}(11) \;=\; [120, 240] \;+\; \{40 \times 11\}^{1.12} \;=\; [1033, 1153],$$

$$N_{required}(13) \;=\; [120, 240] \;+\; \{40 \times 13\}^{1.12} \;=\; [1221, 1341].$$

These rough-and-ready arguments seem to give quite encouraging results by reference to those in (**). Generally we would propose the empirical formula

$$N_{required}(d) \;=\; N_0 \;+\; \{\gamma d\}^{\eta}, \tag{34}$$

where $\eta \leq 2$ and

$$\gamma \;=\; \frac{\rho^2(d)}{[\text{'}failure\ rate\text{'}]^2}.$$

Inverting (6.6) gives us an analogue of Ruelle's formula (1.1):

$$embedding\ dimension \;\leq\; \frac{\sqrt{N} \times (\text{'}failure\ rate\text{'})^2}{normalised\ dynamic\ noise\ variance}. \tag{35}$$

We suggest that the denominator be estimated (albeit roughly) by fitting a linear model of low order first or by the so-called "noise floor" of a principal component analysis. (Broomhead and King, 1986.) Other methods have also been suggested. See, e.g. Szpiro (1993). It is interesting to note that for the annual sunspot numbers, if we are prepared to tolerate a failure rate of $20 - 25\%$ and accept the normalised dynamic noise variance at 15% (-many reported parametric models for the data set have a lower value), then formula (6.7) gives an upper bound of about 6. The cross-validatory choice of order 4 by Cheng and Tong (1992) has a normalised dynamic noise variance of 15% , which is consistent with a failure rate of about 20%.

Using equation (7) of Hannan and Quinn (1979), we may deduce that for the *linear* autoregressive model,

$$\hat{\Delta}(d_0, d_0+1) \approx \frac{\rho^2(d_0) ln ln(N)}{N}. \tag{36}$$

Following the same derivation as (5.7), we get

$$\frac{\rho^2(d_0) ln ln(N)}{N} \approx \frac{\kappa_1}{d_0}. \tag{37}$$

Repeating the same arguments as before we may then obtain an empirical formula analogous to (6.3) but with $\eta = 1$, leading to the formula:

$$order\ of\ linear\ autoregression \leq \frac{N \times (`failure\ rate')^2}{normalised\ dynamic\ noise\ variance}. \tag{38}$$

## 7. Discussions

Our approach to the detection of chaos in real time series in the absence of any substantive theory is quite different from those often found in the literature of the physical sciences. Let us summarise our position briefly. As indicated in Tong (1990), Cheng and Tong (1992) and Chan and Tong (1994), we use the class of nonlinear autoregressive models as our basic framework and start by estimating the order of the autoregressive model for the data at hand by the cross-validation approach of leaving-k-out. We often set k = 1 but for over-sampled data we might prefer to set k much greater than 1. (Cf. Cheng and Tong, 1993b.) We are now convinced that the sample size requirement is not excessive and an empirical formula is available for its evaluation. Indeed, a parsimonious set of stochastic regressors may be identified by using an optimal subset selection based on the cross-validation approach as confirmed in Yao and Tong (1994). We may then perform tests for linearity with respect to the selected regressors using an assortment of those described in Tong (1992, Chapter 5) for example. If linearity is rejected, then the next job is the fitting of an appropriate nonlinear function of the selected regressors. For this a number of techniques are avaiable and they fall roughly into two groups: local function approximation and global function approximation. The former includes the threshold models of Tong (1990) and their younger relatives due to Casdagli *et al* (1991) and Lewis and Stevens (1991), and others; the latter includes the polynomial autoregressive models (See, e.g., Cox, 1977, Chan and Tong, 1994) and others. Casdagli *et al* (1991) gives some details. It is probably fair to say that map reconstruction using noisy data is still an open problem. At present, which function approximation/map reconstruction we use for noisy data is often a trade-off between the curse of dimensionality and subjectivity.

Once a model has been fitted, we may examine its skeleton (=signal) with the aim of detecting chaos by evaluating its correlation dimension, Lyapunov spectrum, etc. Fortunately the curse of dimensionality does not necessarily apply here. In short, we prefer to place chaos detection within the context of signal extraction, the signal being the skeleton which may be a limit point, a limit cycle or a strange attractor.

**References**

Akaike, H., 1974 A new look at the statistical model identification. *IEEE Trans.Autom. Control*, **AC-19**, 716-723.

Broomhead, D.S. and King, G.P.. 1986 Extracting qualitative dynamics from experimental data. *Physica* **D, 20**, 217-236.

Casdagli, M., Jardins, D.D., Eubank, S., Farmer, J.D., Gibson, J., Hunter, N. and Theiler, J. 1991 Nonlinear modeling of chaotic time series: theory and applications. Tech. Rep. LA-UR-91-1637, Los Alamos National Laboratory, U.S.A.

Chan, K.S. and Tong, H. 1994 A note on noisy chaos. *J.Roy.Statist.Soc.,* **B, 56**, 301-11.

Cheng, B. and Tong H. 1992 On consistent nonparametric order determination and chaos. *J.Roy.Statist.Soc.,* **B, 54**, 427-450.

Cheng, B. and Tong, H. 1993a On residual sums of squares in nonparametric autoregression. *Stoch.Proc. and Their Appl.,* **48** 157-74.

Cheng, B. and Tong H. 1993b Nonparametric function estimation in noisy chaos. *Developments in Time Series Analysis*, ed. T. Subba Rao, London: Chapman and Hall, pp. 183-206.

Cox, D.R. 1977 Discussion of paper by Campbell and Walker, Tong and Morris. *J. Roy. stat. Soc.,* **A 140**, 453-4.

Drazin, P.G. and King, G.P. 1992 *Interpretation of Time Series from Nonlinear Systems.* Proc. of IUTAM Symposium and NATO Advanced Research Workshop, Warwick, England. Amsterdam: North Holland

Eckmann, J.-P. and Ruelle, D. 1992 Fundamental limitations for estimating dimensions and Lyapunov exponents in dynamical systems. *Physica* **D, 56**, 185.

Hall, P. and Johnstone, I. 1992 Empirical functionals and efficient smoothing parameter selection (with discussion). *J.Roy.Statist.Soc.,* **B, 54**, 475-530.

Hannan, E.J. and Quinn, B.G. 1979 The determination of the order of an autoregression. *J.Roy.Statist.Soc.,* **B, 41**, 190-195.

Jensen, J.L. 1992 Chaotic dynamical systems with a view towards statistics – a review. Res.Regs. # 245, March 1992, Dept. of Theo.Stat., Univ. of Aarhus, Denmark. (To appear as a chapter in *Proc. of the Séminaire Européen de Statistique*, London: Chapman and Hall.)

Lewis, P.A.W. and Stevens, J.G. 1991 Nonlinear modeling of time series using multivariate adaptive regression splines (MARs). *JASA* **86**, 864-877.

Ruelle, D. 1990 Deterministic chaos: the science and the fiction. *Proc.R.Soc.Lond.* A **427**, 241-248.

Shibata, R. 1980 Asymptotically efficient selction of the order of the model for estimating parameters of a linear process. *Ann Statist.*, **8**, 147-164.

Smith, L.A., 1988 Intrinsic limits on dimension calculations. *Phys.Letters A*, **133**, 283-288.

Szpiro, G.G. 1993 Measuring dynamical noise in dynamical systems. *Physica* **D, 65**, 289-299.

Tong, H. 1990 *Nonlinear Time Series: A Dynamical System Approach.* Oxford: Oxford University Press.

Tong, H. 1994 Akaike's approach can yield consistent order determination. *Frontiers of Statistical Modelling: An Informational Approach*, ed. H. nal Bozdogan, Kleuwer Academic Publication, pp. 91-103.

Tong H. and Smith R.L. 1992 Royal Statistical Society Meeting on Chaos. *J.Roy.Statist. Soc.*, **B, 54**, 301-474.

Yao, Q. and Tong H. 1994 On subset selection in nonparametric stochastic regression. *Statistica Sinica*, **4**, 51-70.

## Appendix

(A1) The existence of bounded first partial derivatives of $f_d$ for each $d$ implies that $\mu = 1$ in condition $(d)$.

(A2) Let $i = i(N) \equiv 1$ and $j = j(N) = N/2$. Then condition $(\ell)$ clearly holds. Also

$$\begin{aligned}(1 + 6e^{1/2}\beta_j^{1/(1+i)})^i &= 1 + 6e^{1/2}\beta_j^{1/2} \\ &= 1 + 0(\beta^{j/2}) \\ &= 1 + 0(\beta^{N/4}) \\ &\to 1 \quad \text{as } N \to \infty\,.\end{aligned}$$

Therefore condition $(k)$ holds.

(A3) $$\beta_j^{\delta/(1+\delta)} = 0(\beta^{\delta j/(1+\delta)})\,.$$

However,

$$j^{2-\varepsilon}\ \beta^{\delta j/(1+\delta)} \le j^2 \beta^{\delta j/(1+\delta)} \to 0 \quad \text{as } j \to \infty\,.$$

Therefore condition $(j)$ holds for any $\varepsilon > 0$ .

(A4) Clearly with $h_{d,N} \in H_{d,N}$, as $N \to \infty$

$$N h_{d,N}^{2d} \to \infty$$

and

$$N h_{d,N}^{2d+2} \to 0\,.$$

Therefore conditions $(m)$ and $(n)$ hold.

(A5) From condition $(i)$, $2d < \theta$. Thus

$$h_{d,N}^{-2d+\theta} \to 0 \quad \text{as} \quad N \to \infty\,.$$

For $h_{d,N} \in H_{d,N}$,

$$N^\varepsilon h_{d,N}^{-2d+\theta} < b^{-2d+\theta} N^{\varepsilon - [1/(2d+1)+\xi](-2d+\theta)}\,.$$

Since condition $(j)$ holds for any $\varepsilon > 0$ , choose

$$\varepsilon < [1/(2d+1) + \xi](-2d + \theta)\,.$$

Then, $N^\varepsilon h_{d,N}^{-2d+\theta} \to$ as $N \to \infty$ and therefore condition $(o)$ holds.

Table 1: Frequencies of estimated order. (true order $d = 4$)

| Estimated order | N=100 | N=120 | N=150 | N=180 | N=300 | N=500 | N=600 |
|---|---|---|---|---|---|---|---|
| 1 | 3 | 2 | 2 | 0 | 0 | 0 | 0 |
| 2 | 0 | 0 | 1 | 1 | 0 | 0 | 0 |
| 3 | 3 | 2 | 0 | 0 | 0 | 0 | 0 |
| 4 | 31 | 37 | 37 | 45 | 58 | 91 | 97 |
| 5 | 8 | 9 | 21 | 14 | 14 | 4 | 3 |
| 6 | 6 | 8 | 5 | 6 | 8 | 1 | 0 |
| 7 | 6 | 5 | 8 | 7 | 5 | 0 | 0 |
| 8 | 9 | 5 | 3 | 3 | 3 | 0 | 0 |
| 9 | 5 | 1 | 0 | 2 | 1 | 0 | 0 |
| 10 | 3 | 4 | 5 | 3 | 0 | 0 | 0 |
| 11 | 4 | 5 | 0 | 3 | 2 | 0 | 0 |
| 12 | 4 | 3 | 3 | 3 | 3 | 1 | 0 |
| 13 | 3 | 6 | 5 | 3 | 4 | 0 | 0 |
| 14 | 0 | 3 | 0 | 0 | 0 | 1 | 0 |
| 15 | 2 | 0 | 2 | 1 | 1 | 0 | 0 |
| 16 | 6 | 3 | 2 | 4 | 1 | 1 | 0 |
| 17 | 1 | 2 | 2 | 2 | 0 | 0 | 0 |
| 18 | 1 | 1 | 2 | 0 | 0 | 0 | 0 |
| 19 | 3 | 3 | 1 | 1 | 0 | 0 | 0 |
| 20 | 2 | 2 | 1 | 2 | 0 | 1 | 0 |

Table 2: Frequencies of estimated order. (true order $d = 8$)

| Estimated order | N=100 | N=250 | N=295 | N=500 | N=750 | N=800 | N=850 |
|---|---|---|---|---|---|---|---|
| 1 | 21 | 4 | 2 | 0 | 0 | 0 | 0 |
| 2 | 8 | 0 | 1 | 0 | 0 | 0 | 0 |
| 3 | 5 | 2 | 2 | 0 | 0 | 0 | 0 |
| 4 | 0 | 0 | 0 | 0 | 0 | 0 | 0 |
| 5 | 2 | 0 | 0 | 0 | 0 | 0 | 0 |
| 6 | 1 | 0 | 0 | 0 | 0 | 0 | 0 |
| 7 | 1 | 0 | 0 | 0 | 0 | 0 | 0 |
| 8 | 23 | 39 | 45 | 60 | 78 | 80 | 81 |
| 9 | 12 | 21 | 22 | 18 | 16 | 14 | 14 |
| 10 | 7 | 9 | 9 | 5 | 4 | 5 | 5 |
| 11 | 6 | 10 | 9 | 5 | 2 | 1 | 0 |
| 12 | 3 | 7 | 5 | 5 | 0 | 0 | 0 |
| 13 | 6 | 3 | 2 | 1 | 0 | 0 | 0 |
| 14 | 0 | 2 | 0 | 2 | 0 | 0 | 0 |
| 15 | 1 | 1 | 0 | 0 | 0 | 0 | 0 |
| 16 | 3 | 0 | 1 | 2 | 0 | 0 | 0 |
| 17 | 1 | 1 | 2 | 2 | 0 | 0 | 0 |
| 18 | 0 | 0 | 0 | 0 | 0 | 0 | 0 |
| 19 | 0 | 1 | 0 | 0 | 0 | 0 | 0 |
| 20 | 0 | 0 | 0 | 0 | 0 | 0 | 0 |

Table 3: Frequencies of estimated order. (true order $d = 11$)

| Estimated order | N=100 | N=300 | N=500 | N=750 | N=1000 |
|---|---|---|---|---|---|
| 1 | 18 | 11 | 0 | 0 | 0 |
| 2 | 4 | 6 | 1 | 0 | 0 |
| 3 | 1 | 3 | 0 | 0 | 1 |
| 4 | 5 | 3 | 0 | 0 | 0 |
| 5 | 2 | 1 | 0 | 0 | 0 |
| 6 | 8 | 1 | 0 | 0 | 0 |
| 7 | 1 | 0 | 0 | 0 | 0 |
| 8 | 1 | 0 | 0 | 0 | 0 |
| 9 | 1 | 0 | 0 | 0 | 0 |
| 10 | 0 | 0 | 0 | 0 | 0 |
| 11 | 20 | 34 | 60 | 61 | 86 |
| 12 | 13 | 15 | 14 | 18 | 11 |
| 13 | 4 | 8 | 14 | 13 | 1 |
| 14 | 2 | 2 | 6 | 4 | 1 |
| 15 | 6 | 5 | 2 | 2 | 0 |
| 16 | 2 | 1 | 1 | 2 | 0 |
| 17 | 1 | 3 | 1 | 0 | 0 |
| 18 | 2 | 1 | 1 | 0 | 0 |
| 19 | 2 | 2 | 0 | 0 | 0 |
| 20 | 7 | 4 | 0 | 0 | 0 |

Table 4: Frequencies of estimated order. (true order $d = 13$)

| Estimated order | N=550 | N=650 | N=750 | N=1000 | N=1250 |
|---|---|---|---|---|---|
| 1 | 3 | 0 | 0 | 0 | 1 |
| 2 | 1 | 0 | 0 | 0 | 1 |
| 3 | 2 | 0 | 0 | 0 | 0 |
| 4 | 0 | 0 | 0 | 0 | 1 |
| 5 | 1 | 0 | 0 | 0 | 0 |
| 6 | 1 | 2 | 0 | 0 | 1 |
| 7 | 1 | 0 | 0 | 0 | 0 |
| 8 | 0 | 0 | 0 | 0 | 0 |
| 9 | 0 | 0 | 0 | 0 | 0 |
| 10 | 0 | 0 | 0 | 0 | 0 |
| 11 | 1 | 0 | 0 | 0 | 0 |
| 12 | 0 | 0 | 0 | 0 | 0 |
| 13 | 56 | 57 | 57 | 64 | 88 |
| 14 | 23 | 26 | 31 | 26 | 7 |
| 15 | 4 | 9 | 8 | 4 | 1 |
| 16 | 5 | 3 | 2 | 4 | 0 |
| 17 | 0 | 3 | 2 | 1 | 0 |
| 18 | 0 | 0 | 0 | 1 | 0 |
| 19 | 2 | 0 | 0 | 0 | 0 |
| 20 | 0 | 0 | 0 | 0 | 0 |

CHAPTER 2

# A THEORY OF CORRELATION DIMENSION FOR STATIONARY TIME SERIES

Colleen D. Cutler
Department of Statistics and Actuarial Science
University of Waterloo
Waterloo, Ontario
Canada N2L 3G1

**Abstract**

We develop a formal theory of correlation dimension for a class of stationary time series that includes both deterministic outputs (functionals of finite-dimensional dynamical systems) and Gaussian processes with continuous paths. This theory enables us to completely analyse correlation dimension in continuous Gaussian processes via spectral methods. In particular, we prove that the theoretical correlation dimension is always $\infty$ (except for those processes which are a superposition of a finite number of harmonics.) Our approach, combined with recent results on the convergence behaviour of the sample correlation integral, is then used to re-examine the behaviour of Gaussian power-law coloured noise. We show that the finite correlation dimension observed by Osborne and Provenzale (Physica D, 1989) is a local quantity due to the nonergodicity of their simulation model; moreover, their observed value corresponds to the correlation dimension of an ergodic component of the simulation process. We also show that nonergodic and weakly ergodic finite-dimensional dynamical systems (such as uncoupled harmonic oscillators and the simple circle map) exhibit a similar phenomenon under the Grassberger-Procaccia algorithm.

## 1. Introduction

In this paper we propose a formal definition and theory of correlation dimension for a class of strictly stationary real-valued time series with continuous paths. This class encompasses those time series generated by smooth functionals of invariant finite-dimensional dynamical systems (the so-called *deterministic outputs,* in which case our theory produces the usual results) but also includes more general time series produced by other means. For example, we verify that our class also includes all stationary Gaussian processes with continuous paths. We examine the Gaussian family in detail, and show how our approach clarifies some points in the *deterministic vs. stochastic* debate.

The chief motivation for this work was the important paper of Osborne and Provenzale (1989) (hereafter called OP), in which it was demonstrated that, for certain random processes with power-law spectra (often called *coloured noise,* and which can be taken to be Gaussian), the sequence of numerical estimates of the correlation dimension (obtained by successively increasing the embedding dimension $d$) converge to a *finite value* which in fact coincides with the fractal dimension of the embedded sample path. This observation (which we call the *OP phenomenon*) was of particular significance because it contradicted a widely-held assumption that correlation dimension estimates obtained from stochastically-generated time series would always increase without bound as the embedding dimension increased. As a consequence of OP, it was clear that convergence of the correlation dimension sequence could no longer be taken as a definitive indicator of an underlying determinism in the mechanism generating the data. This unexpected result has spawned a considerable literature devoted to increasingly sophisticated methods of distinguishing stochastic time series from low-dimensional chaotic dynamics; see, for example, Kaplan and Glass (1992), Kennel and Isabelle (1992), Provenzale *et al.* (1992), Smith (1992), and Theiler *et al.* (1992). Recent related statistical work can be found in Cheng and Tong (1992).

Correlation dimension was initially introduced and defined (Grassberger and Procaccia, 1983, hereafter denoted GP) for $\mathbb{R}^d$-valued data $\boldsymbol{X}_1, \ldots, \boldsymbol{X}_n$ as a double limit:

$$\nu_{\text{GP}} = \lim_{r \to 0} \lim_{n \to \infty} \frac{\log C_n(r)}{\log r} \tag{1.1}$$

where the statistic $C_n(r)$ (which we call the *sample correlation integral*) was defined to be the proportion of pairs of observations no more than distance $r$ apart. That is,

$$C_n(r) = \binom{n}{2}^{-1} \sum_{i=1}^{n} \sum_{j<i}^{n} I_{[\|\boldsymbol{X}_i - \boldsymbol{X}_j\|_d \le r]} \, . \tag{1.2}$$

In this paper the symbols $\| \, \|_d$ always denote the $\ell^\infty$ or "sup" norm:

$$\|\boldsymbol{X} - \boldsymbol{Y}\|_d = \max_{1 \le j \le d} |X_j - Y_j| \tag{1.3}$$

where $\boldsymbol{X} = (X_1, \ldots, X_d)$ and $\boldsymbol{Y} = (Y_1, \ldots, Y_d)$. The statistic $\nu_{\text{GP}}$ in (1.1) is actually independent of the choice of norm, but we will find the sup norm most convenient for computation, as well as the most easily extendable to the case of random continuous functions.

In the case that the data takes the form of a real-valued time series $\{X(t_k)\}_k$ (and hence may either be a realization of a stochastic process or a functional of a finite-dimensional dynamical system evolving in $\mathbb{R}^d$, some $d \geq 1$) the standard practice has been to attempt to reconstruct the underlying dynamics by employing the time-delay embedding method of Packard *et al.* (1980) and Takens (1981). We assume that the reader is fully familiar with this procedure. Its adaptation to the case of correlation dimension, applied to a single realization of a time series, is generally called the *Grassberger-Procaccia (GP) algorithm.* The experimentalist constructs a sequence $\{\nu_{\text{GP}}^{(d)}\}$, $d = 1, 2, \ldots$ of correlation dimensions, where $\nu_{\text{GP}}^{(d)}$ is determined via (1.1) (usually by regressing $\log C_n(r)$ *vs.* $\log r$) after grouping the original observations into vectors $\boldsymbol{X}_k^{(d)} = (X(t_k), X(t_{k+1}), \ldots, X(t_{k+d-1}))$ of length $d$ and computing (1.2). The limit of this sequence

$$\nu_{\text{GP}}^{(\infty)} = \lim_{d\to\infty} \nu_{\text{GP}}^{(d)} \tag{1.4}$$

is often called the *correlation dimension of the time series,* and, prior to the work of OP, was believed to indicate a deterministic output whenever it assumed a finite value. We note that Takens' (1981) embedding theorem is generally regarded as the theoretical basis for the GP algorithm, although Takens' theorem is not a statement about single realizations of a dynamical system.

In this paper we depart from the preceding standard approach in two important ways. First, in contrast to the GP method of definition, we adopt the more traditional statistical viewpoint that it is valuable to define parameters of interest in a manner *independent of data,* so that different methods of estimating the parameter (as well as different methods of obtaining data) can be compared for their ability to reproduce the parameter. Hence our definition of correlation dimension, both for random vectors and time series, is formulated in terms of probability distributions rather than data. Second, we treat a time series as a single object (a random continuous function) and make a definition of correlation dimension which seems appropriate for such objects by extending the formal mathematical approach which can be taken for random vectors in $\mathbb{R}^d$ (see next section). This creates a significant difference in rationale between our approach and the embedding technique — the latter evolved specifically as a method of reconstructing dynamics under the assumption of an underlying deterministic system, whereas our approach makes no such assumption on the time series. Thus our approach leads to a rigorous definition of correlation dimension valid for a larger class of stationary processes. This definition, coupled with recent results on the convergence behaviour of the sample correlation integral (Pesin (1993), Serinko (1993a), Aaronson *et al.* (1993)), provides a simple explanation of the OP phenomenon, and allows us to complete the discussion initiated by Theiler (1991). The main point is that continuous-spectrum Gaussian noise always has infinite correlation dimension and any observed OP phenomenon is *transient* in this case. Moreover, both this transient effect and the behaviour observed by OP in their discrete-spectrum simulation model are simply illustrations of a behaviour common to weakly ergodic and nonergodic systems; in Section 4 we provide examples of finite-dimensional systems exhibiting similar OP phenomena.

## 2. A New Approach to Correlation Dimension in Time Series

A formal mathematical approach to correlation dimension (and other fractal dimensions) for distributions in $\mathbb{R}^d$ has already been taken by various authors; see Cutler (1993) for an overview, also Pesin (1993) for alternatives. Given a probability measure $\mu$ on the Borel sets of $\mathbb{R}^d$, we define the associated *spatial correlation integral* $C_\mu(r)$ by

$$C_\mu(r) \;=\; \mu \times \mu(\{(\boldsymbol{x},\boldsymbol{y}) \in \mathbb{R}^d \times \mathbb{R}^d | \|\boldsymbol{x}-\boldsymbol{y}\|_d \le r\}) \,. \tag{2.1}$$

That is, $C_\mu(r)$ represents the probability that two independent observations from $\mu$ are no more than distance $r$ apart. This is portrayed descriptively by the alternate notation

$$C_\mu(r) \;=\; P([\|\boldsymbol{X}-\boldsymbol{Y}\|_d \le r]) \tag{2.2}$$

where $\boldsymbol{X}$, $\boldsymbol{Y}$ are independent, identically-distributed (i.i.d.) random vectors in $\mathbb{R}^d$, each with distribution $\mu$. We note that the sample correlation integral defined in (1.2) might be considered the "natural estimator" of $C_\mu(r)$, even though the vectors involved in obtaining (1.2) will generally be extracted from the same realization of a time series and hence fail to be independent. The *correlation dimension* of $\mu$ is then defined to be

$$\nu \;=\; \lim_{r\to 0} \frac{\log C_\mu(r)}{\log r} \tag{2.3}$$

assuming this limit exists (more generally, in the absence of a limit, one can define the upper and lower correlation dimensions associated with $\mu$). It is well known that $0 \le \nu \le d$.

This approach extends in the obvious way to the case of random continuous functions with compact time domain. Let $T > 0$ be fixed, and let $\mathcal{C}([0,T])$ denote the space of continuous real-valued functions on $[0,T]$ equipped with the sup norm:

$$\|x-y\|_T \;=\; \sup_{0\le t\le T} |x(t)-y(t)| \,. \tag{2.4}$$

Then, given a probability distribution $\mu$ on the Borel sets of $\mathcal{C}([0,T])$, we define, in analogy with (2.1) and (2.2), the associated spatial correlation integral by

$$C_T(r) \;=\; \mu \times \mu(\{(x,y) \in \mathcal{C}([0,T]) \times \mathcal{C}([0,T]) | \|x-y\|_T \le r\}) \tag{2.5}$$

which has the alternate description

$$C_T(r) \;=\; P([\|X-Y\|_T \le r]) \tag{2.6}$$

where $X$ and $Y$ are i.i.d. random functions in $\mathcal{C}([0,T])$, each with distribution $\mu$. We have used the notation $C_T(r)$, rather than $C_\mu(r)$, in order to emphasize the dependence of (2.5) and (2.6) on the length of the time interval. The corresponding correlation dimension $\nu_T$ is then obtained via (2.3).

In the case of stationary time series, the natural time domain of the process will generally be $[0, \infty)$ rather than $[0, T]$. Thus one might consider evaluating the distance between two i.i.d random functions over all $t \geq 0$, using a uniform metric of the form

$$\text{dist}(x, y) \;=\; \min\left(\sup_{0 \leq t < \infty} |x(t) - y(t)|, 1\right). \tag{2.7}$$

However, this approach seems undesirable, as in most instances we would expect to find that the probability of two random functions staying uniformly close for *all* time is actually zero. This will in fact be the case for most functionals of chaotic dynamical systems (i.e. systems with a positive Lyapunov exponent and extreme sensitivity to initial conditions). Hence we instead study the separation of two independent realizations of the process over bounded time intervals $[0, T]$, and restrict consideration (at least for this paper) to those processes for which the value of the correlation dimension $\nu_T$ is independent of the choice of $T$. For such processes the notion of correlation dimension can be unambiguously defined:

**Definition 2.1:** A process with stationary distribution $\mu$ on the Borel sets of $\mathcal{C}([0, \infty))$ is said to be *time-length invariant* with respect to correlation dimension if, for all $T > 0$, $T' > 0$, we have $\nu_T = \nu_{T'}$. This common value, which we denote by $\nu_\infty$, will be called the *correlation dimension of the process.*

We remark that we do not have any examples of stationary processes on $C([0, \infty))$ which are *not* time-length invariant in the above sense. If such processes do exist, a broader definition of correlation dimension can be given simply by $\nu_\infty = \lim_{T \to \infty} \nu_T$. However, the notion of time-length invariance seems important for practical reasons; it permits alternative sampling mechanisms using finite time intervals (see in particular the last remark at the end of this section).

We also note that we have not taken the approach of trying to introduce an alternative metric (in place of the unsatisfactory (2.7)) to measure distance on $C([0, \infty))$. The reasons here are twofold. The reader will note that the usual topology assigned to $C([0, \infty))$, in which convergence corresponds to uniform convergence of functions over compact time intervals, is close in spirit to Definition 2.1. However, there is no canonical choice of metric generating this topology, and the various metrics that are used do not assign the same dimension values to all sets and measures. (In general, topologically equivalent metrics need not produce similar dimensions.) Moreover, any metric generating the usual topology on $C([0, \infty))$ has the undesirable feature of being nonstationary (distance is measured differently over different time intervals of equal length) which destroys a natural uniformity present in the original problem. Hence we have chosen to take a different route for stationary time series.

There is a second relevant definition we can make which will establish a clear link between our approach and (1.4):

**Definition 2.2:** Let $X(t)$ be a stationary process with invariant distribution $\mu$ on the Borel sets of $\mathcal{C}([0, \infty))$. Given a compact interval $[0, T]$ with partition $0 = t_0 \leq$

$t_1 < \ldots < t_d \leq t_{d+1} = T$, let $\nu_{t_1,\ldots,t_d}$ denote the correlation dimension of the finite-dimensional distribution $\mu_{t_1,\ldots,t_d}$ on $I\!R^d$ corresponding to the vector $(X(t_1),\ldots,X(t_d))$. The process is said to be *accessible* with respect to correlation dimension if, for every $T > 0$, we have $\lim_{\Delta T\to 0} \nu_{t_1,\ldots,t_d} = \nu_T$ where $\Delta T = \max_{1\leq j\leq d+1} |t_j - t_{j-1}|$ is the mesh of the partition over $[0,T]$.

We note that time-length invariance does not necessarily imply accessibility. If X and Y are two independent realizations of the process, the assumed continuity of paths does imply that

$$\lim_{\Delta T\to 0} \sup_{t_1,\ldots,t_d} |X(t_j) - Y(t_j)| = \|X - Y\|_T \tag{2.8}$$

and hence that the corresponding spatial correlation integrals converge appropriately

$$\lim_{\Delta T\to 0} C_{t_1,\ldots,t_d}(r) = C_T(r) \quad \text{for each } r \geq 0 \tag{2.9}$$

where $C_{t_1,\ldots,t_d}(r)$ is the spatial correlation integral of $\mu_{t_1,\ldots,t_d}$. To see this, note that $C_{t_1,\ldots,t_d}(r)$ and $C_T(r)$ are, respectively, the distribution functions of the variables $\sup_{t_1,\ldots,t_d} |X(t_j) - Y(t_j)|$ and $\|X - Y\|_T$; hence by weak convergence of distributions (which follows from the stronger convergence in (2.8)) and the obvious inequality $C_{t_1,\ldots,t_d}(r) \geq C_T(r)$, we get (2.9). However, accessibility further requires that we are able to interchange two limits  i.e. we must assume that

$$\lim_{\Delta T\to 0}\lim_{r\to 0} \frac{\log C_{t_1,\ldots,t_d}(r)}{\log r} = \lim_{r\to 0}\lim_{\Delta T\to 0} \frac{\log C_{t_1,\ldots,t_d}(r)}{\log r}. \tag{2.10}$$

The right hand side of (2.10) is the desired quantity $\nu_T$, while the lim sup of the left hand side will certainly always be bounded above by $\nu_T$.

Accessibility indicates that the entire path behaviour over $[0,T]$ can be approximated in a very strong way by observations at a finite number of time points $t_1,\ldots,t_d$ provided that the partition is sufficiently fine. However, even in the absence of accessibility, a strong link between the finite dimensional distributions and the overall process is provided by (2.9). As with time-length invariance, we do not know the size of the class of accessible processes.

The remainder of this section is devoted to showing that both deterministic outputs and Gaussian processes are time-length invariant and accessible with respect to correlation dimension; moreover, for deterministic outputs, this new definition of correlation dimension coincides with the standard one (discussed below) usually assigned to the underlying dynamical system.

A stationary finite-dimensional dynamical system on a smooth compact $d$-dimensional manifold $M$ is represented by a semigroup of mappings $\{\varphi_t\}_t$, $t \geq 0$, $\varphi_t : M \to M$, and an invariant distribution $\mu^*$ (i.e. $\mu^*\varphi_t^{-1} = \mu^*$ for each $t$) which describes the distribution of initial conditions on $M$. Note that, by the Whitney embedding theorem, there exists a minimal embedding dimension $d^* \leq 2d + 1$ for which $M$ can be viewed as a subset of $I\!R^{d^*}$. In this case, the *correlation dimension of the system* is typically defined to be the correlation dimension $\nu^*$ of $\mu^*$, computed via (2.2) and (2.3) with $d^*$ in place of $d$. In practice, however, one rarely observes the actual system evolving on $M$, but observes instead a real-valued functional $X(t) = h(\varphi_t(\boldsymbol{X}))$ of the system; here $h$ is a smooth mapping of $M$ into $I\!R$ and $\boldsymbol{X}$ denotes a random initial condition (assumed selected according to $\mu^*$). The resulting process $X(t)$ is strictly stationary with invariant distribution $\mu$ on the Borel sets of $\mathcal{C}([0,\infty))$ having finite-dimensional marginals $\mu_{t_1,\ldots,t_k}$ on $I\!R^k$ given by

$$\mu_{t_1,\ldots,t_k}(B_1 \times \ldots \times B_k) = \mu^* \left( \bigcap_{j=1}^{k} \varphi_{t_j}^{-1} h^{-1}(B_j) \right) . \tag{2.11}$$

Let $\nu_T$ denote the correlation dimension of the functional process over $[0,T]$. Then we have:

**Theorem 2.3:** Suppose the semigroup $\{\varphi_t\}_t$ satisfies a uniform Lipschitz condition over each compact time interval $[0,T]$. i.e. for each $T > 0$ there exists $0 \le A_T < \infty$ such that $\sup_{0\le t\le T} \|\varphi_t(\boldsymbol{x}) - \varphi_t(\boldsymbol{y})\|_{d^*} \le A_T \|\boldsymbol{x} - \boldsymbol{y}\|_{d^*}$ for all $\boldsymbol{x}, \boldsymbol{y} \in M$, where $d^*$ is the minimal embedding dimension of $M$.

(a) If $h$ also satisfies a Lipschitz condition over $M$, then $\nu_T \le \nu^*$ for every $T > 0$.

(b) If $h$ and the semigroup further satisfy the hypotheses of the Takens' embedding theorem, then the functional process $\{X(t)\}_t$ is time-length invariant and accessible with $\nu_\infty = \nu^*$.

**Proof:** (a) Suppose $h$ satisfies a Lipschitz condition $|h(\boldsymbol{x}) - h(\boldsymbol{y})| \le B\|\boldsymbol{x} - \boldsymbol{y}\|_{d^*}$ for all $\boldsymbol{x}, \boldsymbol{y} \in M$. If $X(t) = h(\varphi_t(\boldsymbol{X}))$ and $Y(t) = h(\varphi_t(\boldsymbol{Y}))$ are two realizations of the functional process with initial conditions $\boldsymbol{X}$ and $\boldsymbol{Y}$, then $\sup_{0\le t\le T} |X(t) - Y(t)| \le B \sup_{0\le t\le T} \|\varphi_t(\boldsymbol{X}) - \varphi_t(\boldsymbol{Y})\|_{d^*} \le A_T B \|\boldsymbol{X} - \boldsymbol{Y}\|_{d^*}$. Hence, choosing $\boldsymbol{X}$ and $\boldsymbol{Y}$ to be independent initial conditions, each with distribution $\mu^*$, we get $C_T(r) \ge P([\|\boldsymbol{X} - \boldsymbol{Y}\|_{d^*} \le r/(A_T B)]) = C_{\mu^*}(r/(A_T B))$ and the result follows.

(b) Suppose $h$ and the semigroup additionally satisfy the hypotheses of Takens' (1981) embedding theorem. Given any $T > 0$ and time points $0 \le t_1 < \ldots < t_k \le T$, it follows that the mapping $\Phi : M \to I\!R^k$ defined by $\Phi(\boldsymbol{x}) = (h(\varphi_{t_1}(\boldsymbol{x})), \ldots, h(\varphi_{t_k}(\boldsymbol{x})))$ is an embedding for all $k \ge d^*$. Since $\Phi$ is bi-Lipschitz, the correlation dimensions of $\mu^*$ and the image measure $\mu_{t_1,\ldots,t_k}$ coincide i.e. $\nu^* = \nu_{t_1,\ldots,t_k}$ for all $k \ge d^*$. The result now follows from this and (a).

We remark that the Lipschitz condition in the hypothesis of Theorem 2.3 is one that we can reasonably expect to be satisfied by most smooth dynamical systems evolving on compact manifolds. In particular, if $\{\varphi_t\}_t$ arises as the family of solutions to the differential equation $\dot{\boldsymbol{x}} = f(\boldsymbol{x})$ where $f$ satisfies a Lipschitz condition (with coefficient $K$) on $M$, then we may take $A_T = \exp(KT)$ (see Theorem 1.0.4 of Guckenheimer and Holmes (1983)).

We now consider real-valued stationary Gaussian processes. Without loss of generality we can assume that all such processes have mean 0, since (2.6) is unchanged by a shift in location. We immediately note that if the covariance matrix of the components $X(t_1), \ldots, X(t_d)$ is *nonsingular,* then the joint distribution of $X(t_1), \ldots, X(t_d)$ has a bounded density with respect to $d$-dimensional Lebesgue measure and hence $\nu_{t_1,\ldots,t_d} = d$. This shows that any Gaussian process for which *all* finite-dimensional joint distributions are nonsingular is both time-length invariant and accessible with $\nu_\infty = \infty$. Thus we need only consider those cases where singularities arise in some of the joint distributions, and since these distributions are Gaussian, a singularity always corresponds to the existence of a nontrivial linear relation among the components.

It will be useful here (and later when we look at coloured noise) to consider the spectral decomposition of a general complex-valued 2nd-order stationary process (see, for example, Cramér and Leadbetter (1967)). Let $K(t) = E(X(s)\overline{X}(s+t))$ denote the autocovariance function of the process. Under the weak restriction that $K(t)$ be continuous, we have the unique representation

$$K(t) \;=\; \int_{-\infty}^{\infty} e^{i\lambda t}\, S(d\lambda) \tag{2.12}$$

where $S$ is the associated (nonnormalized) spectral measure. The following two lemmas establish the relationship between the spectrum and linear dependencies of a stationary process. We believe that both lemmas are known results, but were unable to locate them in the required form in the literature. Hence proofs are included for completeness.

**Lemma 2.4:** Let $X(t)$ be a complex-valued 2nd-order stationary process with continuous autocovariance $K(t)$. If there exist time points $t_1, \ldots, t_n$ and complex numbers $a_1, \ldots, a_n$ (not all zero) such that $\sum_{j=1}^{n} a_j X(t_j) = 0$ *a.s.*, then the associated spectral measure $S$ is purely discrete.

**Proof:** For any coefficients $a_1, \ldots, a_n$, it follows from (2.12) that

$$\mathrm{Var}\left(\textstyle\sum_j a_j X(t_j)\right) \;=\; \textstyle\sum_j \sum_k a_j \bar{a}_k K(t_k - t_j) \;=\; \int_{-\infty}^{\infty} |\sum_j a_j e^{-i\lambda t_j}|^2\, S(d\lambda)\,. \tag{2.13}$$

Let $h(\lambda) = \sum_j a_j e^{-i\lambda t_j}$ and suppose that the variance on the left hand side of (2.13) is zero. It then follows from the right hand side of (2.13) that $h(\lambda) = 0$ $S$-*a.s.* Let $\mathcal{Z} = \{\lambda | h(\lambda) = 0\}$. If $\mathcal{Z}$ has a cluster point in $\mathbb{R}$, then $h \equiv 0$ by a property of analytic functions; this then forces $a_j = 0$ for every $j$. (Note that if some $a_j \neq 0$, then $h$ is the Fourier transform of a linear combination of Dirac measures and hence, by the uniqueness of Fourier transforms, cannot be identically 0.) This proves that if some $a_j \neq 0$ then $\mathcal{Z}$ has no cluster points (and hence forms a discrete set of points). Since $h(\lambda) = 0$ $S$-*a.s.* it follows that $S(\mathbb{R} \setminus \mathcal{Z}) = 0$ and the lemma is proved.

Now any complex-valued 2nd-order stationary process with mean 0, continuous autocovariance, and purely discrete spectral measure $S$, has the representation

$$X(t) \;=\; \sum_{\lambda_k \in \mathcal{A}} c_k\, e^{i\lambda_k t} Z_k \qquad a.s. \tag{2.14}$$

where $\mathcal{A}$ is the set of atoms of $S$, $c_k^2 = S(\{\lambda_k\})$, and $Z_1, Z_2 \ldots$ are orthogonal (complex-valued) variables with mean 0 and variance 1. In the case of Gaussian processes the $Z_j$'s are complex-valued Gaussian variables. The sum on the right hand side of (2.14) is interpreted as a limit in quadratic mean.

**Lemma 2.5:** Let $X(t)$ be a complex-valued 2nd-order stationary process with mean 0 and purely discrete spectral measure $S$. Let $|\mathcal{A}|$ denote the number of atoms of $S$. If $|\mathcal{A}| \geq n$, then for every $T>0$ and every sequence of partitions $\{\mathcal{P}_m\}_m$ of $[0,T]$ for which the mesh $\Delta T \to 0$, there exists $m_0$ such that, for all $m \geq m_0$, $\mathcal{P}_m$ includes $n$ distinct time points $t_1, \ldots, t_n$ (which may vary with $m$) for which $X(t_1), \ldots, X(t_n)$ are linearly independent.

**Proof:** First suppose $|\mathcal{A}| < \infty$, say $|\mathcal{A}| = n$. Let $t_1 < \ldots < t_n$ be arbitrary time points and let $\boldsymbol{X}_n$ denote the transpose of the vector $(X(t_1), \ldots, X(t_n))$ and $\boldsymbol{Z}_n$ denote the transpose of the vector $(Z_1, \ldots, Z_n)$. Then, by (2.14), we can write $\boldsymbol{X}_n = \boldsymbol{C}_{n,n}\boldsymbol{Z}_n$ where $\boldsymbol{C}_{n,n}$ is the $n \times n$ matrix whose $(r,j)^{th}$ entry is given by $c_j e^{i\lambda_j t_r}$. Since the $Z_j$'s are orthogonal, the number of linearly independent components in $\boldsymbol{X}_n$ is determined by the rank of $\boldsymbol{C}_{n,n}$. It follows by applying elementary column operations that the rank of $\boldsymbol{C}_{n,n}$ is the same as that of $\boldsymbol{C}^*_{n,n}$ where the $(r,j)^{th}$ entry of $\boldsymbol{C}^*_{n,n}$ is $e^{i\lambda_j(t_r - t_1)}$. Now $\boldsymbol{C}^*_{n,n}$ becomes a Vandermonde matrix (see Nering 1970) for every choice of $t_1 > 0$ if we set $t_r = rt_1$ for $r = 2, \ldots, n$. The determinant of $\boldsymbol{C}^*_{n,n}$ is then given by $D = \prod_{1 \leq k < j \leq n}(e^{i\lambda_j t_1} - e^{i\lambda_k t_1})$. Clearly $D \neq 0$ provided we select $t_1$ so that it is not an integer multiple of $2\pi/(\lambda_j - \lambda_k)$ for any $1 \leq k < j \leq n$. Hence in any interval $[0,T]$ we can find suitable $0 < t_1 \leq T/n$ and then take $t_r = rt_1$. Moreover, the continuity of the determinant of $\boldsymbol{C}_{n,n}$ in the $t$ values implies that there will be intervals $I_r \ni t_r$, $r = 1, \ldots, n$, for which the determinant is nonzero for any choice of time points $s_1, \ldots, s_n$ satisfying $s_r \in I_r$, $r = 1, \ldots, n$. Thus a sufficiently small mesh size implies the existence of $n$ timepoints in the partition for which $\boldsymbol{C}_{n,n}$ is of full rank $n$, which proves the lemma for the case $|\mathcal{A}| < \infty$. If $|\mathcal{A}| = \infty$ we may index the atoms by the positive integers and then, by (2.14), write $X(t) = X_n(t) + X^{n+1}(t)$ where $X_n(t) = \sum_{k=1}^{n} c_k e^{i\lambda_k t} Z_k$ and $X^{n+1}(t) = \sum_{k=n+1}^{\infty} c_k e^{i\lambda_k t} Z_k$. Note that $X_n(t)$ has a discrete spectrum of size $n$ and so our earlier analysis applies to $X_n(t)$. Moreover, as $X_n(t_j)$ and $X^{n+1}(t_k)$ are orthogonal for all time points $t_j$ and $t_k$, it follows that $X(t_1), \ldots, X(t_n)$ are linearly independent whenever $X_n(t_1), \ldots, X_n(t_n)$ are linearly independent. This completes the proof.

We see from the preceding lemmas and earlier discussion that the only possible Gaussian candidates for a finite correlation dimension are harmonic processes with a finite number of components. If $X(t) = \sum_{k=1}^{d} c_k\, e^{i\lambda_k t} Z_k$ and $Y(t) = \sum_{k=1}^{d} c_k\, e^{i\lambda_k t} Z'_k$ are two independent realizations of such a process (meaning that $\boldsymbol{Z} = (Z_1, \ldots, Z_d)$ and $\boldsymbol{Z}' = (Z'_1, \ldots, Z'_d)$ are independent vectors), we obtain, for each $T>0$

$$\sup_{0 \leq t \leq T} |X(t) - Y(t)| \leq \left(\sum_{k=1}^{d} c_k\right) \max_{1 \leq k \leq d} |Z_k - Z'_k| = \left(\sum_{k=1}^{d} c_k\right) \|\boldsymbol{Z} - \boldsymbol{Z}'\|_d. \qquad (2.15)$$

Now $P([\|\boldsymbol{Z} - \boldsymbol{Z}'\|_d \leq r]) = C_{\bar{\mu}}(r)$ where $\bar{\mu}$ is the distribution of $\boldsymbol{Z}$ in $\mathbb{C}^d$ ($\mathbb{C}$ denotes the complex plane). Identifying $\mathbb{C}^d$ with $\mathbb{R}^{2d}$, we must have $\nu_T \leq 2d$. However, in the case of a real-valued process, the spectrum is symmetric about 0 and the $Z_k$'s associated with nonzero frequencies occur in conjugate pairs (i.e. $Z_{-k} = \overline{Z}_k$), while the $Z$ variable associated with the zero frequency (should it be present) must be real. This gives $|Z_k - Z'_k| = |Z_{-k} - Z'_{-k}|$ and ensures that $\nu_T$ cannot exceed $d$ (alternately, write $X(t)$ in real form as in (3.6) to arrive at the same conclusion). This plus Lemma 2.5 gives $\nu_T = d$ in the real-valued case.

Combining the above results with the fact that every stationary Gaussian process with continuous paths has a continuous autocovariance (so the lemmas apply), we arrive at the following:

**Theorem 2.6:** Every real-valued stationary Gaussian process with continuous paths is time-length invariant and accessible with respect to correlation dimension. If the spectral measure of the process is purely discrete with exactly $d$ atoms, then $\nu_\infty = d$. In all other cases $\nu_\infty = \infty$.

**Remark 2.7:** We make the following observations:

- We were able to use spectral theory to completely determine correlation dimension in the case of Gaussian processes. This is partly due to the fact that singularities in Gaussian distributions are always reflected in the linear structure, making spectral methods well-suited to the problem. We would not expect such complete success using spectral analysis with general stationary processes.

- Since every bounded Borel measure $S$ is the spectral measure of some stationary Gaussian process, we may choose $S$ to be continuous with a high-frequency cutoff $S(\{\lambda \mid |\lambda| > \lambda_0\}) = 0$. In this case $\nu_\infty = \infty$, but the process has paths which are infinitely differentiable (hence locally predictable) and the Hausdorff dimension of a graph of a realization of the process is *a.s.* 1.

- A Gaussian process comprised solely of $d$ harmonic components has finite correlation dimension $d$, and any single realization of the process is completely predictable once $d$ (linearly independent) observations have been obtained from the realization. However, the solution for one realization is a.s. not valid for any other realization, nor can it be used to predict limiting behaviour and occupation probabilities for other realizations. This situation corresponds to the projection of an (unbounded) $d$-dimensional dynamical system with uncountably many distinct simple attractors. Note that graphs of the process are periodic or almost periodic and *a.s.* have Hausdorff dimension 1.

- If global scaling behaviour (as described by our definition of correlation dimension) is truly the quantity of interest, then our approach suggests that repeated independent small samples of a process (when possible), taken over a short time interval, will generally be preferable to a single large sample taken over a long time period. That is, it is better to obtain $m$ ($m$ large) independent "copies" $\boldsymbol{X}_1, \ldots, \boldsymbol{X}_m$ of a relatively short vector $\boldsymbol{X} = (X(\tau), \ldots, X(k\tau))$ than to obtain a long sequence of observations $X(j\tau)$, $j = 1, \ldots, m$ on a single realization of a time series. Independent copies are generated by running the process repeatedly with different random initial conditions. Once the copies are obtained, the sample correlation integral $C_m(r)$, for a given embedding dimension $d \le k$, is computed over the $d$-vectors $\boldsymbol{Y}_1, \ldots, \boldsymbol{Y}_m$ defined by $\boldsymbol{Y}_j = (X_j(\tau), \ldots, X_j(d\tau))$ (i.e. $\boldsymbol{Y}_j$ consists of the first $d$ components of $\boldsymbol{X}_j$.) This approach to sampling and estimation is the key to circumventing the OP phenomenon; it would correctly identify the correlation dimension as being arbitrarily large (in the above example, at least as big as $k$) for power-law noise.

## 3. Dimension Analysis of Gaussian Power-Law Noise

In this section we resolve the apparent contradiction between the finite correlation dimension observed by OP and the theoretical value of $\infty$ provided by Theorem 2.6. Throughout we employ the real form $S^*$ of the spectral measure $S$ i.e. $S^*([0, \lambda]) = S([-\lambda, \lambda])$. In this case the autocovariance is expressed as

$$K(t) = \int_0^\infty \cos(\lambda t)\, S^*(d\lambda)\,. \tag{3.1}$$

The original intent of OP appears to have been to estimate the correlation dimension of a process with (real) spectral density $f^*(\lambda)$ of the form

$$f^*(\lambda) = \frac{dS^*}{d\lambda} = \lambda^{-\alpha} \qquad \text{for } \lambda > 0 \qquad \text{where } \alpha > 1\,. \tag{3.2}$$

Of course to maintain integrability of $f^*$ we require a low-frequency cutoff $\lambda_0$ on the power law; hence the range in (3.2) should be $\lambda > \lambda_0 > 0$. Any integrable function (including a different power law e.g. $f^*(\lambda) = \lambda^{-\beta}$, $\beta < 1$) can be patched in over the interval $(0, \lambda_0)$.

Since (3.2) corresponds to a continuous spectral measure, Theorem 2.6 of the pre vious section gives the theoretical value $\nu_\infty = \infty$ for any Gauss ian process following this model. OP, however, approached the problem numerically through simulation and the GP algorithm, consistently obtaining the value $\nu_{\mathrm{GP}}^{(\infty)} = 2/(\alpha - 1)$ for $1 < \alpha < 3$. Various explanations of this phenomenon have been attempted, and a particularly insightful discussion can be found in Theiler (1991). Here we examine the problem from a somewhat different point of view, consistent with the approach taken in Section 2. We will conclude that it is important to distinguish the discrete spectrum *simulation* method of OP (discussed below) from the continuous spectrum model (3.2). In the case of continuous spectrum Gaussian processes, the GP algorithm *will* eventually yield the correct answer of $\infty$ (although the sample sizes required to observe such full-dimensional scaling might be extremely large); we will further argue that both low and high frequencies have a role to play in determining the sample size necessary to find a correct scaling region for the sample correlation integral. We then show that the OP simulation model is not ergodic and that the GP algorithm, applied to a single realization of this model, remains trapped in a low-dimensional component or subspace of the process. This leads to a new interpretation of the magic number

$d(\alpha) = 2/(\alpha - 1)$ as the correlation dimension of a local component $\mu^x$ in the discrete spectrum model.

Any strictly stationary process $X(t)$ with continuous paths (this includes deterministic outputs; see Section 2) and invariant distribution $\mu$ can be represented as a functional of a shift dynamical system on $\mathcal{C}([0,\infty))$. (We assume that $\mathcal{C}([0,\infty))$ is equipped with the usual Borel sets generated by the finite-dimensional projection maps.) For each $t \geq 0$, the left-shift operator $L_t : \mathcal{C}([0,\infty)) \longrightarrow \mathcal{C}([0,\infty))$ is defined by

$$L_t(x)(s) \;=\; x(s+t)\,. \tag{3.3}$$

The mappings $\{L_t\}_t$, $t \geq 0$, form a semigroup on $\mathcal{C}([0,\infty))$, and it follows that $\mu L_t^{-1} = \mu$ for every $t \geq 0$ as a consequence of stationarity. Hence $(\{L_t\}_t, \mu)$ defines a stationary dynamical system evolving on $\mathcal{C}([0,\infty))$ with random initial conditions $X \in \mathcal{C}([0,\infty))$ selected according to $\mu$. The original process $X(t)$ is recovered by projecting $L_t(X)$ down to its $0^{th}$ coordinate:

$$X(t) \;=\; L_t(X)(0)\,. \tag{3.4}$$

Since $(\{L_t\}_t, \mu)$ is a dynamical system, the usual definitions of ergodicity and mixing apply. (Note that if $X(t)$ is a functional of a finite-dimensional dynamical system in $I\!R^d$, then $(\{L_t\}_t, \mu)$ will be ergodic whenever the original $d$-dimensional system is ergodic.) The significance of this approach is seen in the following theorem.

**Theorem 3.1:** Let $X(t)$ be a strictly stationary process on $\mathcal{C}([0,\infty))$ with invariant distribution $\mu$.

(a) Suppose that $(\{L_t\}_t, \mu)$ is *ergodic.* Let $\tau > 0$ be fixed, and define the discrete-time vector-valued embedded process $\{\boldsymbol{X}_n\}_n$ in $I\!R^d$ by $\boldsymbol{X}_n = (X(n\tau), X((n+1)\tau), \ldots, X((n+d-1)\tau))$ for $n = 1, 2, \ldots$. Let $C_n(r)$ be the sample correlation integral of $\boldsymbol{X}_1, \ldots, \boldsymbol{X}_n$ as defined in (1.2). Then, for almost all choices of $\tau$, $\lim_{n\to\infty} C_n(r) = C_{\bar{\mu}}(r)$ *a.s.* where $\bar{\mu} = \mu_{\tau,2\tau,\ldots,d\tau}$ is the distribution of the vector $\boldsymbol{X}_1$.

(b) If $X(t)$ is Gaussian, then the dynamical system $(\{L_t\}_t, \mu)$ is ergodic if and only if the real spectral measure $S^*$ of $X(t)$ is purely continuous. (We then say $X(t)$ is ergodic.)

**Proof:** (a) Under the ergodicity assumption, the vector-valued process $\boldsymbol{X}(t) = (X(t), X(t+\tau), \ldots, X(t+(d-1)\tau))$ is also ergodic, for any $\tau > 0$, under the vector-valued shift $\boldsymbol{L}_t$ defined on $C([0,\ infty))^d$ by $\boldsymbol{L}_t(x_1, \ldots, x_d) = (L_t(x_1), \ldots, L_t(x_d))$. This can be seen by noting that $\boldsymbol{L}_t$ is a *factor* of $L_t$ (Cornfeld *et al.* (1982), pg. 230). The discrete-time version (sampled every $\tau^{th}$ unit of time) will also generally be ergodic; however, if the spectrum has a discrete component, $\tau$ should not be a rational multiple of any period. The result then follows directly from the recent work of Pesin (1993), Serinko (1993a) and Aaronson *et al.* (1993), who prove that the sample correlation integral converges to the spatial correlation integral under ergodicity. Serinko's work is more complete in this context as his approach works for all $r \geq 0$ and does not require any continuity assumptions on the joint distributions $\mu_{\tau,\ldots,d\tau}$. (b) is an old result, and can be found in Maruyama (1949) and Grenander (1950).

See also Cramér and Leadbetter (1967) and Cornfeld *et al.* (1982) for discussions.

If $X(t)$ is a Gaussian process with continuous spectral distribution, it follows from Theorem 3.1 that the embedding method, applied to a single realization of the process, yields the desired limit $\lim_{n\to\infty} C_n(r) = C_{\bar{\mu}}(r)$ for each $d \geq 1$. Moreover, we know from Lemma 2.4 that $\bar{\mu} = \mu_{\tau,2\tau,\ldots,d\tau}$ is a nonsingular Gaussian distribution, and hence $C_{\bar{\mu}}(r) \sim cr^d$ for sufficiently small $r$. This is consistent with the theoretical calculations of Theiler (1991), who observed a range of small $r$ with the correct scaling $C_n(r) \sim cr^d$ for very large $n$ in a continuous power-law model of the form (3.2).

**Remark 3.2:** There are two distinct factors influencing the size of $n$ necessary to observe a correct full-dimensional scaling region $C_n(r) \sim cr^d$ in a continuous-spectrum Gaussian model:

• the *statistical effect* (i.e. the slowing of convergence of $C_n(r)$ to $C_{\bar{\mu}}(r)$ due to the dependence structure *between* the embedded vectors $\boldsymbol{X}_1, \ldots, \boldsymbol{X}_n$ of Theorem 3.1(a)) which is a function of the mixing behaviour of the process. Before sufficient mixing occurs, correlation dimension estimates will tend to approximate the dimension of the local subspace spanned by the left-shifts of a single trajectory over a short time interval; the size of this subspace is related to the degree of path smoothness (see rest of this section). In Gaussian processes, a necessary and sufficient condition for ordinary mixing is that $\lim_{t\to\infty} K(t) = 0$. Uniform mixing conditions (see Denker and Keller (1986) and Serinko (1993b) for examples and applications) are likely necessary to get rates near the optimal case of independent $\boldsymbol{X}_1, \ldots, \boldsymbol{X}_n$. Since mixing in Gaussian processes is determined by the tail behaviour of the autocovariance (hence by the behaviour of $S^*$ near 0) we see that the statistical effect increases as the power at very low frequencies increases. In the case of a power-law spectral density, the critical information about mixing behaviour is determined by the size of the low-frequency cutoff $\lambda_0$ and the choice of density over $(0, \lambda_0)$. This phenomenon has been noted by Theiler (1986, 1991) who recommends modifications to the sample correlation integral (such as not counting pairs of vectors which are close in time) in order to improve convergence. In the case of dependencies which persist over long time periods, we suggest that repeated independent sampling of the vector $\boldsymbol{X} = (X(\tau), \ldots, X(d\tau))$ be substituted (when possible) for the GP algorithm.

• the *geometric effect* induced by the degree of path smoothness (i.e. by the degree of dependence between components *within* each vector $\boldsymbol{X}_j$). The smoothness of Gaussian paths depends on the behaviour of $K(t)$ at $t$ near 0 (i.e. on the tail of $S^*$) so the geometric effect is reduced when sufficient power exists at arbitrarily high frequencies. When paths are very smooth, $X(\tau), \ldots, X(d\tau)$ are strongly statistically correlated for small $\tau$, and the distribution $\mu_{\tau,\ldots,d\tau}$ in $\mathbb{R}^d$ appears low-dimensional from a distance. The correct scaling $C_{\bar{\mu}}(r) \sim cr^d$ is observed only for small $r$, and greater measurement precision and larger $n$ are required for $C_n(r)$ to exhibit correct scaling at small $r$, even assuming independent $\boldsymbol{X}_1, \ldots, \boldsymbol{X}_n$. (In practice this effect may be lessened somewhat by a good choice of the time delay $\tau$.) It is worth mentioning that the existence of a power-law *tail* on the spectral density helps, rather than hinders, in the identification of a stochastic process.

Note that the worst-case scenario for a continuous-spectrum Gaussian process corresponds to a spectral measure with small high-frequency cutoff (very smooth paths) and a lot of power near 0. Such processes, based on observation of a single realization, may be almost indistinguishable from smooth deterministic outputs (at least in terms of dimension) until the sample size becomes enormous.

OP attempted to simulate a process with spectral density (3.2) by first choosing a small frequency increment $\Delta\lambda$ and approximating (3.2) by the *discrete* power-law distribution

$$S^*(\{k\,\Delta\lambda\}) \;=\; C\,(k\,\Delta\lambda)^{-\alpha}(\Delta\lambda) \qquad \text{for } k = 1, 2, 3, \ldots \tag{3.5}$$

where $C$ is a normalizing constant chosen to yield unit variance $K(0) = 1$. They then simulated a process with spectral distribution (3.5) by generating random uniform phases (but noted that Gaussian models could be used as well). The real representation of a Gaussian process with general discrete real spectral distribution $S^*$ is given by

$$X(t) \;=\; \sum_{\lambda_k \in \mathcal{A}^*} c_k\,(A_k \cos(\lambda_k t) + B_k \sin(\lambda_k t)) \tag{3.6}$$

where $\mathcal{A}^*$ is the set of atoms of $S^*$, $c_k^2 = S^*(\{\lambda_k\})$, the $A_k$'s, $B_k$'s are i.i.d. real Gaussian variables with mean 0 and variance 1, and the sum in (3.6) converges in quadratic mean. Substituting (3.5) into (3.6) yields a discrete-spectrum Gaussian power-law process

$$X_\alpha(t) \;=\; C^{1/2}(\Delta\lambda)^{1/2} \sum_{k=1}^{\infty} (k\,\Delta\lambda)^{-\alpha/2}(A_k \cos(k\,\Delta\lambda\,t) + B_k \sin(k\,\Delta\lambda\,t))\,. \tag{3.7}$$

We note that $X_\alpha(t)$ is also *a.s.* a continuous periodic function (not just a limit in quadratic mean; see Kahane (1985) for continuity results on random Fourier series) with period $2\pi/(\Delta\lambda)$, and so (3.7) directly defines random elements of $\mathcal{C}([0,\infty))$. OP then sampled the process over the interval $(0, 2\pi/(\Delta\lambda))$ at equally-spaced timepoints $t_k = k\,\Delta t$, $k = 1, \ldots, m$. For simulation purposes, and to avoid the aliasing of high frequencies, the sum in (3.7) was truncated at $k = m/2$.

**Remark 3.3:** OP did not then directly apply the GP algorithm to the sampled trajectory $X_\alpha(t_1), \ldots, X_\alpha(t_m)$. Rather, they generated 15 independent copies of (3.7) in the above manner, then grouped observations vertically over different realizations to produce "embedded" vectors whose components were independent. The GP algorithm was subsequently applied to this sequence of vectors. This method of embedding is not defensible from any theoretical point of view — the dependence structure between components *within* an embedded vector is precisely what we wish to retain from a time series, as it contains the desired information about the underlying dynamics. (It is the dependence between the vectors that we would like to eliminate.) In the OP case the statistical correlations between the vectors are still so dominating that they override the artificial independence introduced by the manner of grouping observations, and one arrives at the same answer as obtained when applying the GP algorithm to a single realization.

In the following we will assume that the GP algorithm has been applied to a single realization of the process. We will need a few further modifications to the OP setup in order that various convergence issues are not confused. First note that (3.7) is not changed in any fundamental qualitative way by taking $C = \Delta\lambda = 1$ (this amounts to rescaling the period) and we do so throughout the following. Thus our expression for $X_\alpha(t)$ simplifies to:

$$X_\alpha(t) = \sum_{k=1}^{\infty} k^{-\alpha/2}(A_k \cos(kt) + B_k \sin(kt)) . \tag{3.8}$$

The phase space then becomes $\mathcal{C}_{2\pi}$, the set of all continuous functions with period $2\pi$. In this context it is natural to endow $\mathcal{C}_{2\pi}$ with the uniform topology, measuring the distance between two functions via the norm $\| \, \|_{2\pi}$ of (2.4). We also note that this is the topology $\mathcal{C}_{2\pi}$ naturally inherits as a subspace of $C([0,\infty))$ under the usual topology. We further assume that the sum in (3.8) has not been truncated, that $\Delta t$ is a small fixed quantity which is *not* a rational fraction of the period $2\pi$ (to avoid sampling a finite cycle), and that we can let the sample size $m$ grow arbitrarily large. This sch eme mimics the usual setup for the GP algorithm with $\Delta t$ interpreted as the ti me delay; for large $m$ it also implies sampling over several periods of the pro cess. We note that OP were careful to sample only a single period of their simul ation model; this practice has been quite justly criticized by Theiler (1991). Examination of the space-time recurrence diagrams of the OP data (see Provenzale *et al.* (1992)) clearly reveals that virtually all spatially close observations are those also close temporally, indicating that "folding" on the attractor (should one exist) has not yet occurred and hence that the sampling range is insufficient to draw a conclusion. However, it is also important to realize that sampling additional periods (with an appropriate value of $\Delta t$ i.e. not a rational fraction of $2\pi$) would not have changed the estimates of correlation dimension (we will see that $d(\alpha) = 2/(\alpha - 1)$ is the correct asymptotic quantity along a single realization); the value of further sampling here lies in the fact that *the discrete nature of the spectrum would have become apparent to the experimentalist.* This last observation would have (perhaps) prevented the experimentalist from erroneously concluding that the results obtained were applicable to the continuous model (3.2).

Now (3.8) possesses a spectrum with an infinite number of distinct atoms; hence Theorem 2.6 gives $\nu_\infty = \infty$ (there is no distinction between the theoretical values for the discrete-spectrum and continuous-spectrum power-law models). However, since the spectrum in (3.8) is discrete, it follows from Theorem 3.1(b) that the invariant distribution $\mu$ is not ergodic. This lack of ergodicity is actually very easy to see in (3.8) because of its simple periodic nature. Therefore, in order to understand the behaviour of the GP algorithm over a single realization of (3.8), it is necessary to decompose $\mu$ into its ergodic components. Theorem 3.1 (a) can then be applied to each component. Note that each initial condition $x$ (here $x$ is a function $x = x(t) \in \mathcal{C}_{2\pi}$) belongs to the basin of attraction (under the left-shift operators $L_t$) of exactly one ergodic component, say $\mu^x$. Since $C_n(r)$ is computed, for each $d$, by embedding over left-shifts of the *same* function $x$, it follows that, given $x$, $\lim_{n\to\infty} C_n(r) = C_{\bar{\mu}^x}(r)$ $\mu^x$-*a.s.*, where $\bar{\mu}^x = \mu^x_{t_1,t_2,\ldots,t_d}$, the finite-dimensional projection of $\mu^x$. Thus, given $x$, we will have

$$\nu^{(d)}_{\mathrm{GP}}(x) \;=\; \lim_{r\to 0}\lim_{n\to\infty} \frac{\log C_n(r)}{\log r} \;=\; \nu^x_{t_1,\ldots,t_d} \qquad \mu^x\text{-}a.s. \tag{3.9}$$

where $\nu^x_{t_1,\ldots,t_d}$ is the correlation dimension of the conditional distribution $\mu^x_{t_1,\ldots,t_d}$. OP observed $\nu^{(d)}_{\mathrm{GP}}(x) \approx d(\alpha) = 2/(\alpha-1)$, independent of $x$, for $1<\alpha<3$ and sufficiently large $d$. Obviously, from (3.9), we require at least $d \geq d(\alpha)$.

We now show that $d(\alpha)$ is $\mu$-*a.s.* the correlation dimension of each ergodic component $\mu^x$ (interpreted in the sense of (2.4) and (2.5) with $T = 2\pi$; we do not know if $\mu^x$ is time-length invariant for $0 < T < 2\pi$); moreover, it is also $\mu$-*a.s.* the Hausdorff dimension of the topological support $M^x$ of $\mu^x$ in $\mathcal{C}_{2\pi}$. For each $x \in \mathcal{C}_{2\pi}$, define $\psi^x : [0,\infty) \longrightarrow \mathcal{C}_{2\pi}$ and $M^x \subset \mathcal{C}_{2\pi}$ by

$$\psi^x(s) \;=\; L_s(x) \qquad \text{and} \qquad M^x \;=\; \psi^x([0,\infty)) \;=\; \psi^x([0,2\pi])\,.$$

Since

$$\|\psi^x(s) - \psi^x(t)\|_{2\pi} \;=\; \sup_{0\leq u\leq 2\pi} |x(s+u) - x(t+u)|\,, \tag{3.10}$$

the uniform continuity of $x$ shows that $\psi^x$ is continuous, and so the image $\psi^x([0,2\pi])$ is compact, hence closed, in $\mathcal{C}_{2\pi}$ in the uniform topology. This closure property is directly related to the lack of ergodicity in the periodic model; the set $M^x$ of all shifts of $x$ is a *closed invariant set* and clearly not dense in the phase space of the total system.

It is quite easy to deduce the form of the ergodic components $\mu^x$. First note that two shifts $\psi^x(s)$ and $\psi^x(t)$ of $x$ $(s \neq t)$ coincide if and only if $x$ has a period of length $|t-s|$. Since we may restrict consideration to those $x$ for which the shortest period is $2\pi$ (equation (3.8) guarantees this property for $\mu$-almost all $x$), we see that $\psi^x$ has period $2\pi$ but is bijective (i.e. 1–1, onto) between $[0,2\pi)$ and $M^x$. An application of the pointwise ergodic theorem then shows that, for $0 \leq s < t < 2\pi$, $\mu^x$ must satisfy

$$\mu^x(\psi^x([s,t])) \;=\; (t-s)/2\pi \tag{3.11}$$

since the shift operator simply cycles through $M^x$ every $2\pi$ units of time. This may also be seen, using probabilistic methods, by transferring from rectangular coordinates $(A_k, B_k)$ to polar coordinates $(R_k, -\phi_k)$ in (3.8). Then $A_k = R_k\cos\phi_k$, $B_k = -R_k\sin\phi_k$, and (3.8) becomes

$$X_\alpha(t) \;=\; \sum_{k=1}^{\infty} k^{-\alpha/2} R_k \cos(\phi_k + kt)\,. \tag{3.12}$$

It is well-known that the $\phi_k$'s are i.i.d. uniform variables over $[0,2\pi)$, independent of the radii $R_k$'s, which have chi distributions (see Cramér and Leadbetter (1967)). We observe from (3.12) that shifting a realization by $s$ units only involves shifting angles

$$L_s(X_\alpha)(t) \;=\; \sum_{k=1}^{\infty} k^{-\alpha/2} R_k \cos((\phi_k + ks) + kt)\,. \tag{3.13}$$

Since any set of shifted angles $\phi_1+s, \phi_2+2s, \ldots, \phi_k+ks, \ldots$ (with addition performed mod $2\pi$) is just as likely as the set $\phi_1, \phi_2, \ldots$ under the i.i.d. uniform model, we conclude that all shifts of $x$ are equally likely and so $\mu^x$ corresponds to a "lifting" of the uniform distribution from the unit circle $[0, 2\pi)$ to $M^x$. This also yields (3.11).

Now from Kahane (1985), pg. 199, the paths of (3.8) $\mu$-*a.s.* satisfy a Lipschitz (or Hölder) condition almost of order $1/d(\alpha)$ for $1<\alpha<3$ i.e. for $\mu$-almost all $x$ we have

$$\|\psi^x(s) - \psi^x(t)\|_{2\pi} \;=\; \sup_{0\le u\le 2\pi} |x(s+u) - x(t+u)| \;=\; O(|t-s|^{(\alpha-1)/2}\sqrt{\log|t-s|^{-1}})\,. \tag{3.14}$$

A slightly less exact result (a Lipschitz condition of order $\beta$ for every $\beta<1/d(\alpha)$) can be deduced from (3.5) and Theorem (ii), pg. 181, of Cramér and Leadbetter (1967). This less exact result is sufficient for our purposes. Applying standard dimension techniques (see Proposition 2.3 and Theorem 16.2 of Falconer (1990)), it follows from (3.14) (or the discussion following it) that

$$\dim(M^x) \;=\; \dim(\psi^x([0,2\pi])) \;\le\; d(\alpha) \quad \mu\text{-}a.s. \tag{3.15}$$

where $\dim(M^x)$ is the Hausdorff dimension of $M^x$ under the sup metric. Similarly, if $\psi^x(S)$ and $\psi^x(T)$ are i.i.d. observations from $\mu^x$ (equivalently, in view of (3.11), if $S$ and $T$ are i.i.d. observations from the uniform distribution $m$ on $[0, 2\pi)$) it follows from (3.14) and (3.11) that, for each $0<\beta<1/d(\alpha)$, there exist $0<C<\infty$ and $0<c<\infty$ such that

$$P^x([\|\psi^x(S) - \psi^x(T)\|_{2\pi} \le r]) \;\ge\; m \times m(\{(s,t)\,|\,C|t-s|^\beta \le r\}) \sim cr^{1/\beta} \tag{3.16}$$

for small $r$. Hence we must have upper correlation dimension $\nu_{2\pi}^x \le d(\alpha)$ $\mu$-*a.s.* The reverse inequalities can be obtained by applying the potential theoretic methods of Frostman; see Appendix.

The above arguments also explain the type of behaviour we will see in *ergodic* processes before we have sampled far enough along a trajectory; we will tend to obtain dimension estimates of $\psi^x([0,T])$, the local subspace spanned by left-shifts of the trajectory over a short time interval $[0,T]$. A Lipschitz or Hölder condition on the paths, such as (3.14), provides an upper bound on this subspace dimension. As we sample further along the trajectory, the sampled points slowly begin to fill in the process attractor (the closure of the set $\psi^x([0,\infty))$) and dimension estimates begin to approach the dimension of this much larger set.

## 4. The OP Phenomenon in Finite-Dimensional Systems

Our discussion in the previous section shows that we will observe the OP phenomenon in every nonergodic system for which the trapping invariant subspaces (or attractors) of the ergodic components have dimensions strictly smaller than the dimension of the total system. This behaviour occurs whether the system is finite-dimensional (see examples below) or infinite-dimensional (as in the case of discrete-spectrum coloured noise). In ergodic systems with poor or no mixing properties (weakly ergodic systems) we may observe a *transient* OP phenomenon; the sample correlation integral may exhibit a significant region of low-dimensional scaling in the $(n, r)$-plane (corresponding to a local subspace dimension) whereas the correct full-dimensional scaling is obtained at small $r$ for sufficiently large $n$. This behaviour is seen in continuous-spectrum coloured noise as well as in the ergodic versions of the examples below.

**Example 4.1:** Consider a system comprised of two harmonic oscillators, first with commensurate frequencies

$$\frac{d^2y}{dt^2} + \lambda^2 y = 0 \quad \text{and} \quad \frac{d^2z}{dt^2} + 4\lambda^2 z = 0\,, \tag{4.1}$$

with amplitudes $R_1 = R_2 = 1$ and random independent phases $\phi_1$, $\phi_2$. Solving (4.1) in 1st-order form yields $y_1(t) = y(t) = \cos(\lambda t + \phi_1)$, $y_2(t) = y'(t) = -\lambda \sin(\lambda t + \phi_1)$, and $z_1(t) = z(t) = \cos(2\lambda t + \phi_2)$, $z_2(t) = z'(t) = -2\lambda \sin(2\lambda t + \phi_2)$. Each oscillator describes an ellipse in $\mathbb{R}^2$ and has invariant distribution absolutely continuous w.r.t. 1-dimensional Lebesgue measure on the ellipse. Hence the total system has invariant distribution absolutely continuous w.r.t. 2-dimensional Lebesgue measure on a torus, but is easily seen to be nonergodic since each realization $\boldsymbol{x}(t) = (y_1(t), y_2(t), z_1(t), z_2(t))$ is periodic and generates a simple closed curve in $\mathbb{R}^4$. As the support of each ergodic component is 1-dimensional, measurements made on any functional of a single realization will always yield local dimension estimates of 1 or less. Takens' theorem is sometimes said to "fail" here, but in fact the failure is that of the GP algorithm (specifically, of the sampling mechanism), not the theorem. Dimension analysis on *independent* embedded vectors, as discussed at the end of Remark 2.7, will yield the correct answer for sufficiently rich $h$ and small $\tau$. If the frequencies in (4.1) are instead chosen to be *noncommensurate,* the system becomes weakly ergodic and a transient OP phenomenon is observed.

**Example 4.2:** The simple circle map $T(x) = (x+\alpha) \bmod 1$ for $0 \le x \le 1$ is an example of a discrete-time dynamical system which exhibits the OP phenomenon for rational $\alpha$ (the nonergodic case) and a transient OP phenomenon for irrational $\alpha$ (the weakly ergodic case). Note that the uniform distribution on $[0, 1]$ is always an invariant distribution for $T$ (for any choice of $\alpha$). In the case that $\alpha$ is rational, the ergodic components are *discrete* uniform distributions over periodic cycles $(x, x+\alpha, \ldots, x+k\alpha)$ of finite length, and the local dimension is 0. Figure 1(a) shows a plot of $\log C_n(r)$ *v.s.* $\log r$ (embedding dimension $d = 1$) for 1000 observations from a realization of this system with $\alpha = 0.1$. In the case of irrational $\alpha$, the system is ergodic but not mixing.

Figure 1(b) shows the analogous log-log plot for $n = 1000$ and $\alpha = 0.101013....$ The plateau in the graph of (b) represents the transient OP phenomenon. This plateau eventually disappears as the number of iterations increases. Of course, for simulation purposes, we used a rational approximation (in this case 0.101013) to the irrational number $\alpha = 0.101013...$; hence, under computer simulation, (b) will collapse to (a) if iteration is carried too far.

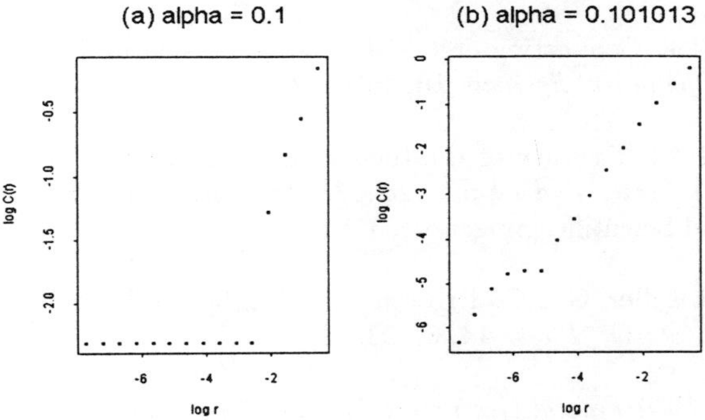


**Figure 1:** Log-log plots of sample correlation integral for the circle map

## Acknowledgements

I am indebted to several people for their contributions to the final version of this paper. My colleague Jiahua Chen participated in many helpful discussions, particularly in resolving (A.19) of the Appendix, and also provided the elegant argument using Vandermonde determinants in Lemma 2.5. Peter Hall suggested the Abel summation formula and the correct rate in (A.19), and Kung-Sik Chan suggested the simple method of proof used in Lemma 2.4. I have also benefitted from conversations with Lenny Smith, Tom Salisbury, and Rege Serinko.

An early draft of this paper was originally presented at the Joint IMS/ASA Statistical Meetings in San Francisco, August 1993. Financial support has been provided by an operating grant from the Natural Sciences and Engineering Research Council of Canada.

## References

Aaronson, J., Burton, R., Dehling, H., Gilat, D., Hill, T., Weiss, B. 1993 Law of large numbers for U-statistics of dependent observations. *preprint.*

Apostol, T. M. 1975 *Mathematical Analysis,* 2nd. Ed. Addison-Wesley: Reading, Mass.

Cramér, H. and Leadbetter, M. 1967 *Stationary and Related Stochastic Processes.* John Wiley & Sons: New York.

Cheng, B. and Tong, H. 1992 On consistent nonparametric order determination and chaos. *J. R. Statist. Soc. B* **54**, 427–449.

Cornfeld, I. P., Fomin, S. V., and Sinai, Ya. G. 1982 *Ergodic Theory.* Springer-Verlag: New York.

Cutler, C. D. 1990 Connecting ergodicity and dimension in dynamical systems. *Ergodic Theory Dynamical Systems* **10**, 451–462.

Cutler, C. D. 1993 A review of the theory and estimation of fractal dimension. In *Nonlinear Time Series and Chaos, Vol. I: Dimension Estimation and Models*, (ed. H. Tong), World Scientific: Singapore, 1–107.

Denker, M. and Keller, G. 1986 Rigorous statistical procedures for data from dynamical systems. *J. Statist. Phys.* **44**, 67–93.

Falconer, K. J. 1990 *Fractal Geometry: Mathematical Foundations and Applications.* John Wiley & Sons: New York.

Frostman, O. 1935 Potentiel d'équilibre et capacité des ensembles avec quelques applications à la théorie des fonctions. *Meddel. Lunds Univ. Mat. Sem.* **3**, 1–118.

Gradshteyn, I. S. and Ryzhik, I. M. 1994 *Table of Integrals, Series, and Products,* 5th Ed. Academic Press: New York.

Grassberger, P. and Procaccia, I. 1983 Characterization of strange attractors. *Phys. Rev. Lett.* **50**, 346–349.

Grenander, U. 1950 Stochastic processes and statistical inference. *Arkiv. Mat.* **1**, 195–227.

Guckenheimer, J. and Holmes, P. 1983 *Nonlinear Oscillations, Dynamical Systems, and Bifurcations of Vector Fields.* Springer-Verlag: New York/Berlin.

Kahane, J.-P. 1985 *Some Random Series of Functions,* 2nd. Ed., Cambridge University Press: Cambridge.

Kaplan, D. T. and Glass, L. 1992 Direct test for determinism in a time series. *Phys. Rev. Lett.* **68**, 427–430.

Kennel, M. B. and Isabelle, S. 1992 Method to distinguish possible chaos from colored noise, and to determine embedding parameters. *Phys. Rev. A* **46**, 3111–3118.

Maruyama, G. 1949 The harmonic analysis of stationary stochastic processes. *Mem. Fac. Sci. Kyusyu Univ.* **A4**, 45–106.

Nering, E. D. 1970 *Linear Algebra and Matrix Theory,* 2nd Ed. John Wiley & Sons: New York.

Osborne, A. R. and Provenzale, A. 1989 Finite correlation dimension for stochastic systems with power-law spectra. *Physica D* **35**, 357–381.

Packard, N. H., Crutchfield, J. P., Farmer, J. D., and Shaw, R. S. 1980 Geometry from a time series. *Phys. Rev. Lett.* **45**, 712–716.

Pesin, Ya. B. 1993 On rigorous mathematical definitions of correlation dimension and generalized spectrum for dimensions. *J. Statist. Phys.* **71**, 529–547.

Provenzale, A., Smith, L. A., Vio, R., and Murante, G. 1992 Distinguishing between low-dimensional dynamics and randomness in measured time series. *Physica D* **58**, 31–49.

Serinko, R. J. 1993a A Glivenko-Cantelli theorem for the Grassberger-Procaccia spatial correlation integral. *preprint.*

Serinko, R. J. 1993b A consistent approach to least squares estimation of correlation dimension in weak Bernoulli dynamical systems. To appear *Ann. Appl. Probab.*

Smith, L. A. 1992 Identification and prediction of low-dimensional dynamics, *Physica D* **58**, 50–76.

Takens, F. 1981 Detecting strange attractors in turbulence. In *Lecture Notes in Mathematics,* No. 898, Springer-Verlag: New York, 366–381.

Taylor, S. J. 1953 The Hausdorff $\alpha$-dimensional measure of Brownian paths in $n$-space. *Proc. Cambridge Philos. Soc.* **49**, 31–39.

Theiler, J. 1986 Spurious dimension from correlation algorithms applied to limited time series data. *Phys. Rev. A* **34**, 2427–2432.

Theiler, J. 1991 Some comments on the correlation dimension of $1/f^{\alpha}$ noise. *Phys.*

*Lett. A* **155**, 480–493.

Theiler, J., Eubank, S., Longtin, A., Galdrikian, B., and Farmer, J. D. 1992 Testing for nonlinearity in time series: the method of surrogate data. *Physica D* **58**, 77–94.

## Appendix

Here we sketch a proof of the reverse inequalities. We employ the potential theory ideas of Frostman (1935); this methodology is described and used in Falconer (1990) to obtain lower bounds on the dimension of Brownian motion paths in $\mathbb{R}^n$ (see also the original paper of Taylor (1953)). We begin by noting that the sup metric imposes a simple structure on $M^x$; specifically, the distance $\|\psi^x(t)-\psi^x(s)\|_{2\pi} = \|\psi^x(t-s)-x\|_{2\pi}$ and depends only on $|t-s|$ mod $2\pi$. Thus $\psi^x(t)$ belongs to the ball $B(\psi^x(s), r)$ if and only if $\psi^x(t-s)$ belongs to $B(x,r)$. This, combined with (3.11) and the translation invariance of the uniform distribution mod $2\pi$, shows that $\mu^x$ assigns identical mass to every ball of radius $r$ i.e. there exists a constant $p(r)$ such that $\mu^x(B(y,r)) = p(r)$ for all $y \in M^x$. This strong uniformity has two important ramifications. First, the *lower pointwise dimension*

$$\sigma^-_{\mu^x}(y) = \liminf_{r\to 0} \frac{\log \mu^x(B(y,r))}{\log r} \tag{A.1}$$

is independent of $y$ and therefore equals a constant $\sigma$. It follows (Cutler (1990)) that $\dim(M^x) = \sigma$. Second, this uniformity implies that $C_{\mu^x}(r) = p(r)$ and so the lower correlation dimension also coincides with $\sigma$ i.e. $\nu^x_{2\pi} = \sigma$. Thus to complete the proof of the reverse inequalities, it suffices to show that $\dim(M^x) \geq d(\alpha)$. Note that this will also prove that $\nu^x_{2\pi}$ exists as a limit (2.3), justifying our use of the same notation $\nu^x_{2\pi}$ for upper and lower correlation dimension.

Given a probability distribution $\nu$ on the Borel sets of a metric space $\mathcal{X}$ with distance $\rho$, the *$\beta$-energy integral* of $\nu$ ($\beta > 0$) is defined to be

$$I_\beta(\nu) = \int_{\mathcal{X}} \int_{\mathcal{X}} \rho(x,y)^{-\beta} \nu(dx)\nu(dy) \tag{A.2}$$

The following lemma is valid in general metric spaces; the proof may be adapted from that of Theorem 4.13(a) of Falconer (1990):

**Lemma A.1.** If $\nu$ is a probability measure on $\mathcal{X}$ and $I_\beta(\nu) < \infty$, then $\dim(\mathcal{X}) \geq \beta$.

Consider now the family of distributions $\mu^x$, $x \in C_{2\pi}$, where $x$ is distributed according to the stationary measure $\mu$ of the process (3.8). The *average $\beta$-energy* of the family is defined to be

$$\begin{aligned} I_\beta = \int_{C_{2\pi}} I_\beta(\mu^x)\, \mu(dx) &= \int_{C_{2\pi}} \int_{M^x} \int_{M^x} \|y-z\|_{2\pi}^{-\beta}\, \mu^x(dy)\, \mu^x(dz)\, \mu(dx) \\ &= \int_{M^x} \int_{M^x} \int_{C_{2\pi}} \|y-z\|_{2\pi}^{-\beta}\, \mu(dx)\, \mu^x(dy)\, \mu^x(dz) \end{aligned} \tag{A.3}$$

where, using the Tonelli theorem, we exchange the order of integration in the last line of (A.3). Note that if $I_\beta$ is finite then $I_\beta(\mu^x)$ must be finite for $\mu$-almost all $x$, yielding $\dim(M^x) \geq \beta$ $\mu$-$a.s.$ via Lemma A.1. Thus to complete our proof it suffices to show that $I_\beta < \infty$ whenever $\beta < d(\alpha)$.

Making the change of variable $y = \psi^x(t)$ and $z = \psi^x(s)$ in the the two outer integrals in the last line of (A.3), we obtain

$$\begin{aligned} I_\beta &= (2\pi)^{-2} \int_0^{2\pi} \int_0^{2\pi} \int_{C_{2\pi}} \|\psi^x(t) - \psi^x(s)\|_{2\pi}^{-\beta} \, \mu(dx)\, dt\, ds \\ &= (2\pi)^{-2} \int_0^{2\pi} \int_0^{2\pi} \int_{C_{2\pi}} \|\psi^x(|t-s| \bmod 2\pi) - x\|_{2\pi}^{-\beta} \, \mu(dx)\, dt\, ds. \end{aligned} \tag{A.4}$$

Setting $h = |t - s| \bmod 2\pi$, the innermost integral in the last line of (A.4) can equivalently be expressed as the expectation

$$E\left( \left( \sup_{0 \leq u < 2\pi} |X_\alpha(u+h) - X_\alpha(u)| \right)^{-\beta} \right) \tag{A.5}$$

where $X_\alpha(t)$ is the process in (3.8). We denote the $h$-increment $X_\alpha(u+h) - X_\alpha(u)$ by $Z_\alpha^h(u)$. From (3.8) it follows that $Z_\alpha^h(u)$ is a stationary mean-zero Gaussian process with autocovariance $K_\alpha^h(|u-w|) = \mathrm{Cov}(Z_\alpha^h(u), Z_\alpha^h(w))$ given by

$$\begin{aligned} K_\alpha^h(|u-w|) &= 2 \sum_{k=1}^{\infty} k^{-\alpha} \, (1 - \cos kh) \cos k|u-w| \\ &= 4 \sum_{k=1}^{\infty} k^{-\alpha} \, \sin^2(kh/2) \cos k|u-w|\,. \end{aligned} \tag{A.6}$$

Now, for any set of timepoints $0 \leq u_1 < \cdots < u_n < 2\pi$, (A.5) is bounded above by

$$E\left( \left( \sup_{u_1,\ldots,u_n} |Z_\alpha^h(u_j)| \right)^{-\beta} \right) \tag{A.7}$$

which in turn is bounded above by

$$n^{\beta/2} \, E\left( \|\boldsymbol{Z}_\alpha^h(\boldsymbol{u})\|_{\mathrm{e}}^{-\beta} \right) \tag{A.8}$$

where $\|\boldsymbol{Z}_\alpha^h(\boldsymbol{u})\|_{\mathrm{e}}$ denotes the usual Euclidean norm of the vector $\boldsymbol{Z}_\alpha^h(\boldsymbol{u}) = (Z_\alpha^h(u_1), \ldots, Z_\alpha^h(u_n))$. It will be sufficient to consider the expectation in (A.8). Now for each $0 < h < 2\pi$ the spectrum of $Z_\alpha^h(t)$ coincides with, or is an infinite subset of, the spectrum of (3.8). Thus the arguments used in Lemma 2.5 show that we may choose $\boldsymbol{u}$ (for arbitrarily large $n$) so that $\boldsymbol{Z}_\alpha^h(\boldsymbol{u})$ is nonsingular for all $0 < h < 2\pi$. In this case the distribution of $\|\boldsymbol{Z}_\alpha^h(\boldsymbol{u})\|_{\mathrm{e}}$ satisfies

$$P([\|\boldsymbol{Z}_\alpha^h(\boldsymbol{u})\|_{\mathrm{e}} \leq z]) = \int_{B(z^2)} \frac{\exp(\frac{-\boldsymbol{v}^T \boldsymbol{K}^{-1} \boldsymbol{v}}{2})}{(2\pi)^{n/2} |\det \boldsymbol{K}|^{1/2}} \, d\boldsymbol{v} \tag{A.9}$$

where $B(z^2) = \{\boldsymbol{v} \mid v_1^2 + \ldots + v_n^2 \leq z^2\}$ and $\boldsymbol{K}$ denotes the $n \times n$ matrix whose $(i,j)^{th}$ entry is given by $K_\alpha^h(|u_i - u_j|)$. For convenience we will initially pretend that the variables $Z_\alpha^h(u_1), \ldots, Z_\alpha^h(u_n)$ are independent (this amounts to choosing independent components as done by OP and discussed earlier in Remark 3.3). In this case we have $\boldsymbol{K} = \sigma_\alpha^2(h)\,\boldsymbol{I}$ where $\boldsymbol{I}$ denotes the identity matrix and $\sigma_\alpha^2(h) = K_\alpha^h(0)$ is the variance of an $h$-increment. Substituting this into (A.9) gives

$$P([\|\boldsymbol{Z}_\alpha^h(\boldsymbol{u})\|_{\mathrm{e}} \leq z]) = \int_{B(z^2)} \frac{\exp(-\sum_{j=1}^n v_j^2/2\sigma_\alpha^2(h))}{(2\pi)^{n/2}(\sigma_\alpha^2(h))^{n/2}}\, dv_1 \ldots dv_n\,. \tag{A.10}$$

Transferring to polar coordinates, with $r^2 = \sum_{j=1}^n v_j^2$ and $dv_1 \ldots dv_n = r^{n-1} dr\, d\theta_1 \ldots d\theta_{n-1}$, it is easy to see that the probability density function of $\|\boldsymbol{Z}_\alpha^h(\boldsymbol{u})\|_{\mathrm{e}}$ is given by

$$f(z) = c\,\sigma_\alpha^2(h)^{-n/2}\, z^{n-1} \exp(-z^2/2\sigma_\alpha^2(h)) \qquad \text{for } z > 0 \tag{A.11}$$

where $c$ depends only on $n$. (This may also be seen by noting that $\|\boldsymbol{Z}_\alpha^h(\boldsymbol{u})\|_{\mathrm{e}}^2/\sigma_\alpha^2(h)$ is chi-square with $n$ degrees of freedom.) It is then straightforward to compute

$$E\left(\|\boldsymbol{Z}_\alpha^h(\boldsymbol{u})\|_{\mathrm{e}}^{-\beta}\right) = c'\,\sigma_\alpha^2(h)^{-\beta/2} \qquad \text{for any } 0 < \beta < n \tag{A.12}$$

where $c'$ does not depend on $h$. This gives us an upper bound for the innermost integral in (A.4). Now from (A.6) we see that $\sigma_\alpha^2(h)$ is bounded away from 0 for $h_0 \leq h \leq \pi$ (for any $h_0 > 0$). Thus the integral in (A.4) is finite (for any $\beta > 0$) over the range $[h_0, \pi]$. We need only concern ourselves with the range $[0, h_0)$. Letting $c(\alpha) = \int_0^\infty \lambda^{-\alpha} \sin^2(\lambda/2)\, d\lambda$ and rewriting $K_\alpha^h(0)$ as a Riemann sum with mesh size $h$, it follows easily that

$$\lim_{h\to 0} \sigma_\alpha^2(h)\, h^{-(\alpha-1)} = c(\alpha) \tag{A.13}$$

and so we may write $\sigma_\alpha^2(h) \approx c(\alpha)h^{\alpha-1}$ for small $h$. Substituting this into (A.12) gives

$$E\left(\|\boldsymbol{Z}_\alpha^h(\boldsymbol{u})\|_{\mathrm{e}}^{-\beta}\right) \approx c''\, h^{-\beta(\alpha-1)/2} \qquad \text{for small } h \text{ and } 0 < \beta < n \tag{A.14}$$

where $c''$ does not depend on $h$. (A.14) is clearly integrable in (A.4) provided $\beta < d(\alpha)$. This completes the proof in the case that we are able to find $n \geq d(\alpha)$ independent $h$-increments.

We now note that the above argument applies directly to (3.8) for $2 \leq \alpha < 3$ because $d(\alpha) \leq 2$ over this range and it is easy to see that $K_\alpha^h(\cdot)$ has a zero in the open interval $(0, \pi)$. Thus there exists $0 < u < \pi$ such that $Z_\alpha^h(u)$ and $Z_\alpha^h(0)$ are independent, validating the above analysis. The range $1 < \alpha < 2$ requires a different argument. We will not have $n \geq d(\alpha)$ independent increments, but rather $n \geq d(\alpha)$ "almost independent" increments because the covariances $K_\alpha^h(u)$ vanish sufficiently quickly as $h \to 0$. We begin by noting that we may write $\boldsymbol{K} = \sigma_\alpha^2(h)\boldsymbol{C}$ where the correlation matrix $\boldsymbol{C}$ has 1's along the diagonal. We then decompose

$$\boldsymbol{K}^{-1} = \sigma_\alpha^2(h)^{-1}\boldsymbol{I} + \sigma_\alpha^2(h)^{-1}(\boldsymbol{C}^{-1} - \boldsymbol{I})\,. \tag{A.15}$$

Substituting the above into (A.9) gives

$$P([\|\boldsymbol{Z}_\alpha^h(\boldsymbol{u})\|_e \le z]) \;=\; \int_{B(z^2)} \phi(\boldsymbol{v})\,\eta(\boldsymbol{v})\,d\boldsymbol{v} \qquad (A.16)$$

where $\phi(\boldsymbol{v})$ is the integrand in (A.10) and $\eta(\boldsymbol{v}) = |\det \boldsymbol{C}|^{-1}\exp((-\boldsymbol{v}^T(\boldsymbol{C}^{-1} - \boldsymbol{I})\boldsymbol{v}) /2\sigma_\alpha^2(h))$ is the correction term due to covariances. Now let $a(h) = |\det \boldsymbol{C}|^{-1}$ and $b(h) = \max_{i,j}|b_{ij}|$ where the $b_{ij}$'s are the elements of the matrix $\boldsymbol{C}^{-1} - \boldsymbol{I}$. It follows that

$$\eta(\boldsymbol{v}) \;\le\; a(h)\exp(nb(h)\textstyle\sum_{j=1}^n v_j^2/\sigma_\alpha^2(h))\,. \qquad (A.17)$$

Now transferring to polar coordinates, an upper bound on the density function of $\|\boldsymbol{Z}_\alpha^h(\boldsymbol{u})\|_e$ is given by the product $f(z)\,g(z)$ where $f(z)$ is the density (A.11) and $g(z)$ is the correction term

$$g(z) \;=\; a(h)\exp(nb(h)z^2/\sigma_\alpha^2(h))\,. \qquad (A.18)$$

We note that the presence of $g(z)$ in the density has no significant effect on an upper bound for $E\left(\|\boldsymbol{Z}_\alpha^h(\boldsymbol{u})\|_e^{-\beta}\right)$ (i.e. we still get (A.12), with a different constant $c'$ which depends on $\boldsymbol{u}$, as an upper bound) provided both $a(h)$ and the ratio $b(h)/\sigma_\alpha^2(h)$ stay uniformly bounded over all $0<h\le\pi$. We now argue that this is so for $1<\alpha<2$. For any small $u_0 > 0$, the autocovariance in (A.6) satisfies

$$|K_\alpha^h(u)| \;=\; O(h^\alpha) \qquad \text{uniformly for } u_0 \le u \le 2\pi - u_0 \text{ and } 1<\alpha<2\,. \qquad (A.19)$$

To see this, we use the Abel summation formula, (9) on pg. 194 of Apostol (1975), to write

$$K_\alpha^h(u) \;=\; -4\sum_{k=1}^{\infty} s_k\,(f(k+1) - f(k)) \qquad (A.20)$$

where $f(k) = k^{-\alpha}\sin^2(kh/2)$ and $s_k = \sum_{j=1}^k \cos ju$. Expanding $f$ in a Taylor series about $y = k$ in each term, we may then write

$$K_\alpha^h(u) \;=\; -4\sum_{k=1}^{\infty} s_k\,(f'(k) + \frac{1}{2}f''(y_k)) \qquad (A.21)$$

where $y_k$ is some point between $k$ and $k+1$. Now $f'(k) = \frac{h}{2}k^{-\alpha}\sin(kh) - (\alpha\sin^2(kh/2)\,k^{-(\alpha+1)})$, and it is easy to show that $|f''(y_k)| \le 6h^2k^{-\alpha}$ by making the substitution $|\sin(yh)| \le yh$ in the expression for $f''$. Thus (A.21) becomes

$$K_\alpha^h(u) \;=\; -2h\sum_{k=1}^{\infty} s_k\,k^{-\alpha}\sin(kh) \;+\; 4\alpha\sum_{k=1}^{\infty} s_k\,k^{-(\alpha+1)}\sin^2(kh/2) \;+\; O(h^2)\,. \qquad (A.22)$$

Now, using **1.342** of Gradshteyn and Ryzhik (1994), we see that $|s_k| \le 2\csc(u/2)$. Also each of the summations in (A.22), with the $s_k$'s ignored, is seen to be $O(h^\alpha)$ by rewriting each as a Riemann sum with mesh size $h$, as done to obtain (A.13). Since $\csc(u/2)$ is bounded for $u_0 \le u \le 2\pi - u_0$, we obtain (A.19) as claimed.

Now, given $1<\alpha<2$ and $n=[d(\alpha)]+1$, let $\boldsymbol{u}=(u_1,\ldots,u_n)$ be such that $\boldsymbol{Z}_\alpha^h(\boldsymbol{u})$ is nonsingular for all $0<h<2\pi$. We consider the behaviour of the correlation matrix $\boldsymbol{C}$ of $\boldsymbol{Z}_\alpha^h(\boldsymbol{u})$. Let $\gamma(h)=\max_{i,j}|\gamma_{ij}|$ where the $\gamma_{ij}$'s are the entries in $\boldsymbol{C}-\boldsymbol{I}$. Note that equivalently $\gamma(h)=\max_{i\neq j}|c_{ij}|$ where $c_{ij}=K_\alpha^h(|u_i-u_j|)/\sigma_\alpha^2(h)$. From (A.13) and (A.19) we see that $\gamma(h)=O(h)$. This shows that $\det\boldsymbol{C}\to 1$ as $h\to 0$, so $a(h)$ is bounded. Moreover, it is easily argued that the quantity $b(h)$ in (A.17) is of the same order as $\gamma(h)$, so we also have $b(h)=O(h)$. Thus the ratio $b(h)/\sigma_\alpha^2(h)=O(h^{2-\alpha})$, which completes the proof.

CHAPTER 3

# ON PREDICTION AND CHAOS IN STOCHASTIC SYSTEMS

Qiwei Yao and Howell Tong
Institute of Mathematics and Statistics
University of Kent
Canterbury, Kent CT2 7NF, U.K.

**Abstract**

We propose a new measure of sensitivity to initial conditions within a stochastic environment and explore its connection with nonlinear prediction and statistical estimation. We use modern statistical developments to construct and illustrate pointwise predictors and predictive intervals/distributions.

## 1. Introduction

It is well known that deterministic chaos is characterized by the sensitivity to initial conditions. However, it is increasingly recognized that a purely deterministic system rarely exists in reality because stochastic noise is ubiquitous; even the computer generation of time series using a purely deterministic chaotic map cannot be free from rounding errors. Accordingly, it is more pertinent to replace the dynamics by the transition probabilities from states to states. Indeed, Chan and Tong (1994) has shown how a deterministic dynamical system which admits a compact attractor can, in a noisy environment, give rise to an ergodic stochastic system. A convenient framework for this stochastic system is the Markov chain over general state space. As a result, nonlinear autoregressive models emerge quite naturally as a realization of this framework for the study of noisy chaos. Within this considerably enlarged stochastic framework, a new notion of sensitivity to initial conditions has to be developed because the distribution of the states should now be more relevant than the positions of the states (in a single realization). This generalized notion should ideally have points of contact with conventional statistical inference, because it is intuitively clear that if we treat the initial value as an unknown parameter, the more information the state variable at a later time point has about the parameter the more sensitively the associated conditional distribution depends on the initial value.

Point predictions are only the first step in any serious study of the subject. The complete picture can only be provided if the predictive distribution is available. It is therefore of substantial practical importance to estimate the interval predictors and the predictive distribution from the observed series and, if possible, to provide indicators of their sensitivity to initial conditions. Being different from the linear case, the nonlinear prediction has three interesting features: (i) the *dependence* of the current position in the state space; (ii) the *sensitivity* to the current state; and (iii) the *non-monotonicity* of the accuracy in multi-step prediction (cf. Yao and Tong 1994).

The rest of the paper is set up as follows. In Section 2, we study chaos in a stochastic environment, discussing noisy chaos and noise amplification. Section 3 develops nonlinear prediction with a view to the estimation, based on the observed time series, of the predictive distribution and measures of initial-value sensitivities. We illustrate our methodology with both simulated data and real data in Section 4. Some technical conditions are relegated to the appendix.

## 2. Stochastic Dynamic System

### *2.1. Noisy Chaos*

It is impossible for us to study nonlinear prediction without touching on *stochastic chaotic systems*, or simply *noisy chaos*. There has been no generally accepted definition of chaos in a stochastic system (or even in a deterministic system for that matter), although the term noisy chaos has appeared in the literature from time to time. It is often implicitly assumed that by a stochastic chaotic system is meant a system with a (deterministically) chaotic skeleton (cf. Tong 1990, and others). Intuitively, this assumption does make sense when the noise is additive and the ratio of the noise to the signal in the system is (very) small. However, it does not apply to non-additive noise systems. Further, it is not always proper even for an additive noise system, because in a stochastic system, the dynamic noise will, by permeating through the system dynamics, interact with the system signal throughout the time evolution. Therefore, a proper description must take account of the effect of the random noise. An extreme case is that if the additive noise tends to be overwhelming, the system would behave like a noise process no matter what the skeleton is.

A discrete-time stochastic dynamical system can be described by the equation

$$X_t = F(X_{t-1}, e_t), \tag{1}$$

for $t \geq 1$, where $X_t$ denotes a state vector in $R^d$, $F$ is a real vector-valued function, and $\{e_t\}$ is a noise process which satisfies the equality $E(e_t|X_0, \ldots, X_{t-1}) = 0$. If the noise is additive, (1) can be written (by an abuse of notation) as follows

$$X_t = F(X_{t-1}) + e_t, \tag{2}$$

To understand the difficulty in giving a general definition of noisy chaos, let us start with deterministic chaos first. It is almost impossible to give a precise mathematical definition of deterministic chaos which encapsulates all that the term implies in the diverse literature. However, it is widely accepted that the sensitive dependence on initial conditions is a typical feature of a (deterministic) chaotic system, and which can be characteristically described in terms of the well-known Lyapunov exponents (cf. Eckmann and Ruelle 1985, Chatterjee and Yilmaz 1992, Berliner 1992, and the references therein). We do not attempt to give a rigorous mathematical definition of chaos for a stochastic system. Instead, *as a working definition, we say that a stochastic dynamic system is chaotic if the (conditional) distribution of the state variable of the system is sensitive to its initial condition.*

The above description is very rudimentary, and it needs further careful exposition. Superficially, it looks similar to the deterministic case. However, in a stochastic system, we would expect that the conditional distribution of $X_m$ given $X_0 = x$ can, under certain conditions, depend sensitively on $x$ for some small or moderate rather than large $m$ because of the accumulation of noise through the time evolution. It would seem unlikely that after a long time, the stochastic system still has a strong memory of its initial conditions. This suggests that asymptotics are unlikely to yield a practically useful characteristic exponent, unless we assume that the different trajectories have the same realization of random noise. Under this assumption, Lyapunov exponents could be defined in a way very much similar to those in deterministic systems, which were initially proposed by Crutchfield *et al.* (1982) and Kifer (1986). (Also see Nychka *et al.* 1992). Obviously, the assumption of the same realization of random noise in different trajectories has an innate drawback in that in practice such trajectories rarely exist.

One way to manifest the sensitivity of the conditional distribution is to use the Kullback-Leibler information. To simplify our discussion, we suppose the system variables are bounded. Let $g_m(y|x)$ denote the conditional density of $X_m$ given $X_0 = x$. We suppose that $g_m(y|x)$ is smooth enough in $x$. For two nearby initial points $x$, $x + \delta \in R^d$, after time $m \geq 1$, the divergence of the conditional distribution of $X_m$ is defined as

$$K_m(x;\delta) = \int \{g_m(y|x+\delta) - g_m(y|x)\} \log\{g_m(y|x+\delta)/g_m(y|x)\} dy. \tag{3}$$

Note that it is well known that the Kullback-Leibler information is invariant under one-one differentiable co-ordinate transformations. Therefore, if $Z_t = \phi(X_t)$ for all $t$, and $\phi(.)$ is a one-one differentiable transformation on $R^d$, then

$$K_m(x;\delta) = K_m^*(\phi(x), \phi(x+\delta) - \phi(x)),$$

where $K_m^*(.;.)$ denotes the divergence of conditonal distribution of $Z_m$, which is defined in the same way as in (3). (Cf. Theorem 2.4.1 of Kullback 1967.)

It is known that for small $\delta$, $K_m(x;\delta)$ has the approximation

$$K_m(x;\delta) = \delta^T I_m(x)\delta + o(||\delta||^2), \tag{4}$$

where

$$I_m(x) = \int \dot{g}_m(y|x)\dot{g}_m^T(y|x)/g_m(y|x)dy, \tag{5}$$

and $\dot{g}_m(y|x)$ denotes $dg_m(y|x)/dx$, $\dot{g}_m^T(y|x)$ denotes its transpose (cf. §2.6 of Kullback 1967). If we treat the initial value $x$ as a parameter vector of the distribution, $I_m(x)$ is the Fisher's information matrix, which represents the information on initial value $X_0 = x$ contained in $X_m$. Roughly speaking, the more information $X_m$ brings, the more sensitively the distribution depends on the initial condition. The converse is also true. Fan, Yao and Tong (1993) has given another measure of sensitivity in the form of an $L_2$ norm. In theory, there is no difficulty in adopting any norm to measure the distance between $g_m(y|x+\delta)$ and $g_m(y|x)$. However, in practice, interpretability may influence the specific choice.

Besides the divergence of distributions, it is also interesting to look at the divergence in some summarizing characteristics, for example the (conditional) means. For $x \in R^d$ and $m \geq 1$, let $F_m(x) = E(X_m|X_0 = x)$. Then for $\delta \in R^d$,

$$F_m(x+\delta) - F_m(x) = \dot{F}_m(x)\delta + o(||\delta||), \tag{6}$$

where $\dot{F}_m(x)$ denotes $dF_m(x)/dx^T$. For the system with additive noise, $F_1(x) = F(x)$, and it follows from (2) that

$$F_m(x) = E\{F(X_{m-1})|X_0 = x\} = E\{F(F(X_{m-2}) + e_{m-1})|X_0 = x\}$$

$$= E\{\ F(\ \ldots\ (\ F(x)\ +\ e_1\ )\ +\ \ldots\ +\ e_{m-1})\ |\ X_0 = x\ \}.$$

By the chain rule of matrix differential, $\dot{F}_m(x)$ can be expressed as

$$\dot{F}_m(x) = E\{\prod_{k=1}^{m} \dot{F}(X_{k-1})\ |\ X_0 = x\}. \tag{7}$$

Roughly speaking, assuming that all the factors on the RHS of (7) are of comparable size, it seems plausible that $\dot{F}_m(x)$ grows (or decays) exponentially with $m$. Let $\nu_m^2(x)$ denote the largest eigenvalue of $\{\dot{F}_m(x)\}^T\dot{F}_m(x)$. It follows from (6) that

$$||F_m(x+\delta) - F_m(x)|| \leq |\nu_m(x)|\,||\delta|| + o(||\delta||),$$

which indicates that the conditional expectation $F_m(x)$ depends on $x$ sensitively when $|\nu_m(x)|$ is large. Since, in practice, the observations will almost certainly be subject to measurement or rounding errors, it seems necessary to take account of this divergence in $m$-step prediction. However, in the context of prediction, the task is usually to predict one component of $X_t$ instead of the whole vector $X_t$. Hence, the above approximation is rather rough. Let $Y_t$ denote the first component of $X_t$. It follows from (2) that

$$Y_t = f(X_{t-1}) + \epsilon_t,$$

where $f(\cdot)$, and $\epsilon_t$ denote respectively the first component of $F(\cdot)$ and the first component of $e_t$. For $m \geq 1$, and $x \in R^d$, let $f_m(x) = E(Y_m|X_0 = x)$. Obviously, $f_1(x) = f(x)$. Then from (6), we have

$$f_m(x+\delta) - f_m(x) = \delta^T \lambda_m(x) + o(\| \delta \|), \tag{8}$$

where $\lambda_m(x) = df_m(x)/dx$. We call $\lambda_m(.)$ the $m$-step *Lyapunov-like index*, or simply the $m$-LI (Yao and Tong 1994). When $d = 1$,

$$\lambda_m(x) = \nu_m(x) = E\{\prod_{k=1}^{m} \frac{d}{dx} f(X_{k-1}) \mid X_0 = x\} = E\{\prod_{k=1}^{m} \lambda_1(X_{k-1}) \mid X_0 = x\}. \tag{9}$$

We will see in the Section 3 that the $m$-LI plays an important role in the pointwise prediction.

So far we have derived some indices which describe the sensitive dependence of the conditional distributions on the initial conditions. However, this falls short of a mathematical definition of noisy chaos. The essential barrier lies with the fact that the 'clear' cutoff between deterministic chaotic systems and deterministic non-chaotic systems becomes vague after stochastic noise comes into play in the systems. To see this, let us consider the following two models

$$\begin{aligned} \text{Model I}: &\quad Y_t = 0.230 Y_{t-1}(16 - Y_{t-1}) + 0.4\epsilon_t; \\ \text{Model II}: &\quad Y_t = 0.222 Y_{t-1}(16 - Y_{t-1}) + 0.4\epsilon_t, \end{aligned} \tag{10}$$

where $\{\epsilon_t\}$ is a sequence of independent random variables with the standard normal distribution truncated in the interval $[-12, 12]$. It is known that the skeleton of Model I is chaotic (cf. Hall and Wolff 1993, for example). However the skeleton of Model II is not chaotic, but is a limit cycle with period 8. The scatter plots of 3000 data points generated from each of the above models are displayed in Figures 1a and 1b, which show that the difference between these two models is more in quantity than in quality. The random noise masks the distinction in the structures of the skeletons.

### *2.2. Noise Amplification*

Another intersting feature of a nonlinear system is noise amplification. We measure the amplification of noise by comparing the conditional variance of the system variables $\{X_t\}$ (given the initial conditions $X_0$) with the variance of the innovations $\{e_t\}$. We will see that the amplification of noise varies with the initial values, and is not necessarily monotonic in time evolution. In fact, the sensitive dependence on the initial values and the noise amplification are related to each other, and both of them are dictated by some functions of the derivatives of $F(.)$. In this sense, we say that a small noise is expected to be amplified rapidly through the dynamics if the system is chaotic.

Deissler and Farmer (1991) studied the noise amplification in a different way. They considered the distance between the state variables in a purely deterministic

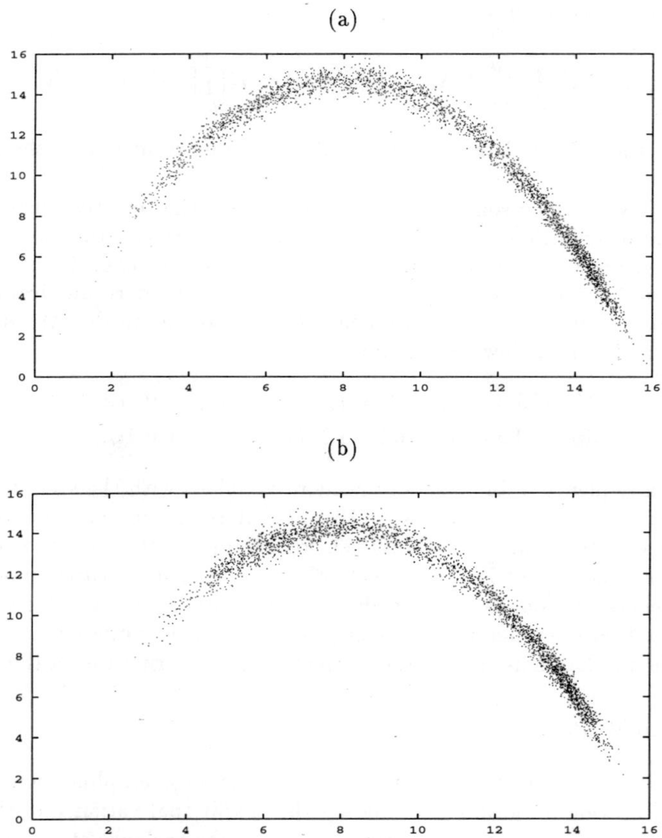


Figure 1: The scatter plots of $Y_{t+1}$ against $Y_t$ for (a) Model I; (b) Model II.

system and the counterparts in the system perturbed by additive system noise. This approach seems improper in the statistical context, because the underlying deterministic skeleton is unknown.

To highlight how nonlinear dynamics amplify noise without going through too much technical detail, we restrict our discussion here to the one-dimensional system with additive noise. Suppose that the process begins at the initial value $Y_0 = x \in R$, and for $t \geq 1$,

$$Y_t = f(Y_{t-1}) + \epsilon_t,$$

where $\{\epsilon_t, t \geq 1\}$ is a noise process with mean value 0 and variance $\sigma_0^2$. Suppose that $\epsilon_t$, is distributed on a bounded set which is independent of $t$. Then for $\sigma_m^2(x) \equiv \mathrm{Var}(Y_m | Y_0 = x)$, it can be proved that as $\sigma_0 \to 0$,

$$\sigma_m^2(x) = \sigma_0^2 \, \mu_m(x) \, (1 + o(1)), \tag{11}$$

where

$$\mu_m(x) \;=\; 1 \;+\; \sum_{j=1}^{m-1} \left\{ \prod_{k=j}^{m-1} \dot{f}[f^{(k)}(x)] \right\}^2 . \tag{12}$$

(Cf. Yao and Tong 1994.) It is easy to see that if the absolute value of $\dot{f}(x)$ is greater than 1 for a large range of values of $x$, $\mu_m(x)$ can be very large for moderate (and even small) $m$. The rapid increase of $\sigma_m^2(x)$ with respect to $m$ is a manifestation of noise amplification. On the other hand, (12) implies that

$$\mu_{m+1}(x) = 1 + \mu_m(x)\{\dot{f}[f^{(m)}(x)]\}^2.$$

Thus, $\mu_{m+1}(x) < \mu_m(x)$ if $\{\dot{f}[f^{(m)}(x)]\}^2 < 1 - 1/\mu_m(x)$. By (11), it is possible that for such $x$ and $m$, $\sigma_{m+1}^2(x) < \sigma_m^2(x)$. This suggests that from the same initial value, the error of an $(m+1)$-step ahead prediction could be smaller than that of an $m$-step ahead prediction in some cases (cf. Tong 1990, Yao and Tong 1994).

## 3. Nonlinear Prediction

By using the ideas developed in Section 2, we study the prediction of nonlinear time series. From (11), we expect that the 'error bounds' of the prediction will vary with the initial value. This is a typical feature of nonlinear (but not necessarily chaotic) model. If the model is stochastically chaotic, (12) indicates that a small noise could be amplified quickly when the system starts at some initial values, which means that the $m$-step prediction based on these initial values could be unreliable even for small $m$. Further, when the system is chaotic, a small change in the initial value $x$ would lead to considerable divergence in the states at time $m$ (cf. (8) and (9)). In this case, it is worth our while to take account of the error in prediction because there is always some measurement error in the initial value.

Since we do not assume any specific form of the model, we choose as our technical tool the nonparametric kernel regression method based on locally linear fit (or simply, locally linear regression, cf. Fan 1992) to estimate both the prediction functions and their derivatives (i. e. the Lyapunov-like indices) simultaneously. However, it does not mean that our results only hold for these particular estimates. In specific practical applications, parametric (nonlinear) models would be more appealing provided that they could be properly justified. Our results can be easily extended to these cases.

### *3.1. Model*

Suppose that $\{Y_t,\ -\infty < t < \infty\}$ is a one-dimensional strictly stationary time series, which has the property that given $\{Y_i,\ i \le t\}$, the conditional distribution of $Y_{t+1}$ depends on $\{Y_i,\ i \le t\}$ only through $X_t$, where $X_t = (Y_t, Y_{t-1}, \ldots, Y_{t-d+1})^T$. Given the observations $\{Y_t,\ -d+1 < t \le n\}$, we shall predict the random variables $Y_{n+m}$ for $m = 1, 2, \ldots$. In fact, the time series model can be considered a special case of a stochastic dynamical system. To see this, let $f(x) = E(Y_1 | X_0 = x)$. Then $Y_t$ can be expressed as

$$Y_t = f(X_{t-1}) + \epsilon_t, \tag{13}$$

where $\epsilon_t = Y_t - f(X_{t-1})$. Define $F(X_{t-1}) = (f(X_{t-1}), Y_{t-1}, \ldots, Y_{t-d+1})^T$, $e_t = (\epsilon_t, 0, \ldots, 0)^T$. Then equation (2) holds. In what follows, the time series model is said to be chaotic if the corresponding stochastic dynamic system is chaotic.

To study the $m$-step prediction, we define

$$f_m(x) = E(\, Y_m \mid X_0 = x\,),$$

for $x \in R^d$ and $m \ge 1$. It follows from (13) that for all $t$, $Y_{t+m}$ can be expressed as

$$Y_{t+m} = f_m(X_t) + \epsilon_{t+m}^{(m)},$$

with

$$E\{\, \epsilon_{t+m}^{(m)} \mid Y_k,\ k \le t\,\} = 0, \qquad \text{a.s.} \tag{14}$$

### *3.2. Point Predictors*

It is easy to see from (14) that the (theoretical) least squares predictor of $Y_{n+m}$ based on $\{Y_t, t \le n\}$ is $f_m(X_n)$, which only depends on the latest vector $X_n = (Y_n, \ldots, Y_{n-d+1})^T$. In what follows, an estimator of the function $f_m(x)$ is constructed by using the locally linear regression method, which also produces an estimator of the $m$-LI $\lambda_m(x) \equiv df_m(x)/dx$. The idea of the locally linear regression is very simple: for a small shift $\delta \in R^d$, the equality (8) holds. Hence, based on the observations $(Y_{-d}, Y_{-d+1}, ..., Y_n)$, the estimation problem can be described as a weighted least-squares problem, namely finding $f_m$ and $\lambda_m$ to minimize

$$\sum_{t=1}^{n-m} \left\{ Y_{t+m} - f_m(x) - \lambda_m^T(x)(X_t - x) \right\}^2 K\left(\frac{X_t - x}{h}\right), \tag{15}$$

where $K(\cdot)$ is a probability density function on $R^d$, and $h = h(n)$ is a bandwidth. Simple calculation yields

$$\hat{f}_m(x) = \{T_0(x) - S_1^T(x)S_2^{-1}(x)T_1(x)\}/\{S_0(x) - S_1^T(x)S_2^{-1}(x)S_1(x) + h^2\}, \quad (16)$$

$$\hat{\lambda}_m(x) = \{S_2(x) - S_1(x)S_1^T(x)/S_0(x)\}^{-1}\{S_1(x)T_0(x)/S_0(x) - T_1(x)\}, \quad (17)$$

where

$$S_0(x) = \frac{1}{n-m}\sum_{t=1}^{n-m} K\left(\frac{X_t - x}{h}\right), \quad S_1(x) = \frac{1}{n-m}\sum_{t=1}^{n-m}(x - X_t)K\left(\frac{X_t - x}{h}\right),$$

$$S_2(x) = \frac{1}{n-m}\sum_{t=1}^{n-m}(x - X_t)K\left(\frac{X_t - x}{h}\right)(x - X_t)^T, \quad (18)$$

and

$$T_0(x) = \frac{1}{n-m}\sum_{t=1}^{n-m} Y_{t+m}K\left(\frac{X_t - x}{h}\right), \quad T_1(x) = \frac{1}{n-m}\sum_{t=1}^{n-m}(x - X_t)Y_{t+m}K\left(\frac{X_t - x}{h}\right). \quad (19)$$

For technical reasons, we add $h^2$ into the denominator in (16), which has little effect for large $n$.

**Theorem 1.** Assume that conditions (A1) – (A7) hold for some $m \geq 1$, which are listed in Appendix.

(i) For $x \in \{p(x) > 0\}$ and $\delta \in R^d$

$$\lim_{n\to\infty} E\{[Y_{n+m} - \hat{f}_m(x)]^2 | X_n = x + \delta\} = \sigma_m^2(x+\delta) + \{\delta^T\lambda_m(x)\}^2 + R_m, \quad \text{a.s.}, \quad (20)$$

where $R_m = o(\| \delta \|^2)$ as $\| \delta \| \to 0$, $\lambda_m(x) = df_m(x)/dx^T$ is the $m$-LI, and $\sigma_m^2(x) = \mathrm{Var}(Y_m | X_0 = x)$.

(ii) For $x \in \{p(x) > 0\}$, as $n \to \infty$, $\hat{\lambda}_m(x)$ converges to $\lambda_m(x)$ in probability.

For the proof, we can easily adapt the arguments in Yao and Tong (1994) which were based on different assumptions on the mixing condition. The first part of Theorem 1 shows that the mean-squared error of the predictor $\hat{f}_m$ at the initial value $x$, which has a small shift from the true but unobservable value $X_n = x + \delta$, can be decomposed into two parts: (i) the conditional variance; (ii) the error due to the small shift at the initial value which is related to the $m$-LI. When $\delta = 0$, i. e. $X_n$ is fully known, equation (20) becomes

$$\lim_{n\to\infty} E\{ [Y_{n+m} - \hat{f}_m(x)]^2 \,|\, X_n = x \} = \sigma_m^2(x) \quad \text{a.s.},$$

which shows that the accuracy of the prediction in a nonlinear (but not necessarily chaotic) model does depend on the initial value $x$, which is strikingly different from the case of a linear prediction. When the measurement error $\delta$ is small but not zero, such as rounding errors in measurement etc., usually the right hand side of (20) is dominated by the conditional variance $\sigma_m^2(x+\delta) = \sigma_m^2(x) + O(\| \delta \|)$. However, for a chaotic system, the $m$-LI $\lambda_m(x)$ can be very large for some values of $x$ (cf. (7) and (8)), in which case the term $\{\delta^T \lambda_m(x)\}^2$ can no longer be ignored. In this sense, we say that the $m$-step prediction is sensitive to the initial values when the model is chaotic. In fact, the asymptotic decomposition (20) dose not depend on the special choice of $\hat{f}_m$. It holds for any estimator which converges to $f_m$ in mean square. Our preference for the locally linear regression stems mainly from the fact that it offers a natural and convenient estimator for $\lambda_m$ by virtue of the weighted least-squares formulation around (15).

In (13), the noise term $\epsilon_t$ is not necessarily homogeneous as indicated in the second expression in (9). However, if it is, $\sigma_1^2(x) \equiv \sigma_1^2$ is a constant. In this case, the variation of the asymptotic mean-squared prediction error is dictated by $\lambda_1(x)$.

Theorem 1 (ii) says that $\hat{\lambda}_m$ is a weakly consistent estimator of the $m$-LI $\lambda_m$. In fact, the estimate of $\lambda_m$ can be improved by using the locally quadratic regression instead of the locally linear regression (cf. Fan et. al. 1993). We use the latter for the simplicity in calculation.

To use (20) in practice, we need to estimate the conditional variance $\sigma_m^2(x)$. In principle, we can use the locally linear regression method to estimate the second conditional moment $E(Y_m^2|X_0 = x)$ by

$$\hat{\zeta}_m(x) = \{V_0(x) - S_1^T(x)S_2^{-1}(x)V_1(x)\}/\{S_0(x) - S_1^T(x)S_2^{-1}(x)S_1(x)\},$$

where $S_k(\cdot)$, $k = 0,1,2$, are as given in (18), and $V_k(\cdot)$, $k = 0,1$, are defined in the same way as $T_k(\cdot)$ with $Y_{t+m}^2$ replacing $Y_{t+m}$ (cf. (19)). Now, we get an estimator for $\sigma_m^2$,

$$\hat{\sigma}_m^2(x) = \hat{\zeta}_m(x) - [\hat{f}_m(x)]^2, \tag{21}$$

where $\hat{f}_m$ is given in (16). However, any smooth regression method would suggest using different bandwidths for the first and second conditional moments. In practice, for the sake of convenience, we tend to adopt the same bandwidth whilst bearing in mind the possibility of misleading results sometimes (cf. Yao and Tong 1994). Note that the positivity of $\hat{\sigma}_m^2(\cdot)$ cannot always be guaranteed even if the same bandwidth is used in estimating the first and second conditional moments.

The discussion in Section 2.2 offers us a tentative way to estimate a 'profile' of $\sigma_m^2(x)$ when the noise terms are small. In the case $d = 1$, it is easy to see from (11) that the variation of $\sigma_m(x)$ is dominated by the variation of the functions $\mu_m(x)$. Equation (12) suggests the following estimator for $\mu_m$,

$$\hat{\mu}_m(x) = 1 + \sum_{j=1}^{m-1} \left\{ \prod_{k=j}^{m-1} \hat{\lambda}_1[\hat{f}_k(x)] \right\}^2,$$

where $\hat{f}_k$ and $\hat{\lambda}_1$ are given in (16) and (17). Simulations show that this estimator is quite good in small-noise experiments (cf. Yao and Tong 1994).

### *3.3. Interval Predictors*

In a stochastic system, an interval predictor is much more relevant than a point predictor, especially in the case of a relatively large noise. A natural way to construct a predictive interval is to estimate the conditional percentiles of $Y_m$ given $X_0$. Specifically, for $\alpha \in [0,1]$, the $100\alpha$-th conditional percentiles of $Y_m$ given $X_0 = x \in R^d$ is defined as

$$\xi_{\alpha,m}(x) \;=\; \arg\min_{|a|<\infty} E\{\, R_\alpha(Y_m - a) \mid X_0 = x \,\},$$

where the loss function

$$R_\alpha(y) = \begin{cases} (1-\alpha)|y| & y \le 0, \\ \alpha|y| & y > 0. \end{cases} \tag{22}$$

It is well known that the relation $\alpha = P\{Y_m \le \xi_{\alpha,m}(x) | X_0 = x\}$ holds. Therefore, given $\{Y_t,\ t \le n\}$, $Y_{n+m}$ will be in the interval $[\xi_{\alpha/2,m}(X_n),\ \xi_{1-\alpha/2,m}(X_n)]$ with probability $1-\alpha$. In fact, the conditional distribution of $Y_{n+m}$ given $X_n$ is determined by the values of $\xi_{\alpha,m}(X_n)$ for $0 \le \alpha \le 1$.

Similar to Section 3.2, we use the locally linear regression to estimate $\xi_{\alpha,m}(.)$ as well as its derivative $\dot{\xi}_{\alpha,m}(.) \in R^d$. More precisely, we use the estimators $\hat{\xi}_{\alpha,m}(x) = \hat{a}$ and $\hat{\dot{\xi}}_{\alpha,m}(x) = \hat{b}$, by setting $(\hat{a},\ \hat{b})$ as the minimizer of the function (with respect to a and b respectively)

$$\sum_{t=1}^{n-m} R_\alpha\{Y_{t+m} - a - b^T(X_t - x)\} K\left(\frac{X_t - x}{h}\right), \tag{23}$$

where $K(.)$ is a probability density function on $R^d$, and $h = h(n)$ is a bandwidth. Unlike (15), (23) does not have an explicit solution for $(\hat{a},\ \hat{b})$. Moreover, since $R_\alpha(y)$ is not differentiable at $y = 0$, either a smooth approximation of $R_\alpha(.)$ or a more complicated software development seems necessary in order to compute the estimates numerically (cf. Bloomfield and Steiger 1983).

An alternative approach is to change the loss function (22) to a quadratic function

$$Q_\omega(y) = \begin{cases} (1-\omega)y^2 & y \le 0, \\ \omega y^2 & y > 0, \end{cases}$$

for $\omega \in [0,1]$, the 100 $\omega$-th conditional expectile of $Y_m$ is defined as

$$\tau_{\omega,m}(x) \;=\; \arg\min_{|a|<\infty}\; E\{\, Q_\omega(Y_m - a) \mid X_0 = x \,\},$$

(cf. Neway and Powell 1987). Obviously, in the case $\omega = \frac{1}{2}$, this definition reduces to the conditional mean $E(Y|X = x)$. Since $Q_\omega(.)$ has a continuous first derivative, $\tau_{\omega,m}(x)$ satisfies the equation

$$E\{\, L_\omega(Y_m - \tau_{\omega,m}(x)) \mid X_0 = x \,\} = 0,$$

where

$$L_\omega(x) = \begin{cases} (1-\omega)y & y \le 0, \\ \omega y & y > 0. \end{cases} \tag{24}$$

Consequently, we have

$$\omega = \frac{E\{\ |Y_m - \tau_{\omega,m}(x)|;\ \ Y_m \le \tau_{\omega,m}(x) \mid X_0 = x\ \}}{E\{\ |Y_m - \tau_{\omega,m}(x)| \mid X_0 = x\ \}}, \tag{25}$$

where $E\{X;\ A\}$ denotes $E\{XI_A\}$, and $I_A$ is the indicator function of the set $A$. Notice that for $\xi_{\alpha,m}(.)$, we can rewrite the relation

$$\alpha = \frac{E\{\ 1;\ \ Y_m \le \xi_{\alpha,m}(x) \mid X_0 = x\ \}}{E\{\ 1\ \mid X_0 = x\ \}}. \tag{26}$$

Comparing the above two expressions, we can see that given $X_0 = x$, the percentile $\xi_{\alpha,m}(x)$ specifies the position below which $100\alpha\%$ of the (probability) mass of $Y_m$ lies; while the expectile $\tau_{\omega,m}(x)$ determines, again given $X_0 = x$, the point such that $100\omega\%$ of the mean absolute distance between it and $Y_m$ comes from the mass below it. Based on this interpretation, $\tau_{\omega,m}(x)$ can also be used to construct a predictive interval: given $\{Y_t,\ t \le n\}$, predict $Y_m$ to lie in the interval $[\tau_{\omega/2,m}(X_n),\ \tau_{1-\omega/2,m}(X_n)]$ with $100(1-\omega)\%$ 'coverage'.

To estimate $\tau_{\omega,m}(.)$, we minimize the function

$$\sum_{t=1}^{n-m} Q_\omega\{Y_{t+m} - a - b^T(X_t - x)\}\, K\left(\frac{X_t - x}{h}\right),$$

and define the estimators $\hat{\tau}_{\omega,m}(x) = \hat{a}, \hat{\dot{\tau}}_{\omega,m}(x) = \hat{b}$. It is easy to see that $\{\hat{\tau}_{\omega,m}(x), \hat{\dot{\tau}}_{\omega,m}(x)\}$ satisfies the following equation

$$\begin{cases} \sum_{t=1}^{n-m} L_\omega\{Y_{t+m} - \hat{\tau}_{\omega,m}(x) - (X_t - x)^T\hat{\dot{\tau}}_{\omega,m}(x)\}K\left(\frac{X_t - x}{h}\right) = 0, \\ \sum_{t=1}^{n-m}(X_t - x)L_\omega\{Y_{t+m} - \hat{\tau}_{\omega,m}(x) - (X_t - x)^T\hat{\dot{\tau}}_{\omega,m}(x)\}K\left(\frac{X_t - x}{h}\right) = 0. \end{cases}$$

Here, $L_\omega(.)$ is the piecewise linear function defined by (24). Based on this relation, it is easy to construct a fast iterative algorithm to computing $\{\hat{\tau}_{\omega,m}(x),\ \hat{\dot{\tau}}_{\omega,m}(x)\}$ (cf. Yao and Tong 1992). Although a predictive interval based on conditional expectiles is convenient to compute, it does not have the conventional probability interpretation in general. However, if we think that the conditional expectation is a good point predictor, $[\tau_{\omega/2,m}(X_n),\ \tau_{1-\omega/2,m}(X_n)]$ could be considered a reasonable interval predictor extended from the conditional expectation. Yao and Tong (1992) has pointed out that, in a special case, the above asymmetric least squares approach can be used to estimate conditional percentiles directly.

**Theorem 2.** Assume that conditions (A3) – (A8) listed in Appendix hold for some $m \ge 1$.

(i) For $x \in \{p(x) > 0\}$,

$$\sqrt{nh^d}\{\hat{\xi}_{\alpha,m}(x) - \xi_{\alpha,m}(x) - h^2\mu_1\} \xrightarrow{d} N(0, \sigma_1^2),$$

$$\sqrt{nh^{d+2}}\{\hat{\dot{\xi}}_{\alpha,m}(x) - \dot{\xi}_{\alpha,m}(x) - h\mu_2\} \xrightarrow{d} N(0, \Sigma_2),$$

where

$$\mu_1 = \frac{1}{2}\sigma_0^2 \mathrm{tr}\{\ddot{\xi}_{\alpha,m}(x)\} + o(1), \qquad \mu_2 = \frac{1}{2\sigma_0^2}\int uu^T\ddot{\xi}_{\alpha,m}(x)uK(u)\,du + o(1),$$

$$\sigma_1^2 = \frac{\alpha(1-\alpha)\int K^2(u)du}{p(x)[g_m(\xi_{\alpha,m}(x)|x)]^2}, \qquad \Sigma_2 = \frac{\alpha(1-\alpha)\int uu^TK^2(u)du}{p(x)\sigma_0^2[g_m(\xi_{\alpha,m}(x)|x)]^2}.$$

(ii) For $x \in \{p(x) > 0\}$,

$$\sqrt{nh^d}\{\hat{\tau}_{\omega,m}(x) - \tau_{\omega,m}(x) - h^2\mu_3\} \xrightarrow{d} N(0, \sigma_3^2),$$

$$\sqrt{nh^{d+2}}\{\hat{\dot{\tau}}_{\omega,m}(x) - \dot{\tau}_{\omega,m}(x) - h\mu_4\} \xrightarrow{d} N(0, \Sigma_4),$$

where

$$\mu_3 = \frac{1}{2}\sigma_0^2 \mathrm{tr}\{\ddot{\tau}_{\omega,m}(x)\} + o(1), \qquad \mu_4 = \frac{1}{2\sigma_0^2}\int uu^T\ddot{\tau}_{\omega,m}(x)uK(u)\,du + o(1),$$

$$\sigma_3^2 = \frac{\int K^2(u)du\mathrm{Var}\{\dot{Q}_\omega(Y_m - \tau_{\omega,m}(x))|X_0 = x\}}{p(x)\gamma^2},$$

$$\Sigma_4 = \frac{\int uu^TK^2(u)du\mathrm{Var}\{\dot{Q}_\omega(Y_m - \tau_{\omega,m}(x))|X_0 = x\}}{p(x)\sigma_0^2\gamma^2},$$

and $\gamma = 2\omega P\{Y_m \le \tau_{\omega,m}(x)|X_0 = x\} + 2(1-\omega)P\{Y_m > \tau_{\omega,m}(x)|X_0 = x\}$.

Using the convexity lemma (cf. Pollard 1991), Yao and Tong (1992) proved Theorem 2 (ii) in the special case $d = 1$. The multidimensional case is technically more involved, but contains no fundamentally new ideas for the current version. Theorem 2 (i) can be proved in a similar way (also see Fan, Hu and Truong 1992).

Theorem 2 gives the asymptotic normality of the the estimators for the conditional percentiles, expectiles and their derivatives. Notice that $\hat{\tau}_{1/2,m}(x) = \hat{f}_m(x)$ and $\hat{\dot{\tau}}_{1/2,m}(x) = \hat{\lambda}_m(x)$. Therefore, Theorem 2 (ii) also includes the asymptotic normality of the point estimators as a special case. As shown in the theorem, the 'asymptotic bias' is of the order of $h^2$ for the estimators $\hat{\xi}_{\alpha,m}$ and $\hat{\tau}_{\omega,m}$, and order of $h$ for the estimators of their derivatives; they come from the error in the local approximation of the underlying curve by a linear function. A locally quadratic fit will improve the estimation for the derivatives (cf. Fan et. al. 1993). However, it creates further complications in practical implementation.

We use the following two kinds of intervals to predict $Y_{n+m}$ from $\{Y_t,\, t \le n\}$

$$[\hat{\xi}_{\alpha/2,m}(X_n),\;\; \hat{\xi}_{1-\alpha/2,m}(X_n)], \tag{27}$$

$$[\hat{\tau}_{\omega/2,m}(X_n),\;\; \hat{\tau}_{1-\omega/2,m}(X_n)]. \tag{28}$$

The interval (27) is based on the conditional percentiles, which has the conventional probability interpretation. The interval (28) does not have any probability meaning. Relations (25) and (26) show that the interval (28) is constructed in the same way as (27) except that we use the mean distance instead of the probability mass.

From Theorem 1, we have learned that in nonlinear prediction, we should take account of two kinds of error: the error caused by the stochastic noise, which varies over the state space, and the error caused by a small shift in the initial value. In the context of interval prediction, the first kind of error can be tentatively represented by the width of the interval. To monitor the second kind of error, i.e. the sensitivity of the predictive intervals, we can use the estimates of the derivatives of the conditional percentiles or expectiles. The estimates presented in the next subsection also offer measures for the sensitivity.

### *3.4. Estimates of $I_m(x)$*

To estimate the Fisher information $I_m(x)$ as given in (5), we use the method proposed by Fan, Yao and Tong (1993). For simplicity, let us discuss the first order case, i.e. $d = 1$, noting that the generalisation to the higher order cases is straightforward. Notice that

$$I_m(x) = 4 \int \left\{ \frac{d\sqrt{g_m(y|x)}}{dx} \right\}^2 dy.$$

We first construct the estimators for $\sqrt{g_m(y|x)}$ and its derivative first. Let $q_m(x,y)$ denote $\sqrt{g_m(y|x)}$.

For given bandwidths $h_1$ and $h_2$, let

$$C_m(X_i, Y_i) = \#\{(X_t, Y_t), 1 \le t \le n \;:\; ||X_t - X_i|| \le h_1 \text{ and } |Y_{t+m} - Y_{i+m}| \le h_2\},$$

$$C_m(X_i) = \#\{X_t, 1 \le t \le n, \;:\; ||X_t - X_i|| \le h_1\},$$

for $1 \le i \le n$. Then

$$Z_t \equiv \sqrt{C_m(X_t, Y_t)/\{C_m(X_t)\, h_2\}}$$

is a natural estimate of $q_m(x,y)$ at $(x,y) = (X_t, Y_t)$. Fitting it into the context of locally linear regression, we estimate $q_m(x,y)$ and its derivative with respect to $x$, denoted by $\dot{q}(x,y)$, by using $\hat{q}_m(x,y) = \hat{a}$ and $\hat{\dot{q}}_m(x,y) = \hat{b}$, where $(\hat{a}, \hat{b})$ are the minimizer of the function

$$\sum_{t=1}^{n-m} \{Z_t - a - b^T(X_t - x)\}^2 K\left(\frac{X_t - x}{h_1}, \frac{Y_t - y}{h_2}\right),$$

$K$ being a probability density function on $R^{d+1}$. Consequently, we estimate $I_m(x)$ by

$$\hat{I}_m(x) = 4 \int \{\hat{\dot{q}}_m(x,y)\}^2 \, dy.$$

For further discussion of this estimation, we refer to Fan, Yao and Tong (1993).

## 4. Examples

We have shown, via asymptotics, that the performance of nonlinear prediction is influenced by the initial values. In this section, we use two simulated models and two real data sets to illustrate the finite-sample behaviour. We use Gaussian kernel in our estimation.

### *4.1. Logistic Map*

We begin with the simple one-dimensional model (10). In fact, its skeleton is a transformed logistic map with the coefficient 3.68(=16 $\times$ 0.23). We adopt the transformation in order to enlarge the dynamic range of the model. A sample of 1200 is generated from model (10). Note that $\sigma_1^2(x) \equiv 0.16$ (also cf. Fig. 1(a)); therefore the one-step prediction is uniformly good for different initial values. Hence, the case is not reported here. The scatter plots of $Y_{t+m}$, for $m = 2, 3, 4$, against $Y_t$ are displayed in Fig. 2, which show obvious change of the variability of $Y_{t+m}$ with respect to the different values of $Y_t$. For example, in the case $m = 3$, the variability of $Y_{t+m}$ is at its largest when $Y_t$ is around 8, and at its smallest when $Y_t$ is about 5.6 and 10.4 (see Fig. 2(b)). We use the first 1000 observations to estimate the unknown functions. The last 200 observations are used to demonstrate the quality of prediction. The predicted values for those 200 observations together with their absolute prediction errors and estimated conditional variance $\hat{\sigma}_m^2(x)$ (cf. (21)) are plotted in Fig. 3 for the cases of two, three, and four steps ahead. Since rounding errors in the calculation are below $10^{-6}$, the accuracy is dominated by the conditional variance. For example, Fig. 3(b) shows that the three-step-ahead prediction is at its worst when the initial value is around 8, and at its best when the initial value is near 5.6 or 10.4, which is in agreement with the observation from Fig. 2(b). Similar remarks apply to the two-step and the four-step predictions.

To see how a small shift in the initial values affects the prediction, we round the initial value $x$ to the nearest value from amongst $[x], [x] + 0.5$, and $[x] + 1$, where $[x]$ denotes the integer part of $x$. Hence, $|\delta| \le 0.5$. Fig. 4 shows that for $m = 1, 2$, the absolute prediction error increases as $|\hat{\lambda}_m(x)|$ increases, which is consistent with the asymptotic conclusion presented in Theorem 1. There, $\hat{\lambda}_m(.)$ is estimated by using (17).

Fig. 5 presents the predictive intervals constructed by the estimated conditional percentiles $\hat{\xi}_{\alpha,m}(.)$, together with 200 real values. To estimate the conditional percentiles, we use the multidimensional *downhill simplex method* (cf. §10.5 of Press et. al. 1992). Fig. 5 shows that the width of the interval varies with respect to the initial

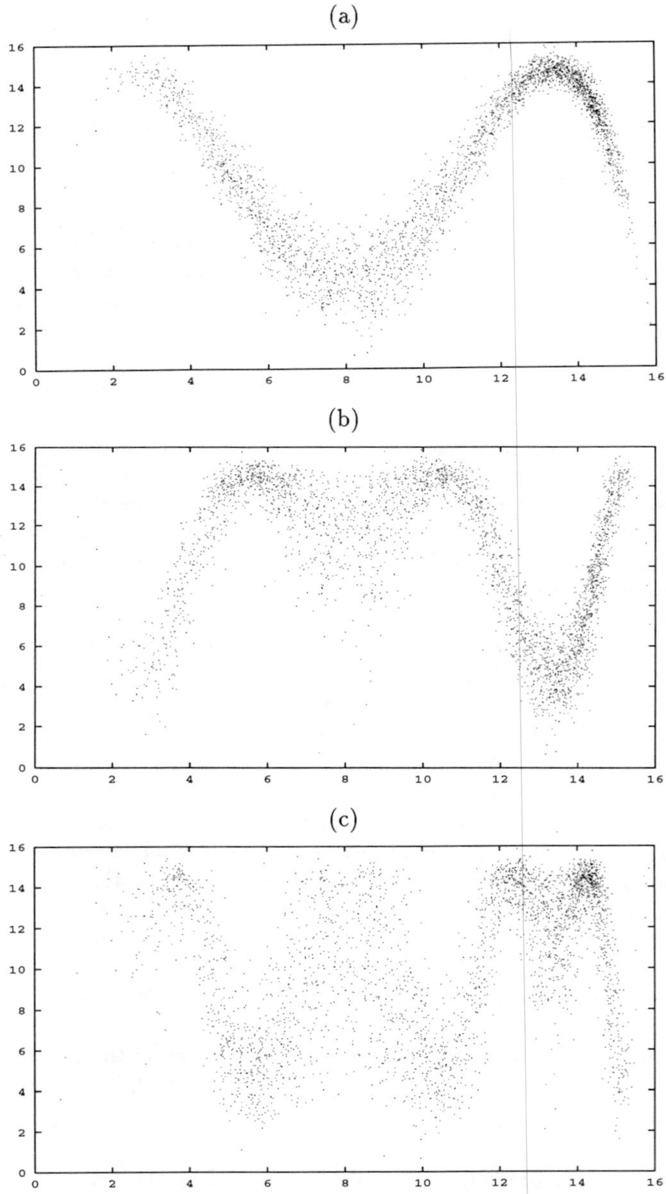


Figure 2: The scatter plots of $Y_{t+m}$ against $Y_t$ for Logistic map: (a) $m = 2$; (b) $m = 3$; (c) $m = 4$

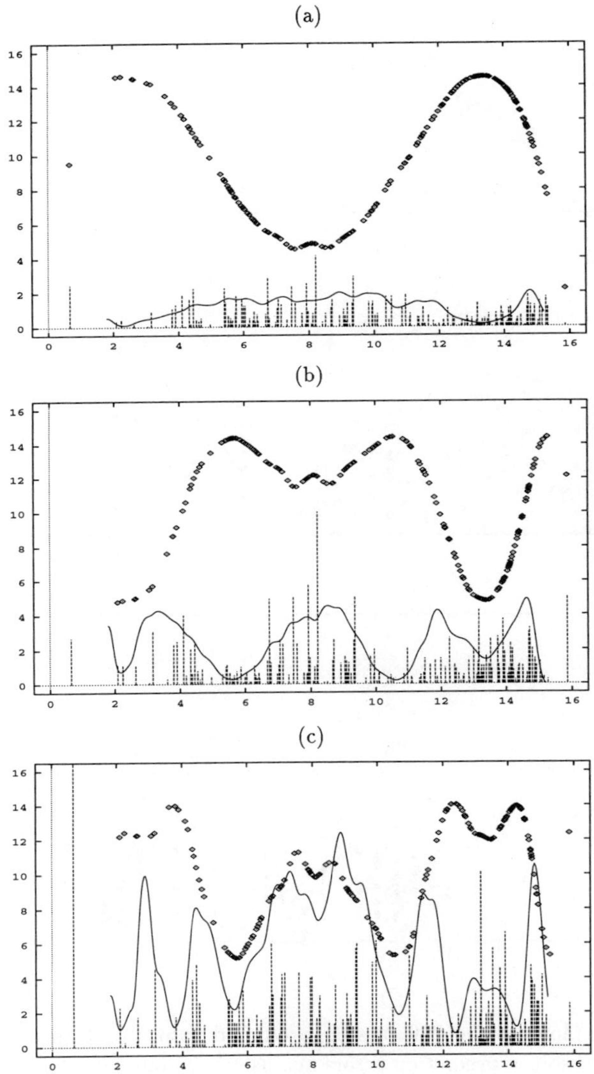


Figure 3: The plots of the 200 $m$-step predicted values of Logistic map, and the corresponding absolute prediction errors against their initial values, as well as the estimated conditional variance $\hat{\sigma}_m^2(x)$: (a) $m = 2$ ($h = 0.25$); (b) $m = 3$ ($h = 0.2$); (c) $m = 4$ ($h = 0.18$). Diamonds — predicted values; impulses — absolute prediction errors; solid curve — $\hat{\sigma}_m^2(x)$.

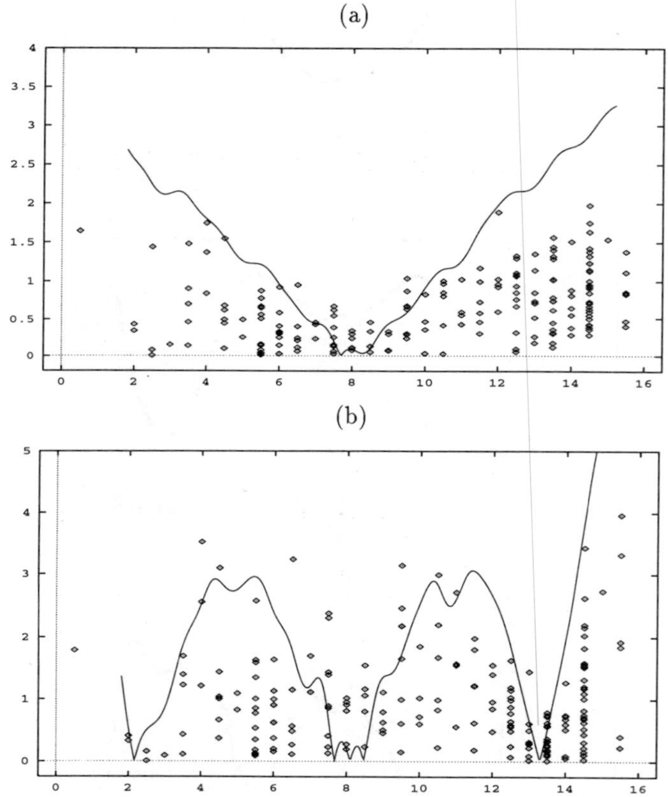


Figure 4: The plots of the 200 $m$-step predicted values of Logistic map, and the corresponding absolute prediction errors against their rounded initial values, and the estimated function $|\hat{\lambda}_m(x)|$; (a) $m = 1$ ($h = 0.32$); (b) $m = 2$. Diamonds — absolute prediction errors; solid curve — $|\hat{\lambda}_m(x)|$.

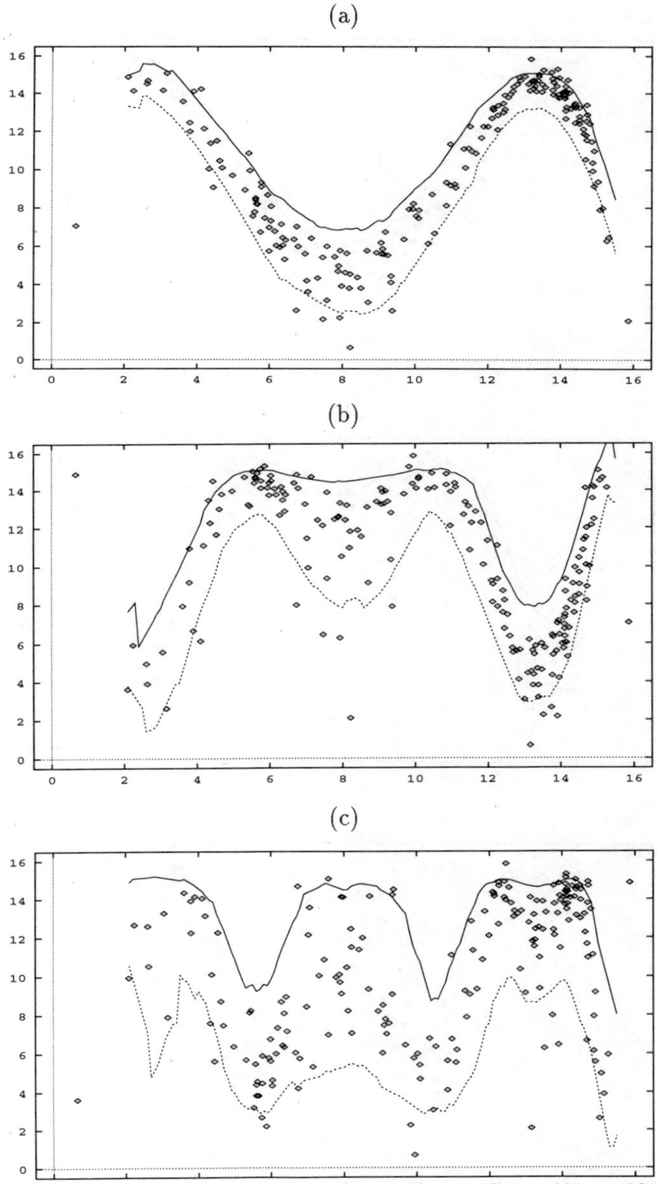


Figure 5: The predictive interval $[\hat{\xi}_{0.05,m}(x),\ \hat{\xi}_{0.95,m}(x)]$, and 200 real values for Logistic map. (a) $m = 2$ ($h = 0.5$); (b) $m = 3$ ($h = 0.42$); (c) $m = 4$ ($h = 0.37$). Solid curve — $\hat{\xi}_{0.95,m}(x)$; dotted curve — $\hat{\xi}_{0.05,m}(x)$; diamonds — real values.

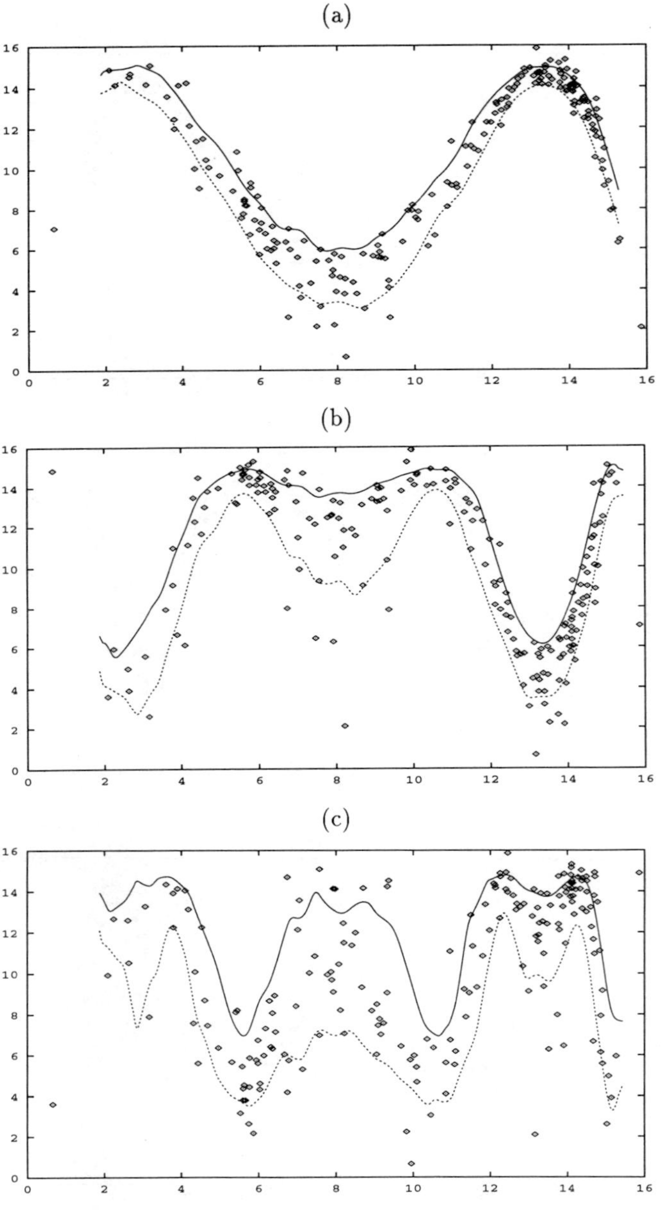


Figure 6: The predictive interval $[\hat{\tau}_{0.05,m}(x),\ \hat{\tau}_{0.95,m}(x)]$, and 200 real values for Logistic map. (a) $m = 2$ ($h = 0.25$); (b) $m = 3$ ($h = 0.2$); (c) $m = 4$ ($h = 0.18$). Solid curve — $\hat{\tau}_{0.95,m}(x)$; dotted curve — $\hat{\tau}_{0.05,m}(x)$; diamonds — real values.

(a)

(b)

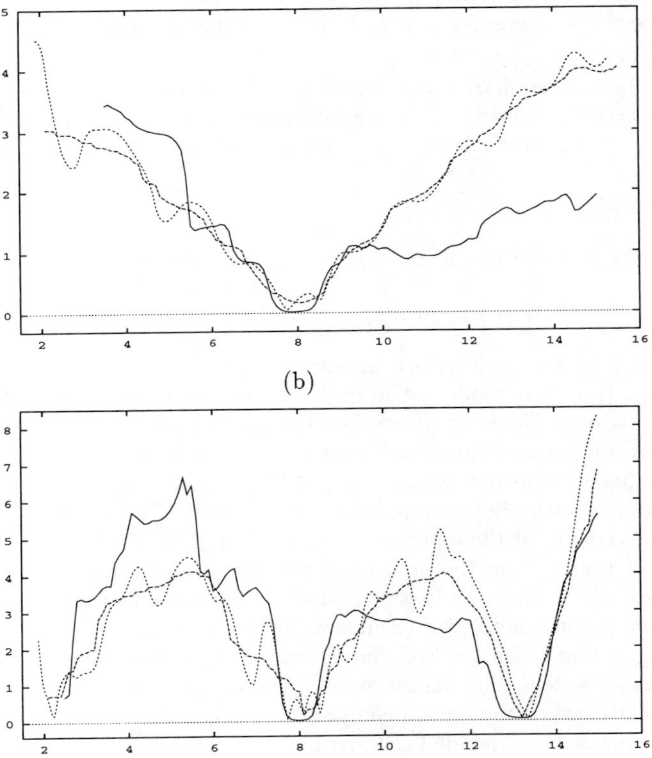


Figure 7: The estimated Fisher information $\hat{I}_m(x)$, and the derivatives of conditional percentiles and expectiles for Logistic map. (a) $m = 1$ ($h_1 = 0.61,\ h_2 = 0.24$ for $\hat{I}_1(x)$); (b) $m = 2$ ($h_1 = 0.57,\ h_2 = 0.22$ for $\hat{I}_2(x)$). Solid curve — $\hat{I}_m(x)$; dashed curve — $\{(\dot{\hat{\xi}}_{0.05,m}(x))^2 + (\dot{\hat{\xi}}_{0.95,m}(x))^2\}^{1/2}$; dotted curve — $\{(\dot{\hat{\tau}}_{0.05,m}(x))^2 + (\dot{\hat{\tau}}_{0.95,m}(x))^2\}^{1/2}$.

value. For example, in the case of $m = 3$, the width attains its maximum around $x = 8$, and its minimum about $x = 5.6$ and 10.4 (cf. Fig. 5(b)). Notice that the presented intervals are supposed to contain the predicted values with probability 0.9; we notice that about 10% of the 200 samples lie outside the intervals. The predictive intervals constructed by the estimated conditional expectile $\hat{\tau}_{\omega,m}(.)$ are displayed in Fig. 6.

To monitor the sensitivity of the predictive interval to the initial value, we plot the three sensitive measures in Fig. 7. It shows that the profiles of the Fisher information $I_m(x)$, $\{(\dot{\hat{\xi}}_{0.05,m}(x))^2 + (\dot{\hat{\xi}}_{0.95,m}(x))^2\}^{1/2}$ and $\{(\dot{\hat{\tau}}_{0.05,m}(x))^2 + (\dot{\hat{\tau}}_{0.95,m}(x))^2\}^{1/2}$ are generally quite similar. Comparing Fig. 7(b) with Fig. 5(a), or Fig. 6(a), we can see that where the sensitive measures are large, small shift will lead to a relatively large change in the intervals (e.g. $x$ near 5 and 11), and vice versa. (Also see Fig. 4.)

### *4.2. Hénon Map*

We clothe a Hénon map with dynamic noise to obtain

$$Y_t = 6.8 - 0.19Y_{t-1}^2 + 0.28Y_{t-2} + 0.2\epsilon_t \quad t \geq 1, \tag{29}$$

where $\epsilon_t$, $t \geq 1$, are as described in equation (10). A sample of 1200 observations is generated from this model. The first 1000 observations are used for estimation, and the remaining 200 observations for checking the prediction. Although there are two components for each initial value (or rather initial vector), we only plot the data against its first component, namely $Y_{t-1}$ of $(Y_{t-1}, Y_{t-2})$. Fig. 8 reports the predicted values together with the corresponding true values. The estimated values of the conditional variance at these points are shown in Fig. 9, which indicate the accuracy of the prediction. (Note the occasional negative estimates as discussed in Section 3.1.) For example, when the first component of the initial value is near $-6.8$ or 6.5, the two-step prediction is good (compare Fig. 8 (a) with Fig. 9 (a)). It can also be seen in Fig. 8 that when the first component of the initial value is near 0, the curve has two branches depending on the signs of the second component. The prediction is evidently better when the second component is negative.

In Fig. 10, we have rounded the first half of the checking sample in the same way as in Fig. 4. (Using the complete checking sample would clutter the figure with too many points.) Note that $\hat{\lambda}_m$ is a two-dimensional vector now and we plot $\| \hat{\lambda}_m \|$ instead of $\hat{\lambda}_m$.

Table 1 reports the two-step and three-step ahead predictive intervals for the first 10 checking sample. Here we use the interval predictor $[\hat{\xi}_{0.05,m}, \hat{\xi}_{0.95,m}]$, which contains the predicted variable with probability 0.90 theoretically. The bandwidth is chosen as 0.53 for two-step prediction and 0.48 for three-step prediction. It has so happened that in both cases of $m = 2$, and 3, 1 out of 10 true values lies outside the estimated interval. The results also show that the width of the interval varies considerably with respect to the initial values. For example, for the case $m = 2$, the shortest interval in the table is [4.27, 4.76] with the width 0.49, and the largest interval is [-7.76, -

(a)

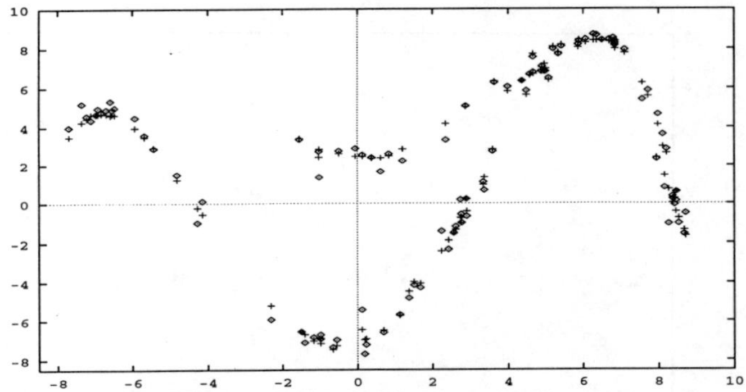


(b)

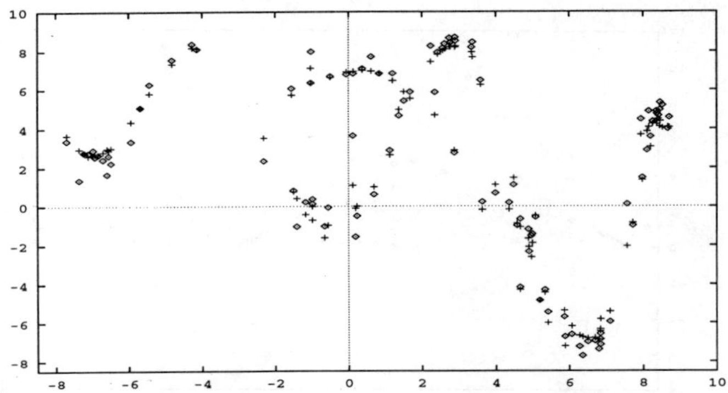


Figure 8: The plots of the 200 $m$-step predicted values for Hénon Map, and the corresponding true values against the first component of their initial values: (a) $m = 2$ ($h = 0.47$); (b) $m = 3$ ($h = 0.45$). Diamonds — predicted values; crosses — true values.

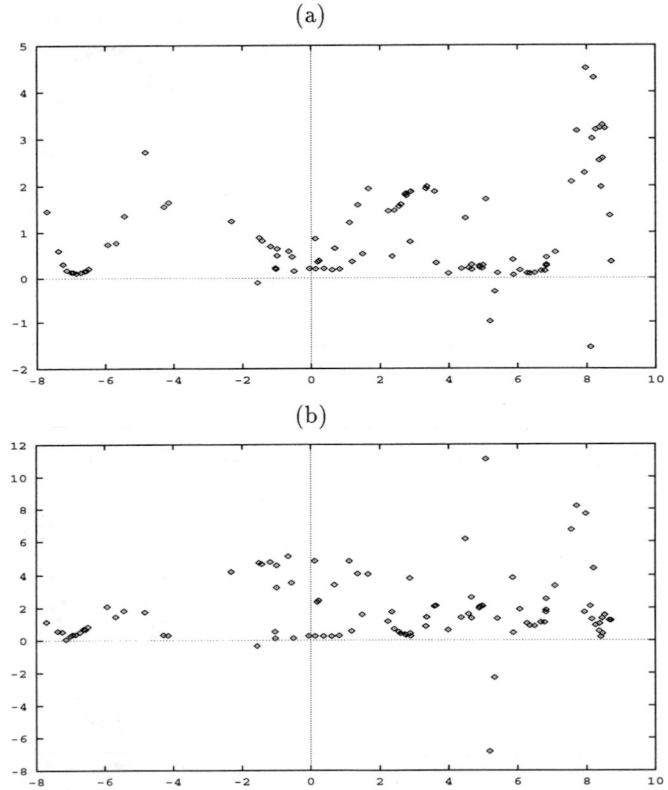


Figure 9: The plots of the 200 estimates values of $\hat{\sigma}_m^2$ against the first component of their initial values for Hénon Map: (a) $m = 2$; (b) $m = 3$.

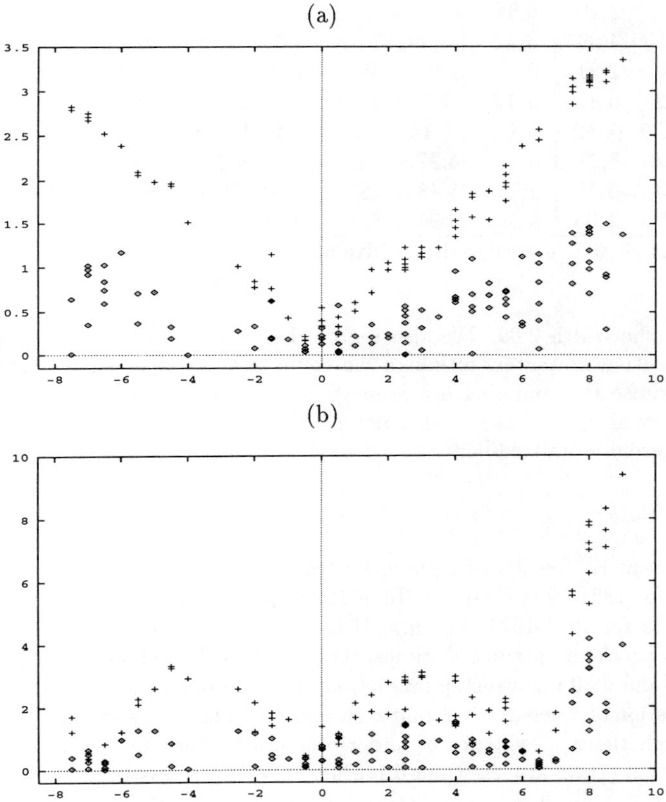


Figure 10: The plots of the 100 absolute prediction errors and the corresponding estimated values $\| \hat{\lambda}_m \|$ against the first component of their first (rounded) initial values for Hénon Map: (a) $m = 1$ ($h = 0.5$); (b) $m = 2$. Diamonds — predicted errors; crosses — $\| \hat{\lambda}_m \|$. (Note that some of the initial values, after rounding, may be coincident. This leads to fewer crosses than diamonds in some columns.)

Table 1: Interval prediction for Henon map

| Initial values | | Two-step ahead prediction | | | Three-step ahead prediction | | |
|---|---|---|---|---|---|---|---|
| $Y_t$ | $Y_{t-1}$ | $Y_{t+2}$ | PI | SM | $Y_{t+3}$ | PI | SM |
| -6.49 | 8.12 | 4.99 | [3.80, 5.00] | 1.00 | 2.24 | [2.36, 4.27] | 2.56 |
| 1.19 | -6.49 | 2.24 | [2.27, 3.51] | 2.01 | 6.84 | [5.63, 7.16] | 3.00 |
| 4.99 | 1.19 | 6.84 | [6.45, 7.26] | 2.44 | -1.42 | [-2.57, -0.42] | 6.80 |
| 2.24 | 4.99 | -1.42 | [-3.39, -1.44] | 4.09 | 8.27 | [6.27, 8.28] | 2.55 |
| 6.84 | 2.24 | 8.27 | [7.06, 8.38] | 1.68 | -7.12 | [-7.08, -3.96] | 4.82 |
| -1.42 | 6.84 | -7.12 | [-7.76, -5.67] | 2.57 | -1.03 | [-2.65, 2.94] | 6.07 |
| 8.27 | -1.42 | -1.03 | [-0.15, 1.52] | 5.44 | 4.36 | [2.29, 4.96] | 2.44 |
| -7.12 | 8.27 | 4.36 | [4.27, 4.76] | 2.04 | 2.74 | [1.96, 3.17] | 1.94 |
| -1.03 | -7.12 | 2.74 | [1.48, 3.28] | 3.90 | 6.35 | [5.98, 8.03] | 4.31 |
| 4.36 | -1.03 | 6.35 | [5.94, 6.81] | 2.54 | 0.19 | [-1.44, 0.83] | 6.15 |

PI: predictive interval; SM: sensitive measure.

5.67] with the width 2.09. We also report the values of $\{||\dot{\hat{\xi}}_{0.05,m}||^2 + ||\dot{\hat{\xi}}_{0.95,m}||^2\}^{1/2}$ as our sensitive measures, although we could not see their effects in the reported results because the rounding errors in the calculation are below $10^{-6}$. We can also use conditional expectiles to construct predictive intervals. The results are similar and are therefore omitted here.

### *4.3. Lynx Data*

We present the results of pointwise prediction for $m = 1$ and 2 for the Canadian lynx data for 1821-1934 (listed in Tong 1990) in Table 2. Here, we choose $d = 4$. We use the data for 1821-1924 (i.e. $n = 104$) to estimate $f_m(\cdot), \lambda_m(\cdot)$ etc., and the last 10 data to check the predicted values. The bandwidth is chosen as 0.55 for one-step prediction and 0.50 for two-step prediction. The column under $\hat{\sigma}_2^2$ is not complete due to the omission of a negative estimate. Roughly speaking, the prediction is reasonably good though there is evidence of under-prediction. For the case of one-step ahead, the prediction errors are less than 0.1 when $\| \hat{\lambda}_1(x) \|$ is less than 1. They tend to be larger when $\| \hat{\lambda}_1(x) \|$ is 'large'. Occasionally (e.g. in 1934) the error is small even though $\| \hat{\lambda}_1(x) \|$ is 'large'. For the two-step prediction, $\hat{\sigma}_2^2$ and $\| \hat{\lambda}_2 \|$ provide some indication of the prediction reliability. Typically, in 1927 the values of both $\hat{\sigma}_2^2$ and $\| \hat{\lambda}_2 \|$ are large, and the error of the prediction is also large.

We also perform the interval prediction using conditional percentiles to this data set. The results are reported in Table 3. The bandwidth is chosen as 0.57 for one-step prediction and 0.51 for two-step prediction. We use the predictor $[\hat{\xi}_{0.05,m}, \hat{\xi}_{0.95,m}]$. In the case $m = 1$, two predictive intervals (out of the ten) do not cover the true values. In the case $m = 2$, although all the intervals contain the corresponding true values, the widths of the intervals are considerably larger than those in the case $m = 1$.

Table 2: Point prediction of the Canadian lynx data (on natural log scale)

| Year | True value | error ($\hat{f}_1$) | $\parallel \hat{\lambda}_1 \parallel$ | error ($\hat{f}_2$) | $\hat{\sigma}_2^2$ | $\parallel \hat{\lambda}_2 \parallel$ |
|---|---|---|---|---|---|---|
| 1925 | 8.18 | -0.05 | 0.58 | -0.13 | 0.08 | 0.77 |
| 1926 | 7.98 | -0.23 | 2.67 | -0.39 | 0.69 | 1.04 |
| 1927 | 7.34 | -0.16 | 2.49 | -0.60 | 1.99 | 4.21 |
| 1928 | 6.27 | 0.22 | 3.12 | 0.13 | 1.60 | 2.30 |
| 1029 | 6.18 | -0.43 | 1.94 | -0.45 | 0.61 | 3.42 |
| 1930 | 6.50 | -0.28 | 2.34 | -0.60 | – | 3.38 |
| 1931 | 6.91 | -0.19 | 1.23 | -0.46 | 0.37 | 2.35 |
| 1932 | 7.37 | 0.02 | 0.70 | -0.21 | 1.17 | 1.43 |
| 1933 | 7.88 | -0.26 | 1.21 | -0.22 | 0.08 | 0.59 |
| 1934 | 8.13 | -0.07 | 2.28 | -0.22 | 0.51 | 2.02 |

Table 3: Interval prediction of the Canadian lynx data (on natural log scale)

| Year | True value | Predictive interval | |
|---|---|---|---|
| | | $m = 1$ | $m = 2$ |
| 1925 | 8.18 | [7.88, 8.67] | [7.84, 8.36] |
| 1926 | 7.98 | [7.35, 8.27] | [6.89, 8.47] |
| 1927 | 7.34 | [6.48, 7.88] | [5.92, 7.58] |
| 1928 | 6.27 | [5.68, 8.09] | [4.77, 8.47] |
| 1929 | 6.18 | [4.97, 6.35] | [4.76, 7.29] |
| 1930 | 6.50 | [5.75, 6.43] | [5.31, 6.53] |
| 1931 | 6.91 | [5.99, 6.97] | [6.28, 7.41] |
| 1932 | 7.37 | [7.04, 7.63] | [6.65, 7.87] |
| 1933 | 7.88 | [7.07, 7.83] | [7.31, 8.07] |
| 1934 | 8.13 | [7.55, 8.40] | [7.22, 8.32] |

Table 4: Point prediction of the sunspot numbers

| Year | True value | error ($\hat{f}_1$) | $\| \hat{\lambda}_1 \|$ | error(MARS) |
|---|---|---|---|---|
| 1979 | 155.4 | - 8.88 | 2.64 | -21.67 |
| 1980 | 154.7 | -47.26 | 6.32 | -9.24 |
| 1981 | 140.5 | - 5.83 | 2.98 | -10.55 |
| 1982 | 115.9 | -32.0 | 12.59 | -12.58 |
| 1983 | 66.6 | 2.80 | 1.10 | 15.76 |
| 1984 | 45.9 | 1.01 | 0.96 | -2.59 |
| 1985 | 17.9 | 17.94 | 1.16 | 2.75 |
| 1986 | 13.4 | -2.57 | 0.64 | -7.66 |
| 1987 | 29.2 | -19.73 | 0.92 | -2.82 |
| 1988 | 100.2 | -53.67 | 3.92 | -24.27 |
| 1989 | 157.6 | 35.56 | 8.27 | -9.32 |
| 1990 | 142.6 | 34.51 | 9.84 | 4.38 |
| 1991 | 145.7 | -11.63 | 3.08 | -27.90 |
| 1992 | 94.3 | -10.40 | 11.35 | 12.98 |

*4.4. Sunspot Data*

In many respects, the Wolf's annual sunspot numbers are known to be quite a challenging set of data (see e.g. Tong 1990). We use the data for 1700-1978 (i.e. $n = 279$) to estimate the predictor function and its related functions, and the data for 1979-1992 (i.e. 14 points) to check the prediction reliability as monitored by $\| \hat{\lambda}_1 \|$. In the fitting, we adopt $d = 4$ and $h = 6.43$. The results are summarized in Table 4. The overall impression is that $\| \hat{\lambda}_1 \|$ tends to be small (around 1 say) for the 'trough-years' and large for the 'peak-years'. With the exception of 1985 and 1987, the prediction reliability is fairly closely monitored by reference to $\| \hat{\lambda}_1 \|$.

We also fit this data set by using the MARS method with linear base functions (cf. Friedman 1991, Lewis and Stevens 1991). Using the sunspot data in the period 1700-1978, the selected model is

$$\begin{aligned} Y_t &= 181.67 - 0.23Y_{t-3} + 0.86(Y_{t-1} - 190.2)^- - 1.93(Y_{t-2} - 79.7)^- \\ &\quad -0.0023(Y_{t-2} - 79.7)^-(Y_{t-9} - 190.2)^- - 0.0081(Y_{t-1} - 190.2)^-(Y_{t-2} - 92.6)^- \\ &\quad -0.0043Y_{t-1}Y_{t-3}(Y_{t-2} - 21.3)^-, \end{aligned}$$

where $x^- = x$ if $x < 0$, and 0 otherwise. The prediction errors of this model for the sunspot data in the period 1979-1992 are reported in the last column of Table 4. Again, we have better prediction in the 'trough-years' than the 'peak-years'.

## Appendix: The Regularity Conditions

To discuss the asymptotic properties of the estimators, we need the following assumptions.

(A1) All second partial derivatives of $f_m(x)$ are bounded and continuous.

(A2) The conditional variance $\sigma_m^2(x) = \text{Var}(Y_m|X_0 = x)$ is bounded and continuous.

(A3) The joint density of distinct elements of $(X_1, Y_1, X_k, Y_k)$ is bound by a constant independent of $k$.

(A4) $X_t$ has the probability density function $p$, and $|p(x) - p(y)| \leq C \parallel x - y \parallel$ for any $x, y \in R^d$.

(A5) The precess $\{Y_t\}$ is $\rho$-mixing, i.e., $\rho_j = \sup_{U \in \Im_{-\infty}^0, V \in \Im_j^\infty} \text{Corr}(U, V) \to 0$, as $j \to \infty$, where $\Im_i^j$ is the $\sigma$-field generated by $\{Y_k,\ k = i, \ldots, j\}$. Further, we assume that $\sum_{k=1}^\infty \rho_k < \infty$.

(A6) $K(\cdot)$ is a continuous density function with a bounded support in $R^d$, and $\int xK(x)dx = 0$, $\int xx^T K(x)dx = \sigma_0^2 I_d$, where $I_d$ denotes the $d \times d$ identity matrix.

(A7) The bandwidth $h \to 0$, $nh^{2+d} \to \infty$, and $(\log n)/(nh^d) \to 0$.

(A8) For any compact subset $B \in R^d$, there exists a constant $c$ such that for any $x,\ y \in B$, $|\int z^2 g_m(z|x)dz - \int z^2 g_m(z|y)dz| \leq ||x - y||$, where $g_m(y|x)$ denotes the conditional density of $Y_m$ given $X_0$.

## References


Berliner, L.M. 1992 Statistics, probability and chaos. *Statistical Science*, **7**, 69-122.

Bloomfield, P. and Steiger, W.L. 1983 *Least Absolute Deviations.* Birkhäuser, Boston.

Chan, K.S. and Tong, H. 1994 A note on noisy chaos. *J. R. Statis. Soc.* **B**, **56**, 301-311.

Chatterjee, S. and Yilmaz, M.R. 1992 Chaos, fractals and statistics. *Statistical Science*, **7**, 49-121.

Crutchfield, J.P., Farmer, J.D. and Huberman, B.A. 1982 Fluctuations and simple chaotic dynamics. *Phys. Rep.*, **92**, 45-82.

Deissler, R.J. and Farmer, J.D. 1989 Deterministic noise amplifiers. Tech. Rep., LA-UR-89-4236, Los Alamos Laboratory, USA.

Eckmann, J.P. and Ruelle, D. 1985 Ergodic theory of chaos and strange attractors. *Rev. Modern Physics*, **57**, 617-656.

Fan, J. 1992 Design-adaptive nonparametric regression. *J. Ameri. Statis. Assoc.*, **87**, 998-1004.

Fan, J., Hu, T.C. and Truong, Y.K. 1992 Robust nonparametric function estimation.

Technical Report 035-92, Math.Science Research Inst., Berkeley.

Fan, J., Gasser, T., Gijbels, I., Brockmann, M., and Engel, J. 1993 Local polynomial fitting: a standard for nonparametric regression. Technical Report, University of North-Carolina.

Fan, J., Yao, Q., and Tong, H. 1993 Estimation of conditional densities and sensitivity measures in nonlinear dynamical systems. Technical Report, Univ. of Kent.

Friedman, J.H. 1991 Multivariate adaptive regression splines. *Ann. Statist.*, **19**, 1-50.

Hall, P. and Wolff, R.C.L. 1993 Properties of invariant distributions and Lyapunov exponents for chaotic logistic maps. Technical Report, Australian National University.

Kifer, Y. 1986 *Ergodic Theory of Random Transformations.* Birkhäuser, Basel.

Kullback, S. 1967 *Information Theory and Statistics.* Dover Publi., New York.

Lewis, P.A.W. and Stevens, J.G. 1991 Nonlinear modeling of time series using multivariate adaptive regression splines. *J. Ameri. Statis. Assoc.*, **86**, 864-877.

Neway, W.K. and Powell, J.K. 1987 Asymmetric least squares estimation and testing. *Econometrica*, **55**, 819-847.

Nychka, D., Ellner, S., Gallant, A.R. and McCaffrey, D. 1992 Finding chaos in noisy systems. *J. R. Statis. Soc.* **B**, **54**, 399-426.

Pollard, D. 1991 Asymptotics for least absolute deviation regression estimators. *Econometric Theory*, **7**, 186-198.

Press, W.H., Flannery, B.P., Tenkolsky, S.A., and Vetterling, W.T. 1992 *Numerical Recipes.* Cambridge Univ.Press, Cambridge.

Tong, H. 1990 *Non-Linear Time Series: A Dynamical System Approach.* Oxford University Press, Oxford.

Yao, Q. and Tong, H. 1992 Asymmetric least squares regression estimation: a nonparametric approach. Technical Report, Univ. of Kent.

Yao, Q. and Tong, H. 1994 Quantifying the influence of initial values on nonlinear prediction. *J. R. Statis. Soc.* **B**, **56**, 701-725.

CHAPTER 4

# LOCALLY OPTIMIZED PREDICTION OF NONLINEAR SYSTEMS: STOCHASTIC AND DETERMINISTIC

Leonard A. Smith
Mathematical Institute
University of Oxford
Oxford OX1 3LB, U.K.
email: lenny@maths.ox.ac.uk

**Abstract**

Forecasting the behavior of chaotic dynamical systems tends to be more difficult at some times than at others. This arises for two basic reasons: first, that the intrinsic predictability of a chaotic system usually varies with the state of the system, and second, the information we have about different regions of state space varies as well. The intrinsic variability implies, for instance, that Lyapunov exponents need not reflect any practical "limit of prediction." A method of locally optimizing the parameters of prediction schemes, which takes advantage of variability from both sources, is introduced and illustrated on both stochastic and deterministic nonlinear systems; difficulties in distinguishing the two are noted. It is also demonstrated that even very good (but not perfect) predictors will leave information in the time series of the residual prediction errors from chaotic signals: it is not possible to reduce the residuals to a sequence of independent, identically distributed random variables. A straightforward method to search for this residual predictability is demonstrated.

## 1. Introduction

Variability and unpredictability are two of the hallmarks of deterministic chaos; it is our aim here to show how the first can be exploited to reduce the second. We begin by examining this variability in the Lorenz attractor (Lorenz 1963), and considering the types of uncertainty in initial condition that give rise to prediction error, even when a perfect model is at hand. In section 3, we cast the prediction problem as one of interpolation in state space; recalling previous work on local linear prediction, we illustrate a weakness in this approach. We also show that irreducible approximation error will, in general, leave information in the residual prediction errors from chaotic systems. The predictability of a chaotic system varies with the location of the initial condition; the main result of this paper is to introduce a prediction scheme which embraces the variability both of dynamics and geometry. This approach is used to optimize local prediction in section 4, where it is illustrated on time series data from a low dimensional, presumably deterministic laser system, and a nonlinear stochastic model for sunspots. In these examples, local linear prediction is optimized by allowing the number of points used in constructing the predictor to vary with location in state space. Next, in section 6, we demonstrate why a time series from a chaotic system (defined as a deterministic system with at least one positive Lyapunov exponent) need not be difficult to predict all the time; the issue here is one of the uniformity of stretching around state space. The penultimate section contains a discussion of residual predictability, proposing a simple test to detect systematic prediction error, which indicates that further improvement in prediction accuracy is possible. The test is straightforward to apply, and has been found very effective in practice. Our conclusions, along with a pointer toward ensemble predictions and the limit of predictability, are given in section 8. In short, accurate prediction of a nonlinear system from limited data requires sensitivity to the variation of the system's properties in state space; examining this variability can cast new light on the "limits of predictability" as well as individual predictions.

## 2. Variations in Predictability with Initial Condition

The cover illustration shows the error-doubling time for a variety of initial conditions on the Lorenz attractor (Smith 1994; Ziehmann-Schlumbohm 1994). The colors reflect the minimum time required for an infinitesimal uncertainty to double; the red points double within one time unit, the orange within two, the yellow three and so on. This illustrates the main points of this paper: that the predictability of the flow is both variable and highly organized. Points with short error doubling times will tend to be more difficult to predict than those with long ones: the predictability of a chaotic system varies with initial condition (Benzi and Carnevale 1989; Doerner *et al.* 1991; Nese 1989; Palmer 1993; Smith 1992; Ziehmann-Schlumbohm 1994). By exploiting this variability, we can significantly improve the reliability of our predictions. Moreover, arguments based on uniform error growth are misleading. The argument linking the largest Lyapunov exponent to the "prediction horizon" is a good example; Lyapunov exponents need not reflect practical limits of prediction.

Before diving into the details of predicting chaotic dynamical systems, we consider two more general forecasting issues: the types of initial uncertainty we are likely to encounter and the implications of the statistic adopted to evaluate a given set of predictors. The short term error growth we observe will depend on the nature of the initial uncertainty in our observation. In the case of a perfect model, error arises solely from uncertainty in the initial conditions. When dealing with data on a strange attractor, the growth of uncertainty is often quantified through the divergence of nearest neighbors, whose likely orientation varies systematically around the attractor. A different structure is observed when this uncertainty is oriented by the largest (global) Lyapunov exponent (as on the cover), which will differ yet again from cases where it is determined either by the locally fastest growing direction, or by a random displacement. In systems like the Earth's atmosphere, we must consider systematic errors both in the observations themselves and from any preprocessing required by the model (as discussed by Palmer *et al.* 1994 and in this volume). In addition, arguments based on perfect predictors overlook the systematic effects of optimal, yet imperfect predictors, an issue examined in section 3 below. Finally, we note that the vast majority of uncertainty growth studies concentrate exclusively on infinitesimal initial uncertainties. The quantities of interest when determining the limitations of prediction must, by definition, consider the regime where the magnitude of the uncertainty becomes comparable with the range of the observable.

To evaluate a set of predictors, we must choose a criteria for comparison. For nonlinear systems, the result will be much more sensitive to the particular statistic chosen than an intuition based on independent, identically distributed (IID) gaussian residuals would suggest. Both the root mean square (RMS) error and the average absolute error may be dominated by a few large errors (Benzi and Carnevale 1989). Using the median absolute error or throwing out a small fraction of the worse predictions (Casdagli *et al.* 1992; Farmer and Sidorowich 1987), reduces this sensitivity, but clouds interpretation of the results. In practice, the predictor which is most frequently the most accurate may be wildly wrong when it is inaccurate. Indeed, it is the application for which the predictions are made that determines, for example, whether the prediction which is most frequently optimal is preferred over the one which is least frequently horrid. While it is obvious that the cost function will play a central role in determining the "best" predictor, the sensitivity to this choice in nonlinear systems is observed to be much greater than in linear systems. Prediction errors from chaotic systems are neither gaussian nor independent; they are correlated both in time and state space, and, as we shall see, they are usually chaotic themselves.

## 3. Prediction as Nonlinear Interpolation

The recent success of prediction methods for many chaotic dynamical systems (Abarbanel *et al.* 1993; Eubank and Farmer 1990; He and Lapedes 1993; Sauer 1993; Smith 1992; Sugihara and May 1990; Tong 1990; Ziehmann-Schlumbohm 1994; Ziehmann-Schlumbohm *et al.*1995) is due, in large part, to the successful translation of an extrapolation problem to an interpolation problem (Eckmann and Ruelle 1985). Ideally, these methods consider a point in the state space of a deterministic physical system, where the future of each initial condition is uniquely determined and, if the equations of motion are known, may be estimated. Alternatively, *if* enough data are available to provide a sufficient number of recurrent trajectories passing near the initial condition in question, then we can interpolate the future trajectory given only the observations. Yet the functions involved in chaotic dynamical systems are, at

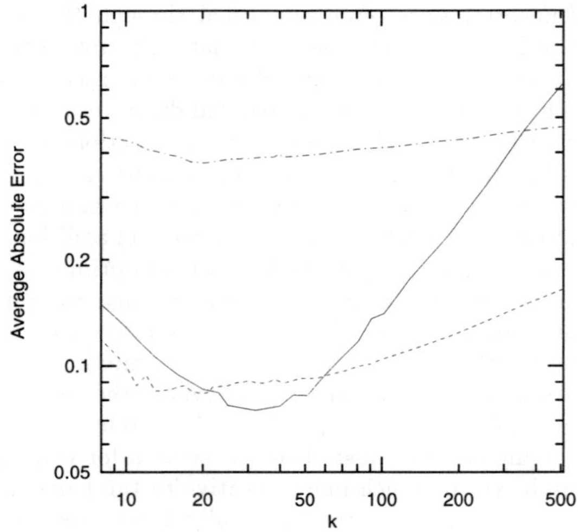


Fig. 1. Average absolute forecast error, $E$, as a function of the number of near neighbors, $k$, used for out-of-sample prediction of (solid) $x_i$ of the Ikeda map, (dashed) the laser system, and (dot-dashed) the stochastic sunspot model. $E(k)$ is normalized by the average deviation.

their simplest, nonlinear, and the data are typically distributed on a strange attractor. We are faced with a high dimensional ($> 2$) interpolation of a complicated function sampled on an inhomogeneous distribution. With noise.

The two basic approaches to this problem consider either global interpolation functions for the entire state space, or restrict attention to local regions. The global approach requires a complicated interpolation scheme (*e.g.* radial basis functions or neural nets). Here we will consider the "local" approach (Farmer and Sidorowich 1987; Sugihara and May 1990; Tong 1990), which allows much simpler interpolation schemes. Following Casdagli *et al.* (Casdagli 1992; Casdagli *et al.* 1992), we can test for nonlinearity by evaluating a series of local predictors based on the $k$ nearest neighbors and determining the value, $k_c$, at which the observed prediction error is smallest. For linear stochastic systems, $k_c$ should correspond to the largest $k$, while for noise free deterministic systems, $k_c$ should be of the order of the dimension of the system, given enough data. As illustrated below, for both deterministic systems with noise and nonlinear stochastic systems, a minimum at "moderate" values of $k$ is to be expected.

As a concrete example, consider the chaotic, 2 dimensional Ikeda map

$$\mathbf{D}_{\text{ikeda}}(x,y) = (1.0 + \mu\ [x\ \cos(t) - y\ \sin(t)], \mu\ [x\ \sin(t) + y\ \cos(t)]) \qquad (1)$$

where $t = 0.4 - \frac{6.0}{x^2+y^2+1}$ and $\mu = 0.90$. We will make one step ahead predictions of the value of $x$, given 1024 observations and base points $(x, y)$ with gaussian noise*

*For simplicity, we consider additive white noise, which is represented through an independent and

($\sigma = 0.125$) added to the observations. The solid line in figure 1 shows the out-of-sample, normalized average absolute error, $E(k)$ indicating $k_c \approx 32$. Yet if we consider only the 5 predictors with $k = 8$, 16, 32, 64 and 128, then we find the $k = 32$ predictor is the most accurate just 27% of the time (the distribution being 19, 23, 27, 23, and 8%, respectively). This illustrates a shortcoming of selecting a predictor with this approach: a global choice of $k$ fails to account for the variation in the length-scales of either the dynamics or the data distribution. These two effects are shown schematically in figures 2 and 3. Figure 2 shows a one-dimensional local linear approximation to a polynomial curve. The circle shows the radius at which the expected value of the noise is equal to the error introduced by the linear approximation of the true curve. The optimal local radius within which data should be used, $r_{\text{opt}}$, depends upon the statistics of the noise, the data density and the local curvature; and this is our point: that even with uniformly distributed data, $r_{\text{opt}}$ will change with the local curvature of the function estimated. Figure 3 illustrates additional complications due to the nonuniform distribution of data; these effects change not only with location in state space, but also at the same location with different data sets. In a given realization, there are likely to be regions where the local data density is too low to accurately estimate the local component of the dynamics and, simultaneously, other regions where the data density is sufficient to obtain a very good estimate. The (locally) optimal neighborhood to fit will vary between these regions, regardless of whether it is parameterized by $k$ or $r$. Our goal is to develop a scheme which adjusts $k$ to minimize the expected prediction error when the local curvature is not known. We will return to local linear predictors shortly, but first we consider the general question of irreducible approximation error in optimal predictors.

### *3.1. Imperfect Predictions: Colorfast chaos*

Let $\mathbf{D}(\mathbf{x})$ define a deterministic dynamical system at a point $\mathbf{x}$ in state space. For simplicity, we consider a scalar measurement function $D(\mathbf{x})$ of $\mathbf{D}(\mathbf{x})$ representing the future property of $\mathbf{D}(\mathbf{x})$ which we wish to predict. The problem is then one of function approximation; we wish to approximate $D(\mathbf{x})$ given a particular family of predictors, $F$, with parameters $\lambda$. We may divide $D$ heuristically into two parts:

$$D(\mathbf{x}) = F(\lambda, \mathbf{x}) + E_F(\mathbf{x}) \tag{2}$$

where $F(\lambda, \mathbf{x})$ represents an optimal fit to $D$ and $E_F(\mathbf{x})$ represents the deterministic structure in $D(\mathbf{x})$ orthogonal to $F(\lambda, \mathbf{x})$. $F(\lambda, \mathbf{x})$ is optimal in the sense that the remaining approximation error is due to the structure of $F$ itself. For example, if $F$ is a linear model, it reproduces the linear behavior of $D(\mathbf{x})$ exactly; in this case $E_F(\mathbf{x})$ would consist of the quadratic and higher order terms in $D(\mathbf{x})$.

The main point of this section is that, in general, we expect $E_F(\mathbf{x})$ to be a good measurement function: when Taken's theorem (Sauer *et al.* 1991; Takens 1981)

identically distributed random variable added to each observation. The case of red (autocorrelated) noise can be handled in a similar manner, when the time scales of the noise are shorter than the recurrence time of the underlying dynamics.

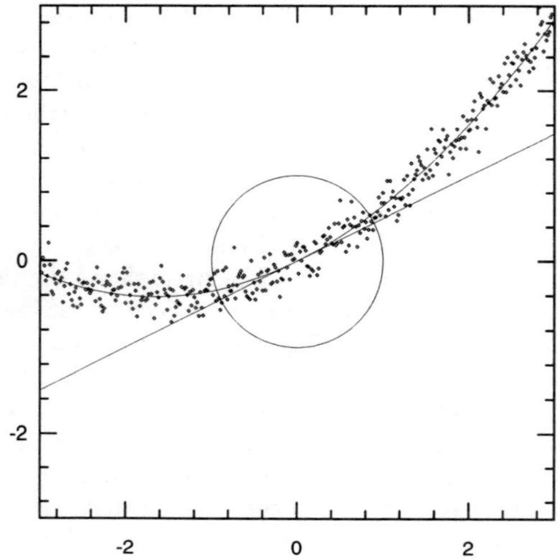


Fig. 2. Schematic diagram showing the need to balance the noise level against the local curvature in a linear fit to a polynomial function. The circle shows the radius at which the deviation due to the linear approximation equals the expected value of the noise.

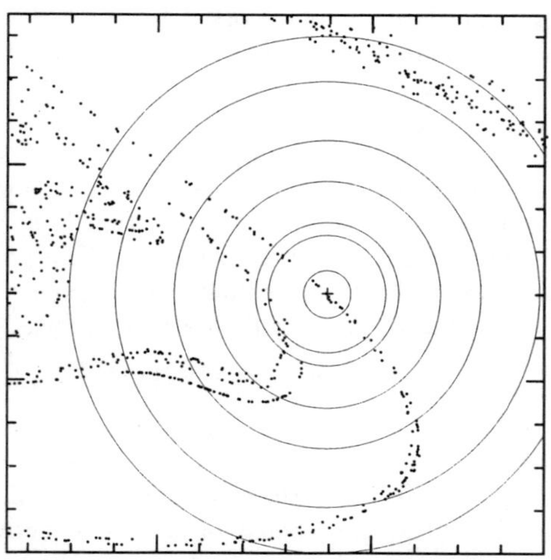

Fig. 3. Circles containing 9, 17, 33, 65, 129 and 257 points near the point to be predicted (+). The data is not evenly distributed within the circles, tending to be almost linear for $k = 8$ and remaining completely skewed to the one side of the prediction point until $k = 256$.

applies to a chaotic data stream, it will also apply to the time series of residuals from our prediction scheme. This implies that the residuals will be chaotic, albeit with more complicated macroscopic structure and a smaller signal to noise ratio, and indicates the impossibility of truly bleaching chaotic data (Theiler and Eubank 1994); chaos is colorfast in the sense that the residuals of non-perfect predictors will, in general, be chaotic. Removing the global linear structure of a chaotic signal (*i.e.* "pre-whitening" or "bleaching") will not result in IID residuals (Brock *et al.* 1991), although it will obscure the results of a nonlinear analysis, as stressed by Theiler and Eubank (Theiler and Eubank 1994). Similar results hold for more complex filters (Broomhead *et al.* 1992; Sauer *et al.* 1991), where the filter or predictor influences the particular value that we observe, but the dynamics are determined by the underlying system.[†] Ultimately, the goal of nonlinear prediction is to reduce residuals to an IID sequence; we now examine obstacles to such a reduction even in the absence of noise.

As an explicit example of the separation in (2), consider local polynomial prediction near a point $\mathbf{x_0}$. We expand $D(\mathbf{x})$ about $\mathbf{x_0}$; setting $r = ||\mathbf{x} - \mathbf{x_0}||$ and denoting angular orientation by the vector $\mathbf{\Theta}$ yields

$$D(\mathbf{x}) = \underbrace{D(\mathbf{x_0}, 0) + a_1(\mathbf{x_0}, \mathbf{\Theta})\, r}_{F(\lambda, \mathbf{x})} + \underbrace{a_2(\mathbf{x_0}, \mathbf{\Theta})\, r^2 + a_3(\mathbf{x_0}, \mathbf{\Theta})\, r^3 + \dots}_{E_F(\mathbf{x})} \qquad (3)$$

where we have shown $F(\lambda, \mathbf{x})$ and $E_F(\mathbf{x})$ for local linear prediction. For higher order polynomial predictors, additional terms would be shifted to $F(\lambda, \mathbf{x})$. For other nonlinear predictors (*e.g.* a particular radial basis function fit or a specific neural net) this decomposition may be difficult to write down, but holds in principle as long as $D(\mathbf{x})$ is not a member of the family of functions which may be approximated exactly by $F$. If $D(\mathbf{x})$ is such a function, then $F(\lambda, \mathbf{x}) = D(\mathbf{x})$, and $F$ is called a perfect predictor.

For local linear prediction of systems in which $a_2(\mathbf{x}, \mathbf{\Theta})$ varies significantly with $\mathbf{x}$, quantities like the optimal radius will be a function of $\mathbf{x}$ even when the learning data is uniformly distributed: $r_{\text{opt}}$ will tend to be smaller where $a_2(\mathbf{x}, \mathbf{\Theta})$ is large. Typically, $a_2(\mathbf{x}, \mathbf{\Theta})$ does vary; figure 4 reflects this variation in the case of the Ikeda map. Here the value of $a_2(\mathbf{x})$, the absolute value of $a_2(\mathbf{x}, \mathbf{\Theta})$ averaged over $\mathbf{\Theta}$, is shown. Variations in data density will enhance this effect, particularly on strange attractors where the distribution is typically fractal and extremely inhomogeneous.

In the noise-free case, $k_c$ is the minimum number of neighbors required to solve for $F(\lambda, \mathbf{x})$ (assuming the data is not distributed in a degenerate way). In the presence of noise, the number of neighbors (or equivalently, the optimal radius, $r_{\text{opt}}$) will depend on the noise level and the local curvature of $D$ at $\mathbf{x}$, as well as the local data distribution. It is this balance between curvature and noise which is illustrated in figure 2. This approach is applicable to all local prediction schemes; it is the aim of local optimal prediction to vary the parameters of $F$ (*e.g.* the value of $k$) in order to both balance the local structure of $E_F$ against that of the noise, and adapt to the details of the data distribution, in cases where the analytic structure of $D$ is unknown.

[†]In contrast to the series obtained by repeatedly iterating a non-perfect predictor.

A similar approach might be applied to global schemes as well, but given that global schemes are more costly to construct and, usually, quite easy to evaluate, it might be preferable to stack (Wolpert 1992) global predictors or even to choose the best reconstruction as a coarse function of location in state space (Smith 1992), and then employ ensemble prediction (Palmer 1992; Palmer 1993; Smith 1994)

Returning to local predictors, we note that the effects of variation in curvature can be shown by analyzing, for a given $\mathbf{x}$, an ensemble of data sets, each consisting of uniformly distributed learning data with gaussian noise. By determining the value of the radius which provides the best fit for each realization and observing this distribution for different points on the attractor, the $a_2(\mathbf{x})$ dependence of $r$ can be seen, for example, by plotting the distribution of $r$ against the local value of $a_2(\mathbf{x})$ (not shown).

Thus we have seen that the optimal number of neighbors will vary with the local curvature of $D(\mathbf{x})$ in state space, the local data density, (which will itself vary "strangely" for dissipative chaotic attractors) and, of course, the details of the noise. This last observation implies not only that the details will be system dependent and vary with location in state space, but they may also change at a given point $\mathbf{x}$ from realization to realization; if, by chance, a number of low noise data points are the near neighbors of the point of interest, the optimal radius for this realization may be much smaller than that expected from the general arguments above; local optimal prediction adapts to these special cases.

## 4. Locally Optimizing Linear Prediction

We now present an algorithm to determine the optimal $k$ from the data alone. In each of the cases discussed below, the data set is divided into a learning set from which the predictors are constructed, and a test set for out-of-sample evaluation. The learning set is analyzed to estimate the global parameters $k_c$ and $k_{\max}(>> k_c)$, the largest neighborhood to be considered. The basic idea is to consider a small number of points in the learning set close to the point to be predicted, $\mathbf{x_0}$, and then employ local "drop-one-out" predictors with various $k$ to predict the (known) image of each of these points in turn (Smith 1993). The best value of $k$ for this $\mathbf{x_0}$ is then determined from the observed errors in predicting these points from the learning set.

More specifically, to determine the best value of $k$ at the point $\mathbf{x_0}$, determine the $k_{\max}$ nearest neighbors of $\mathbf{x_0}$ in the learning set. From this subset, select the $N_{\text{drop}}(\approx 8)$ points nearest to $\mathbf{x_0}$ with the requirement that these test points are well separated in time; this requirement is crucial to avoid picking consecutive points from the same segment of the trajectory, which leads to highly correlated (and misleading) estimated errors.[‡] In the example shown in figure 3, these $N_{\text{drop}}$ points lie within the smallest circle. For several values of $k$ (corresponding to the circles in figure 3), construct $N_{\text{drop}}$ distinct predictors by dropping out, in turn, each of the $N_{\text{drop}}$ test points and predicting the point omitted. Finally, combine these results for each $k$ to

[‡]A similar phenomena is known to bias dimension estimates (Theiler 1986)

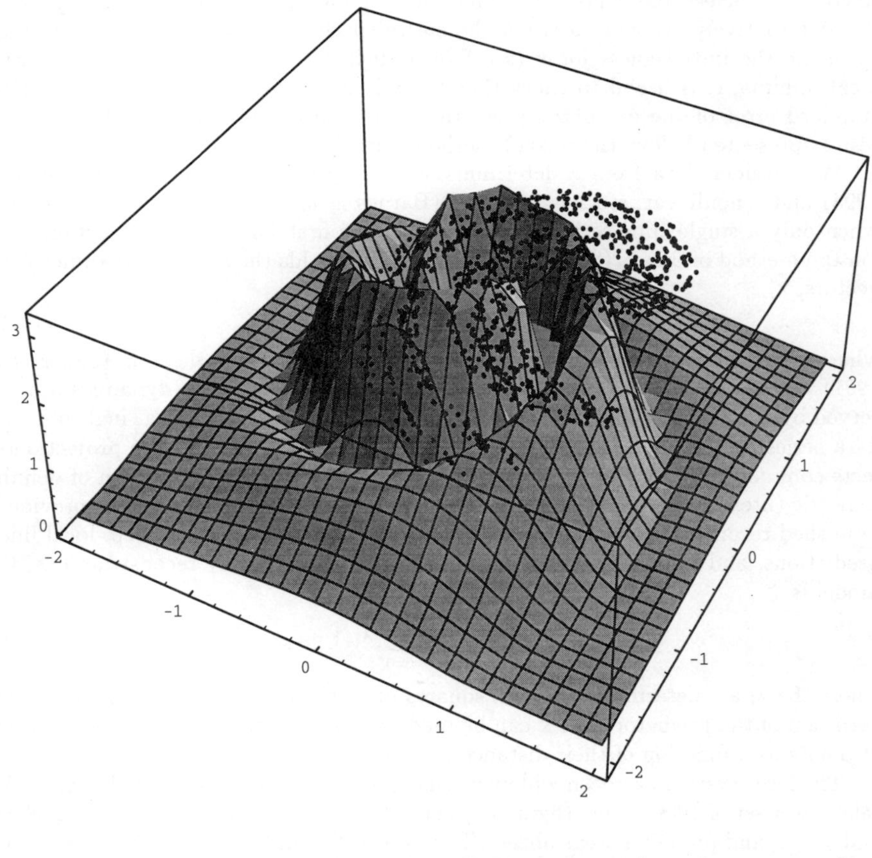


Fig. 4. This surface reflects the variation in the quadratic component for perfect, one step ahead prediction of the $x$ value of the Ikeda map. The height corresponds to the absolute value of the quadratic term, averaged over angle. The dots show the location of the attractor.

obtain an estimated prediction error at $\mathbf{x_0}$ for a $k$ neighbor predictor, and determine the optimal radius at $\mathbf{x_0}$ by comparing these estimates.

The choice of optimal radius is non-trivial, and detailed results will be presented elsewhere (Smith and Drysdale 1994). Frequently, choosing the value of $k$ with the minimum estimated error improves the predictor a large fraction of the time, but *increases* the overall RMS error by making a few bad choices. A second approach is to consider the estimated error as a function of $k$ and find the local minimum nearest $k_c$. Alternatively, we can partition the entire space, and use the partition element to define the initial guess for $k$ (see Smith 1992; Smith 1993). When considering local minima, it is useful to check that there is not a significantly (in terms of the standard error of the estimated prediction error) deeper minima elsewhere. In the results presented below, the second method is used.

We consider data from a deterministic laser system (Weigend and Gershenfeld 1993) and a nonlinear, stochastic model (Barnes *et al.* 1980). To apply these ideas when only a single time series, $s_i$, is available, we first form a reconstruction space via the method of delays (Sauer *et al.* 1991). This yields the series of $M$-dimensional vectors, $\mathbf{x}_i$

$$\mathbf{x}_i = (s_i, s_{i-j}, \ldots, s_{i-j(M-1)}) \tag{4}$$

where $j$ is the delay time. Under ideal circumstances, Taken's theorem (Sauer *et al.* 1991) assures us that many properties from the true state space dynamics are preserved by this reconstruction. Of course, this theorem suggests that a single observed data series can be used, but the use of multivariate data can remove projection effects commonly found in finite series, although it introduces the question of defining a metric (literally, an apples and oranges question) . In each case, we use previously published reconstruction parameters $M$ and $j$ to make direct, fixed step, local linear predictions, and evaluate the models out-of-sample. For delay reconstructions, this model is

$$F(\lambda, \mathbf{x}) = \lambda_0 + \sum_{\ell=0}^{M-1} \lambda_{\ell+1}\, s_{i-j\ell} \tag{5}$$

where the $\lambda_i$ are determined by least squares fit to the $k$ nearest neighbors of $\mathbf{x}$. The variation of the predictor with $\mathbf{x}$ can be made smooth by weighting the contributions of points as a function of their distance from $\mathbf{x}$.

The laser system has been widely discussed (Weigend and Gershenfeld 1993). We take a test set of 1000 points (figure 5a) and a learning set of 8192 points with $M = 4$ and $j = 1$, and predict 4 steps ahead. The dashed line in figure 1 shows the expected behavior of the out-of-sample prediction error with $k$ with a minimum at $k = 20$. Using the drop-one-out scheme of the previous section to allow $k$ to vary with $\mathbf{x}$ improves the average absolute error, $E$, relative to the $k_c$ predictor by 6% (median absolute error by 10%). If we suppress the worst 10% predictions of each predictor, following Casdagli *et al.*(Casdagli *et al.* 1991), this becomes a 6.5% reduction of $E$; omitting these points also reduces the RMS error for the $k_c$ predictor by an order of magnitude. When we examine, after the fact, how often each predictor from the set $k$ = 8, 11, 16, 23, 32, 45 and 64 was the most accurate out-of-sample, we find they were

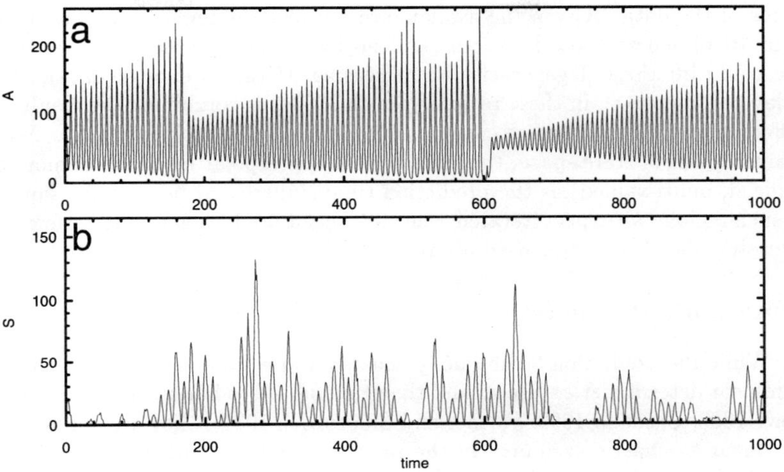


Fig. 5. Time series data from (a) a laser experiment and (b) the stochastic sunspot model.

best 20, 14, 14, 12, 11, 11 and 18% of the time, respectively; the optimal $k$ clearly varies with location.

We observe that large errors are likely to occur when exploring new regions of state space (*i.e.* the macroscopic structure of the attractor not represented in the learning set); this is to be expected, in as much as, in new regions of state space, our procedure for prediction by interpolation reverts to an extrapolation problem. In regions of very low data density, reverting to the "zeroth order" method of simply taking the image of the nearest neighbor (often called analog prediction (Lorenz 1969)) can significantly improve the overall prediction error. An advantage of drop-one-out schemes is that, even when extrapolating, they provide an estimate of the expected error which differs from that available from previous predictions in the same neighborhood (Smith 1992). Local optimal prediction can also outperform fixed $k$ prediction in data dense regions; small fixed $k$ local predictors, like iterated predictors, get lost in data-dense regions where the dynamics are simple, but the series is slowly varying and nearest neighbors are determined by chance (the noise). We expect significant improvement using local optimal prediction, for example, in the chemical experiments presented by Prof. Olsen (Olsen *et al.* 1994).

While we do not know how to construct the analog of figure 4 for this four dimensional reconstruction of experimental data, we exploit the quantization of the data to quickly search for "exact" returns in the reconstruction space, and then examine the images of these points. This reveals that the majority of exact returns have very similar images, but also shows several cases of exact recurrence (*i.e.* to the digital resolution of the data), which have vastly different images (comparable to the range of the data). Given the available resolution, the function we are interpolating is multi-valued at these locations, and no interpolation scheme can assure success. We could lift these degeneracies by increasing $M$ (given enough data), altering the delay parameter, $j$ , in these regions (Smith 1993), or combining the results of several different prediction schemes (Sauer 1993; Smith 1993; Wolpert 1992). Note that in many regions of state space, the surface we are interpolating becomes more contorted (indeed, multi-valued) as the prediction time is increased beyond the sampling rate; in such regions we expect iterated, one step ahead prediction to out perform the direct four step ahead predictor used above.

## 5. Aleatoric Dynamics

While the prediction techniques discussed above were developed in the context of nonlinear deterministic systems and then generalized to handle noise (Broomhead and Lowe 1988; Casdagli 1992; Farmer and Sidorowich 1987), they are applicable in some nonlinear stochastic systems. In the past, successful nonlinear prediction has been interpreted as evidence of deterministic dynamics. This interpretation is incorrect for a class of nonlinear, but fundamentally stochastic systems which do **not** meet the criteria of Laplacian determinism; the current state of such a system does not define a unique future (or past). We call such systems "aleatoric" since, while the underlying driving mechanism appears not to be deterministic, the dynamics are governed in large part by deterministic laws, as with a roll of the dice. The boundaries of what should be included under this heading are not clear, save that the dynamics should be nonlinear (and not trivially made linear), and the random contribution to the dynamics should be essential. A chaotic system with dynamic noise (*i.e.* noise which effects the state of the system, not just the observation of that state) falls into this category, but the concept of aleatoric systems was motivated by stochastic systems with no chaotic analog. One such system was proposed by Barnes *et al.*(Barnes *et al.*1980) as a model for annual mean sunspot numbers, $Y$ (figure 5b). Based on an ARMA(2,2) model with nonlinear modifications to ensure that $Y$ remains positive and tends to increase more rapidly than it decreases, the model is

$$Z_n = \phi_1 Z_{n-1} + \phi_2 Z_{n-2} + a_n - \theta_1 a_{n-1} - \theta_2 a_{n-2} \tag{6}$$

$$Y_n = Z_n^2 + \alpha(Z_n^2 - Z_{n-1}^2)^2 \tag{7}$$

where $\phi_1 = 1.90693, \phi_2 = -0.98751, \theta_1 = 0.78512, \theta_2 = -0.40662$, $\alpha = 0.03$ and the $a_n$ are IID gaussian random variables with zero mean and $\sigma = 0.4$. This system was used by the author and Weiss (Weiss 1990) as a surrogate process (Smith 1992; Theiler *et al.*1992) to illustrate the difficulty of estimating the correlation dimension of the observed sunspot record. It also illustrates the advantage of high data densities with an "incorrect" model, over low data densities from the true model: forecast errors of local linear predictors fit to 6000 years of model data can outperform the same predictor fit to the first half of the observed sunspot series, when the evaluation is based on the RMS error of the residuals of 1 year ahead predictions of the second half of the observed series.

To avoid any bias from our knowledge of the underlying system, the reconstruction parameters of Casdagli *et al.*(Casdagli *et al.*1992) ($M = 3$, $j = 1$) for the observed sunspot series were used to make one year ahead predictions. The dot-dashed curve in figure 1 traces the out-of-sample average absolute error as a function of $k$ showing a minimum at $k = 20$, a behavior similar to that expected from a deterministic system.[§] Further, the data density and sensitivity of the dynamics vary tremendously in different regions of reconstruction space, so local optimal prediction improves prediction; in addition, the time series of prediction errors is far from IID. While Casdagli *et al.* (Casdagli *et al.* 1992) explicitly note that nonlinear stochastic systems may display this behavior, they imply that one can distinguish a minimum at "medium" $k$ from one at "small" $k$. In practice, such a distinction may be invisible (*e.g.* in figure 1), and attempts to distinguish stochastic aleatoric dynamics from deterministic chaotic dynamics may require huge data sets, in order to get good statistics on very near returns in reconstruction space and to quantify the decay of predictability more directly (Casdagli *et al.*1992; Sugihara and May 1990). While of philosophical interest, the distinction may have few, if any, practical implications, and the ultimate resolution of this question requires the evaluation of limits that cannot be determined from any quantity of finite precision measurements. In contrast, the question of whether a given system is better modeled by a stochastic approach or a deterministic approach can be profitably addressed.

## 6. Lyapunov Exponents and Limits on Predictability

We now return to deterministic chaos and support our earlier claim that Lyapunov exponents can provide a misleading indication of predictability. Lyapunov exponents represent a growth rate averaged over the attractor. In a simple example, we contrast this average with the average time taken for an initial uncertainty to double; when the rate of uncertainty growth is not uniform, the two may be very different. We then consider the implications this holds for iterated and direct prediction schemes.

### *6.1. Baker's Apprentice Maps*

In the Baker's Map (see *e.g.* Tong 1990 and references therein), the unit square is stretched by a factor of 2 in the $x$ direction, compressed by a factor of 2 in $y$, and the resulting rectangle is then cut and stacked to form an area preserving map. In this uniform case, the largest Lyapunov exponent, $\lambda_1$, is 1 bit per iteration. The error doubling time for a given initial condition is simply the minimum time required for an initial uncertainty to increase by a factor of two. In this case, all initial uncertainties in the expanding direction double on each iteration, hence the average error doubling time, $\tau_{\text{double}}$, for the Baker's Map is equal to one iteration. Contrast this situation with a class of generalized Baker's Maps, the family of Baker's Apprentice Maps (Smith

[§]Indeed, this system is in a sense "low-dimensional," as discussed in reference (Smith 1992).

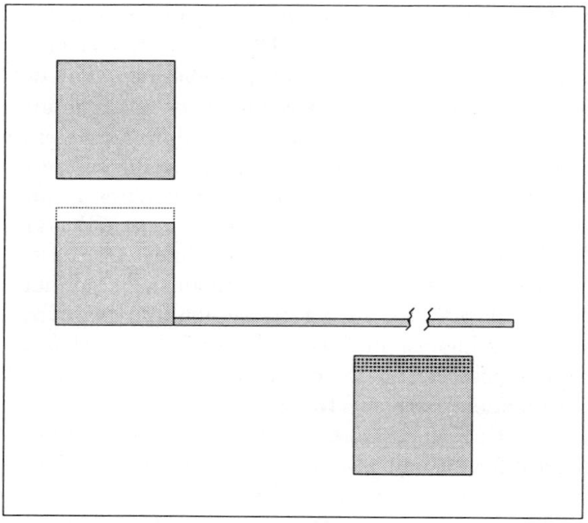

Fig. 6. Schematic representation of a Baker's Apprentice Map showing how a small portion of the unit square is stretched a great deal, then cut and stacked, while the majority of initial conditions are displaced only slightly.

1993), given by

$$x_{i+1} = \begin{cases} \frac{1}{\alpha} x_i & \text{if} \quad 0 \leq x_i < \alpha \\ \beta\ (x_i - \alpha) \mod 1 & \alpha \leq x_i < 1 \end{cases}$$

$$y_{i+1} = \begin{cases} \alpha y_i & \text{if} \quad 0 \leq x_i < \alpha \\ \alpha + \frac{1}{\beta}(\lfloor \beta(x_i - \alpha) \rfloor + y_i) & \alpha \leq x_i < 1 \end{cases} \tag{9}$$

where $\alpha = \frac{2^n - 1}{2^n}$, $\beta = 2^{2^n}$ and $\lfloor z \rfloor$ denotes the greatest integer less than or equal to $z$. A schematic diagram illustrating the action of these maps is shown in figure 6 of the "dough" is stretched a great deal ($2^{2^n}$) before being cut and stacked, while the majority of the initial conditions are displaced only slightly (a factor of $(\frac{1}{\alpha}) \approx 1$ ). For each positive integer $n$, equations (9) define an area preserving map whose positive Lyapunov exponent, $\lambda_1$ $(= 1 - \alpha \log_2 \alpha)$ is greater than 1 bit per iteration. Thus the largest Lyapunov exponent of each of these maps is *greater* than that of the Baker's Map, yet for the majority of initial conditions, the Apprentice Maps are much more predictable.

To quantify this predictability, we compute the doubling time for infinitesimal uncertainties in the expanding direction of an Apprentice map; points of equal doubling time fall in vertical bands, reminiscent of those seen in the Lorenz system on the cover. In particular, initial conditions with $x$ in the range $\alpha \leq x < 1$ will double after one iteration, those in the preimage of this band (*i.e.* $\alpha^2 \leq x < \alpha$) after two iterations, and so on until we reach the number of iterations, call it $k$, required for the uncertainty to double without having reached the region of extreme stretching. But $k$ is then just the smallest integer such that $\left(\frac{1}{\alpha}\right)^k \geq 2$; initial conditions in the band $0 \leq x < \alpha^k$ will have doubling time $k$. Hence the doubling time for the $n^{th}$ Apprentice Map is

$$\tau_{\text{double}} = k\alpha^k + \sum_{i=1}^{k} i\,[(1-\alpha)\alpha^{i-1}] = \sum_{i=0}^{k-1} \alpha^i = \frac{1-\alpha^k}{1-\alpha}$$

where $k = \lceil -\frac{1}{\log_2(\alpha)} \rceil$, and $\lceil z \rceil$ is the smallest integer greater than or equal to $z$. For large $n$, $\tau_{\text{double}} \approx 2^{n-1}$. And this happens fairly quickly: at $n = 4$, $\tau_{\text{double}} = 8.133$ while by $n = 8$, $\tau_{\text{double}} = 128.45$. The chaotic maps corresponding to large $n$ are easily predicted (most of the time), although their largest Lyapunov exponent is always greater than that of the Baker's Map.

Predictability is reflected in the error doubling time, but not by the Lyapunov exponents. The extent to which this occurs in other systems will depend on how localized (and densely populated) the regions of maximum stretching are; here we note that there are extended regions within which all uncertainties *decrease* with time for chaotic attractors in the Lorenz, Moore-Spiegel and Rössler systems (Smith *et al.*1994; Ziehmann-Schlumbohm 1994; Ziehmann-Schlumbohm *et al.*1995). In such regions, any initial uncertainty will be damped, an unexpected occurrence in these paradigm chaotic attractors. This not only eases the prediction of chaotic systems, but may also be of practical importance, for example, in determining to best times at which to perturb a system one is attempting to manipulate. In addition, characterizing the decay of predictability by the growth of errors will be complicated by the lack of independence between consecutive errors. In fact, the correlation of errors in state space provides the basis for a test for residual predictability, which we will consider in the next section.

### *6.2. Modes of prediction : Direct and iterative forecasts*

Variations in predictability will also influence the choice between constructing iterated predictors (which forecast a short time $\Delta T$ ahead and then are iterated $T/\Delta T$ times to get a period $T$ forecast) and direct predictors (for which $\Delta T = T$). Farmer and Sidorowich (Farmer and Sidorowich 1988) argue that, in the limit of good predictors, iterative forecasts are preferred. As they note, the argument assumes a simple relation between the error growth and Lyapunov exponents; it relies on uniformity in the stretching in state space and fails, for example, in data dense regions where the magnitude of the predicted change in one time step is small compared to the observational noise. The choice between iterative and direct prediction will vary in a manner quite like that of the optimal neighborhood size; given sufficient data the choice of $\Delta T$ should also be made locally.

## 7. Detecting Residual Predictability

We began by examining the structure of error doubling times on a strange attractor and observed organization in this structure. To conclude, we suggest adopting the same approach to look for residual predictability, but using a "color code" based on the (signed) prediction error, rather than the doubling time. Organization in these errors means improved prediction is possible. Figure 8 provides an example based on 1024 test points on the Ikeda attractor and a global radial basis function prediction of the next value of $x$; those points at which the prediction error is positive are denoted by a plus, while those with negative errors are marked with dot. The large scale structure in the distribution of prediction error is clear (although the magnitude may, of course, be quite small). In higher dimensions, where visualization is more difficult, we test for structure in the residuals by considering nearest neighbor pairs in the reconstruction space, and simply count the number of pairs where both of the associated prediction errors have the same sign. If the residuals are independent and identically distributed (IID), then the expected number of pairs with like signs is easily computed. As an IID sequence should remain IID under any general regrouping (Dawid 1984; von Mises 1957), contrasting the expected and observed number of pairs with like signs provides an immediate, quantitative evaluation of the prediction scheme. There are, of course, more powerful tests for structure in the residuals; one example being the BDS test, which considers delay reconstructions of the series of the residuals themselves (Brock *et al.*1991). The major advantages of the test presented below are its simplicity and the use of the delay space coordinates of the original data, allowing immediate access to the small length scales in state space where traces of predictability are most likely to be found.

The test is based on the observation that errors in the predictions of nonlinear systems are likely to be correlated in state space due to systematic biases (*e.g.* $E_F$). To determine whether nearest neighbor pairs in reconstruction space have independent residuals:

*i )* Consider original delay reconstruction coordinates as a collection of points in $M$ dimensional space.

*ii )* Classify each point into one of two groups depending on whether the (signed) prediction error at that location was positive or negative (alternatively, if it is above or below its median value).

*iii )* Determine unique nearest neighbor pairs, and count the number pairs in which both neighbors belong to the same group.

*iv )* Test the resulting distribution (number of pairs the same, number different) against the null hypothesis that the points are randomly classified.

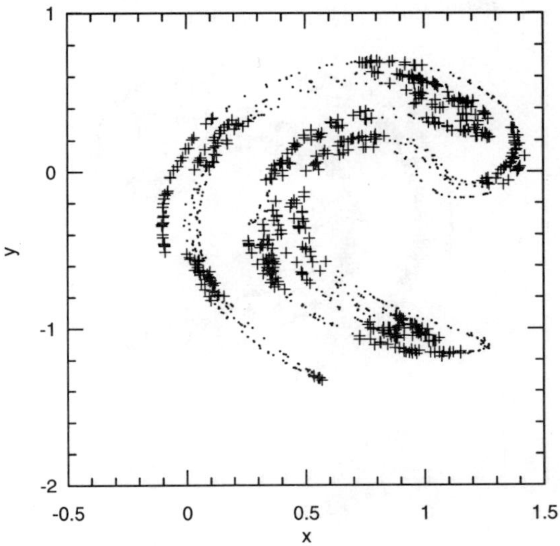


Fig. 7. Spatial organization in the residuals of the Ikeda map for a global radial basis function predictor. Points marked "+" have positive residuals, while those marked "." are negative. Note the large scale structure within which points tend to cluster into a fairly small number of regions of like type.

When coherence is detected, the test may be repeated for larger collections of points (triples,...) to determine the scale of the correlations. Figures 8 and 9 show the results for a global radial basis function fit and a local quadratic predictor of the Ikeda map, respectively. The correlation in the signs of the errors is apparent by eye. This large scale variation in the residuals of global fits is commonly observed when the embedding dimension is small enough to view the fit. The correlation lengths in the local quadratic predictor are smaller, but the null hypothesis of independent errors can still be easily rejected. Rejection of the null hypothesis is, of course, sufficient to show that significant residual predictability remains in a reconstruction, but it is **not** a necessary condition. While less powerful than the BDS test, it is simpler to implement and interpret. Our test is most likely to err on the side of underestimating the remaining predictability, indicating that a given prediction scheme is doing better than it truly is; we have not yet found such "false positive" results to be a problem. Indeed, we detect residual predictability in almost all of the chaotic systems we have examined.

## 8. Conclusions

The difficulty of predicting the evolution of a nonlinear system varies with the state of the system. Fortunately, this variation is highly organized, and we may exploit it

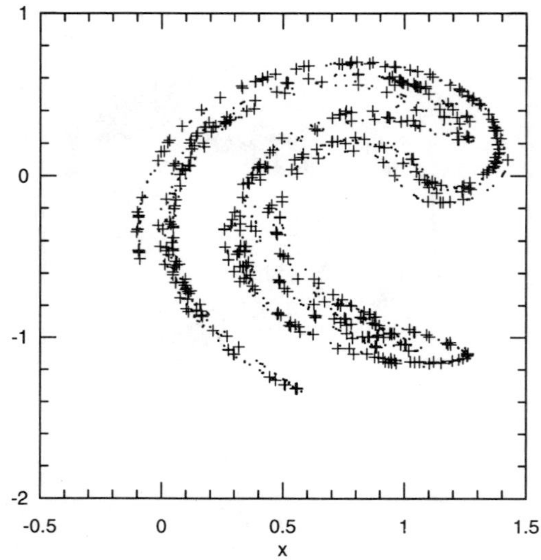


Fig. 8. Spatial organization in the residuals of the Ikeda map for 8 neighbor, local quadratic prediction. Once again, points marked "+" have positive residuals, while those marked "." correspond to negative residuals. In contrast to figure 7, the points with positive and negative error are much better mixed, but still tend to cluster into groups of like type.

both to improve our predictions and our estimate of the degree to which we believe them. This organized variability also requires careful consideration when attempting to distinguish aleatoric and chaotic signals, and when interpreting global "bounds" on predictability and the statistics of error growth; the average error doubling time may paint a very different picture of predictability than the time at which the average error has doubled.

In this paper, we have discussed the growth of small, usually infinitesimal, uncertainty in an initial condition. In practice, any functional definition of a "limit of predictability" should allow an uncertainty to attain a magnitude comparable with the range of observed values. Consider the evolution of an ensemble of initial conditions on the Lorenz attractor, all initially within a region of state space which makes them *indistinguishable*, given a series of three, 8 bit measurements. The behavior of such ensembles reveals a wealth of phenomena (Palmer 1993; Smith 1994; Smith *et al.*1994), we draw attention to only three here. First, the distribution may initially sharpen, indicating that in the short term, initial uncertainties decrease with time (Smith *et al.*1994; Tong and Moeanaddin 1988; Ziehmann-Schlumbohm 1994; Ziehmann-Schlumbohm *et al.*1995). Second, the distribution will become multimodal fairly quickly; this will cause difficulty if we tune our predictors by minimizing the squares of the prediction error. The "best" least squares predictor will predict the mean of the distribution, which may not reflect the value of any element in the ensemble. Third, even after large times when the distribution is spread over the entire range of observable values, as long as the distribution can be distinguished from the projection of the invariant measure onto our measurement function, then meaningful conclusions can be drawn from the distribution. The limit of predictability is reached when the ensemble can no longer be distinguished from this projection of the invariant measure (Smith 1994). Needless to say, this limit will vary with the location of the initial conditions.

The goal of nonlinear prediction is to remove all recognizable structure from the time series. For chaotic systems, it is not possible eliminate all structure in the series of out of sample prediction errors without a perfect predictor; to detect this structure in the residuals, it is useful to work within the framework of the original reconstruction space of the observations. Ultimately, our goal is to exploit the structure in the dynamics to minimize structure in the residuals.

## Acknowledgements

The research presented in this paper has greatly benefited from many exchanges with C. Ziehmann-Schlumbohm. It is also a pleasure to acknowledge both technical assistance and mathematical discussions with D. Drysdale, A. Fowler, K. Fraedrich, and M. Muldoon. I am grateful to S. Ellner, B. LeBaron, B. Seifert, R. Smith and H. Tong for both suggestions and directions through the statistical literature. Issues related to residual predictability have been enlightened by conversations with D. Broomhead and J. Huke and the observations of R. Jones. This work was supported by a Senior Research Fellowship at Pembroke College.

## References

Abarbanel,H. D. I., Brown, R., Sidorowich, J. J. and Tsimring, L.S.1993 The analysis of observed chaotic data in physical systems. *Rev Mod Phys*, **65(4)**, 1331–1392.

Barnes, J. A., Sargent, H. H. and Tryon, P. V. 1980 Sunspot cycle simulation using random noise. In *The Ancient Sun* (ed. R.O. Pepin, J.A. Eddy, and R.B. Merrill), pp. 159 - 163, New York, Pergamon.

Benzi, R. and Carnevale, G. F. 1989 A possible measure of local predictability. *Journal of the Atmospheric Sciences*, **46**, 3595-3598.

Brock, W. A., Hsieh, D. and LeBaron, B. 1991 *Nonlinear Dynamics, Chaos, and Instability: Statistical Theory and Economic Evidence.* MIT Press, Cambridge, MA.

Broomhead, D. S. and Lowe, D. 1988 Multivariable functional interpolation and adaptive networks. *J. Complex Systems*, **2**, 321-355.

Broomhead, D. S., Huke, J. P. and Muldoon, M. R. 1992 Linear filters and nonlinear systems. *J. R. Stat. Soc.*, **B(54)**, 373-382.

Casdagli, M. 1992 Chaos and deterministic *versus* stochastic non-linear modeling. *J. R. Statist. Soc. B*, **54(2)**, 303-328.

Casdagli, M., *et al.* 1992 Nonlinear modeling of chaotic time series. In *Applied Chaos* (ed. J. H. Kim and J. Stringer), pp. 335-380, New York, John Wiley.

Dawid A. P. 1984 Statistical theory: The prequential approach. *J. R. Statist. Soc. A*, **147(2)**, 278-292.

Doerner, R., *et al.* 1991 Predictability portraits for chaotic motinons. *Chaos, Solitons and Fractals*, **1**, 553-571.

Eckmann, J. P. and Ruelle, D. 1985 Ergodic theory of chaos and strange attractors. *Rev. Mod. Phys.*, **57**, 617-656.

Eubank, S. and Farmer, D. 1990 An introduction to chaos and randomness. In *Proc. SFI Summer School* (ed. E. Jen), Addison-Wesley.

Farmer, J. D. and Sidorowich, J. 1987 Predicting chaotic time series. *Phys. Rev. Lett.*, **59**, 8.

Farmer, J. D. and Sidorowich, J. 1988 Exploiting chaos to predict the future and reduce noise. In *Evolution, Learning, and Cognition* (ed. Y. C. Lee), pp. 277. World Scientific.

He, X. and Lapedes, A. 1993 Nonlinear modeling and prediction by successive approximation using radial basis functions. *Physica D*, **70**, 289-301.

Lorenz, E. N. 1963 Deterministic nonperiodic flow. *J. Atmos. Sci.*, **20**, 130.

Lorenz, E. N. 1969 Atmospheric predictability as revealed by naturally occurring analogues. *J. Atmos. Sci.*, **26**, 636-646.

Nese, J. M. 1989 Quantifying local predictability in phase space. *Physica D*, **35**, 237-250.

Olsen, L. F., Valeur, K. R., Geest, T., Tidd, C. W. and Schaffer, W. M. 1994 Nonlinear forecasting of non-uniform chaotic attractors in an enzyme reaction. *Phil. Trans. R. Soc. Lond.* **A, 348(1688)**, 421-430.

Palmer, T. N. 1992 Ensemble prediction. *Meteorological*, **(58)**, 5-15 (June).

Palmer, T. N. 1993 Extended-range atmospheric prediction and the Lorenz model. *Bull. Amer. Meteo. Soc.*, **74(1)**, 49 (June).

Palmer, T. N., Buizza, R., Molteni, F., Chen, Y. C. and Corti, S. 1994 Singular vectors and the predictability of weather and climate. *Phil. Trans. R. Soc. Lond.* **A, 348(1688)**, 459-475.

Sauer, T. 1993 Time series prediction using delay coordinate embedding. In *Predicting the Future and Understanding the Past: A Comparison of Approaches* (ed. A. Weigend and N. Gershenfeld), **volume XV** of *SFI Studies in Complexity*, pp. 175-194, New York, Addison-Wesley.

Sauer, T., Yorke, J. A. and Casdagli, M. 1991 Embedology. *J. Stat. Phys.*, **65**, 579-616.

Smith, L. A. 1992 Identification and prediction of low-dimensional dynamics. *Physica D*, **58**, 50-76.

Smith, L. A. 1993 Do Lyapunov exponents limit predictabilty? Preprint.

Smith, L. A. 1993 Does a meeting in Santa Fe imply chaos? In *Predicting the Future and Understanding the Past: A Comparison of Approaches* (ed. A. Weigend and N. Gershenfeld), **volume XV** of *SFI Studies in Complexity*, pp. 323-344, New York, Addison-Wesley.

Smith, L. A. 1994 Visualising predictability with chaotic ensembles. In *Advanced Signal Processing: Algorithms, Architectures and Implementations* (ed. F.T. Luk), **volume 2296**, Bellingham, WA, SPIE. to appear.

Smith, L. A., Ziehmann-Schlumbohm, C. and Fraedrich, K. 1994 The limits of predictability: The decrease of uncertainty and return of skill in the lorenz system. Pre-print.

Smith, L. A. and Drysdale, D. 1994 Local optimal prediction of low dimensional dynamics. In preparation.

Sugihara, G. and May, R. M. 1990 Nonlinear forecasting as a way of distinguishing chaos from measurement error in a time series. *Nature*, **344**, 734-741.

Takens, F. 1981 Detecting strange attractors in fluid turbulence. In *Dynamical Systems and Turbulence* (ed. D. Rand and L. S. Young), **volume 898**, pp. 366, New York, Springer-Verlag.

Theiler, J. 1986 Spurious dimension from correlation algorithms applied to limited time-series data. *Phys. Rev. A*, **34(3)**, 2427-2432.

Theiler, J. and Eubank, S. 1994 Don't bleach chaotic data. *Chaos*, To appear.

Theiler, J., Eubank, S., Longtin, A., Galdrikan, B. and Farmer, J. D. 1992 Testing for nonlinearity in time series: the method of surrogate data. *Physica D*, **58**, 77.

Tong, H. 1990 *Non-Linear Time Series Analysis.* Oxford Univ. Press, Oxford.

Tong, H. and Moeanaddin, R. 1988 On multi-step non-linear least squares prediction. *The Statistician*, **37**, 101-110.

von Mises, R. 1957 *Probability Statistics and Truth.* George Allen and Unwin, London.

Weigend, A. and Gershenfeld, N. (editors 1993) *Predicting the Future and Understanding the Past: A Comparison of Approaches*, **volume XV** of *SFI Studies in Complexity.* Addison-Wesley, New York.

Weiss, N. O. 1990 Periodicity and aperiodicity in solar magnetic activity. *Phil. Trans. R. Soc. Lond.*, **A 330**, 617-625.

Wolpert, D. 1992 Stacked generalization. *Neural Networks*, **5**, 241-259.

Ziehmann-Schlumbohm, C. 1994 *Vorhersagestudien in chaotischen Systemen und in der Praxis - Anwendung von Methoden der nichtlinearen Systemanalyse.* PhD. Thesis, *Freie Universität Berlin. Meteorologische Abhandlungen N.F. Serie A Monographien.*

Ziehmann-Schlumbohm, C., Fraedrich, K., and Smith, L. A. 1995 Ein internes vorhersagbarkeitsexperiment im lorenz-modell. *Meteorologische Zeitschrift N.F.*, **14**.

CHAPTER 5

# A POISSON DISTRIBUTION FOR THE BDS TEST STATISTIC FOR INDEPENDENCE IN A TIME SERIES

Rodney C. L. Wolff
Department of Statistics
University of Glasgow
Glasgow. G12 8QW, U.K.

**Abstract**

This paper examines a statistical method derived from chaos theory. The correlation integral was proposed over a decade ago as a way of detecting chaos in a possibly partial realisation of a dynamical system, because it depends on the spatial arrangement of the reconstructed attractor of the system. We exploit geometrical properties of an embedded time series in order to establish a test of independence in the original time series. Earlier efforts here have employed the Central Limit Theorem to obtain Normality as the null distribution; however, the testing procedure was, to an extent, *ad hoc.* By making moderately weak assumptions about the marginal distribution of the given series, we obtain a Poisson law for the correlation integral under the null hypothesis of independence, and use non–parametric methods to specify the test precisely. We compare the size and power of the present test with its predecessor and with other non–parametric tests for serial dependence.

## 1. Introduction

### *1.1. Motivation*

In classical time series analysis, in which one utilises linear models incorporating Gaussian innovations, questions of independence are often dealt with by examining correlations. However, if there is departure from Normality and linearity, then independence and lack of correlation need no longer be equivalent. A series with zero autocorrelation at each lag need not necessarily comprise independent observations; see, for example, Hall & Wolff (1994b). The existence of independence in a time series is of much interest in exploratory analyses and modelling, as well as in model–checking, since one typically fits models incorporating independent and identically distributed (i.i.d.) innovations (noise). Prediction and forecasting (Kendall & Ord, 1990, p. 122) are also a concern in time series analysis. There exist many methods of forecasting which resort to using fitted parametric models, non–parametric approaches, and other *ad hoc* ideas, such as exploiting the 'momentum' of a series. In general, all such approaches are governed by dependence structures in the series, and precision of a forecast may be better in a correlated series than in another one of independent observations with identical marginal properties to those of the former. Hence, there is the need to examine dependence structures in time series without resorting to statistical correlations or other linear features. The correlation integral provides a useful medium. Moreover, the use of the correlation integral in assessing independence in a time series is a fine example of the use of chaos theory in the context of stochastic, non–chaotic processes.

*1.2. The correlation integral*

We give a brief introduction to the correlation integral. Accompanying papers in this issue by Dr Cutler and Dr Smith, as well as the comprehensive review articles by Berliner (1992) and Isham (1993), provide more details.

A deterministic dynamical system can be specified in terms of a map, $f$, operating on a manifold, $M$, such that $f : M \to M$. The system may operate either in discrete time, in which case $f$ is applied iteratively on some initial conditions, or in continuous time, in which case $f$ specifies a collection of differential equations which produce a flow beginning at the initial conditions. Suppose that $f$ produces a discrete time trajectory, namely $\{x_0, f(x_0), f^2(x_0), \ldots\}$, where $x_0 \in M$ is the initial condition and $f^k$ denotes the $k$–fold iterate of $f$. The trajectory may be regarded as a time series. The long–run behaviour of a dynamical system, its equilibrium distribution, is embodied in the attractor. Therefore, an investigation of a dynamical system, given only a realisation of it, often commences with an investigation of the nature of its attractor, and especially its dimension. The attractor of a chaotic dynamical system can be Cantor–like, and so may be summarised using a fractional dimension.

We can define a fractional dimension as follows. Choose a point $x_0$, specified in $d$ coordinates, from the attractor. If $\Delta$ is the dimension of the attractor, then, heuristically, the number of points, $N(h)$, of a 'typical' trajectory of length $n$ of the dynamical system within distance $h$ of $x_0$ should satisfy $N(h) \propto h^{\Delta}$. This leads to the definition of the information or correlation dimension,

$$\Delta = \lim_{n\to\infty} \lim_{h\to 0+} \frac{\log N(h)}{\log h}.$$

If the limit exists then it need not be an integer.

The observed realisation of a dynamical system may be a cross–section of the true realisation, for example, a projection into a lower–dimensional space or the realisation of the first–return map to a lower–dimensional subspace. Takens (1981) proves that such a cross–section of a generic dynamical system with attractor of dimension $\Delta$ can be embedded in a Euclidean space of dimension at most $2\Delta + 1$. Grassberger & Procaccia (1983) provide an algorithm to obtain the embedding dimension, given a time series $Y_1, \ldots, Y_n$. They use the correlation integral

$$C(n,p,h) = \binom{n-p+1}{2}^{-1} \sum_{i=1}^{n-p} \sum_{j=i+1}^{n-p+1} I\Big(||V_i - V_j|| \le h\Big), \tag{1}$$

where the vectors

$$V_i = \Big(Y_i, \ldots, Y_{i+p-1}\Big)' \qquad (i = 1, \ldots, n-p+1) \tag{2}$$

constitute the embedding of the time series into $p$–dimensional Euclidean space, and where $I(A)$ denotes the indicator function of event $A$. Assuming that the underlying series is completely random, in the sense of having no deterministic structure, then the number of pairs of points which are within distance $h$ of each other should be proportional to the volume containing such points, $h^p$. Complete Spatial Randomness in spatial point patterns is an example of this (Diggle, 1983). If the series is deterministic or chaotic, then vectors of length $p_0$ or more, for some $p_0$, will contain functionally dependent entries, and so the power law will saturate at $h^{p_0}$ when $p \ge p_0$. Suitable log–log plots can be used to check for saturation in the power law. Smith (1988) discusses limitations in this regard.

### *1.3. A test for independence (Brock et al., 1986)*

Brock, Dechert & Scheinkman (1986) apply the Central Limit Theorem to the correlation integral and deduce its asymptotic distribution. Their definition of the correlation integral uses the $L^\infty$ norm in equation (1) for comparing distances, $h$, between points of the embedded series. For convenience, we use the $L^2$ norm, as in Grassberger & Procaccia (1983), but we conjecture that the choice of norm will not substantially affect the procedure, as $h \to 0^+$, if the underlying series is independent. Brock *et al.* prove that, under the null hypothesis that $Y_1, \ldots, Y_n$ constitute a series of i.i.d. observations, $C(n,p,h)$ converges to $\Big\{\mathrm{pr}\Big(|Y_i - Y_j| \le h\Big)\Big\}^p$ as $n \to \infty$, with probability 1 — recall that they use the $L^\infty$ norm in creating this expression. Further, they show that $\alpha_1 C(n, p_1, h) + \alpha_2 C(n, p_2, h)$, suitably scaled, has an asymptotic normal distribution, the variance of which is given explicitly in terms of $\mathrm{pr}\Big(|Y_i - Y_j| \le h\Big)$ and $\mathrm{pr}\Big(|Y_i - Y_j| \le h \;\&\; |Y_j - Y_k| \le h\Big)$ $(i, j, k$ distinct). There is a numerical study, which we shall discuss more fully along with our results, in which they observe that the actual type I error exceeds the nominal significance when $p \ge 3$. Type II error is calculated for three specific parametric alternative hypotheses.

Brock *et al.* do not place any conditions on the order of magnitude of the comparison distance, $h$. If $h$ becomes large then the correlation integral equals unity with probability 1: it is obvious that this can occur for finite $h$. For fixed $n$, as $h \to 0^+$ there will be fewer and fewer points of the embedded series deemed sufficiently close and therefore $C(n,p,h) \to 0$. We suspect that their simulated type I error would converge to the nominal significance level if $h$ were chosen in accordance with their rate of convergence in distribution to the Normal of the test statistic.

### *1.4. Some other tests for independence in a time series*

Skaug & Tjøstheim (1993a) consider estimating the dependence measure proposed by Hoeffding (1948) for the case of two random variables, which consists of calculating the squared error between the bivariate cumulative distribution function (c.d.f.) and the product of the marginals. Skaug & Tjøstheim consider the time series case by estimating Hoeffding's measure *via* the empirical bivariate c.d.f. The bivariate c.d.f. for a time series is defined to be that of observations at successive epochs. In particular, they obtain the empirical marginal c.d.f.s directly from the bivariate empirical c.d.f., and they note the relationship of their test statistic with the Cramér–von Mises functional for test of fit. Skaug & Tjøstheim (1993b) similarly construct a test using kernel estimates of the bivariate and marginal densities.

Some classical tests for independence are based on the distributions of partial autocorrelations, as described in Hannan (1970, Ch. 3). The Durbin–Watson statistic (Durbin & Watson, 1950, 1951, 1971) enjoys popularity in testing for serial correlation.

### *1.5. Outline of paper*

We proceed as follows. Assuming a technical condition on the marginal distribution of the observed time series, we treat the unnormalised correlation integral as a sum of dependent Bernoulli variables with possibly different parameters. Given the null hypothesis of independence amongst the time series values, we calculate the departure, in law, of the count from a Poisson distribution, and find a proportionality connecting $n$, $h$ and $p$ which guarantees convergence to a Poisson distribution as the sample size increases. We then indicate how the constants of the said proportionality may be calculated, and relate such calculations to computing power. We conclude with a comparative numerical study of this and other tests of independence.

## 2. Poisson convergence

### *2.1. The Chen–Stein method*

Chen (1975), using methods established by Stein (1972), proved that convergence to a Poisson distribution, for the number of occurrences of dependent events, can often be deduced by computing no more than the first and second moments. Arratia, Goldstein & Gordon (1989, 1991) review this so-called Chen–Stein method and discuss extensions to the multivariate case and to approximations by Poisson processes, along with applications. See also Barbour, Holst & Janson (1992). We shall use the following result.

Let $\mathcal{I}$ be an index set and let $X_k$ $(k \in \mathcal{I})$ be Bernoulli random variables with $0 \le p_k = \text{pr}\Big(X_k = 1\Big) = E(X_k) < 1$. Let

$$D = \sum_{k \in \mathcal{I}} X_k, \qquad \lambda = E(D) = \sum_{k \in \mathcal{I}} p_k,$$

such that $0 < \lambda < \infty$. Further, for each $k \in \mathcal{I}$, define

$$\mathcal{I}_k \;=\; \mathcal{I} \Big\backslash \Big\{\ell\,;\; X_\ell \text{ is independent of } X_k\Big\},$$

which Arratia *et al.* call a neighbourhood of dependence. Put

$$b_1 = \sum_{k \in \mathcal{I}} \sum_{\ell \in \mathcal{I}_k} p_k p_\ell, \qquad b_2 = \sum_{k \in \mathcal{I}} \sum_{\ell \in \mathcal{I}_k \backslash \{k\}} p_{k\ell},$$

$$b_3 = \sum_{k \in \mathcal{I}} E\left| E\Big\{X_k - p_k \,\Big|\, \sigma\Big(X_\ell\,;\; \ell \in \mathcal{I}\backslash\mathcal{I}_k\Big)\Big\}\right|,$$

where $p_{k\ell} = E(X_k X_\ell)$ and $\sigma$ denotes the sigma field generator. All of $b_1$, $b_2$ and $b_3$ are non-negative. Let $P$ have a Poisson distribution such that $E(P) = E(D)$, and let

$$\text{TVD}(P, D) = 2 \sup_{A \subseteq \mathbf{Z}} \left| \text{pr}\Big(P \in A\Big) - \text{pr}\Big(D \in A\Big) \right|$$

denote the total variation distance between the distributions of $P$ and $D$. Arratia *et al.* (1989) show that $\text{TVD}(P, D) \le 2(b_1 + b_2 + b_3)$, and that, if $X'_k$ has the same distribution as $X_k$ $(k \in \mathcal{I})$ such that $\{X'_k\}$ are mutually independent, then

$$\text{TVD}\Big(\{X_k\}_{k \in \mathcal{I}}, \{X'_k\}_{k \in \mathcal{I}}\Big) \;\le\; 2(2b_1 + 2b_2 + b_3) + 4 \sum_{k \in \mathcal{I}} p_k^2.$$

*2.2. Proof of Poisson convergence of the correlation integral*

Berman & Eagleson (1983) obtain Poisson limit results for a family of incomplete symmetric statistics, of which the correlation integral is a member. However, we give details of our proof here, as some aspects will provide useful details for the testing procedure.

Let the time series $Y_1, \ldots, Y_n$ be i.i.d. with a bounded, continuous probability density function (p.d.f.) $g$. With $n' = n - p + 1$, define an embedding of the original time series in a phase space of dimension $p$, as in (2). The binary-valued variables

$X_{ij} = I\{V_i - V_j \in \mathcal{B}_p(0,h)\}$ $(1 \le i < j \le n')$ will play the rôles of the Bernoulli variables in the Chen–Stein theorems. Note that $\mathcal{B}_p(0,h)$ denotes the ball of radius $h$ in $p$–dimensional Euclidean space; implicitly, we are using the $L^2$ norm when comparing distances between pairs of vectors, $V_i$ and $V_j$. Call

$$D(n,p,h) = \sum_{i=1}^{n'-1} \sum_{j=i+1}^{n'} X_{ij}$$

the correlation count. Clearly, it is $\frac{1}{2}n'(n'-1)$ times the value of the correlation integral, given in (1). We can write

$$D(n,p,h) = \sum_{i=1}^{n'-1} \sum_{j=1}^{p-1} X_{i,i+j} + \sum_{i=1}^{n'-1} \sum_{j=p}^{n'-i} X_{i,i+j},$$

and $\mathcal{S}_j = \{X_{i,i+j}\}_{i=1}^{n'-i}$ is a sequence of $M_{p,j}$–dependent random variables, such that $M_{p,j} = p - jI(1 \le j \le p-1)$, and where the sequence $\{U_i\}$ of random variables is said to be $M$–dependent if $U_i$ is independent of $\sigma\Big(U_{i+M+1}, U_{i+M+2}, \ldots\Big)$ $(i = 1, 2, \ldots)$. When $M = 0$, $M$–dependence corresponds to independence.

For $i < j$,

$$E(X_{ij}) \;=\; \mathrm{pr}\Big(X_{ij} = 1\Big) \;=\; \mathrm{pr}\Big\{V_i - V_j \in \mathcal{B}_p(0,h)\Big\} \;\sim\; h^p v_p g_{ij}(0) \qquad (3)$$

as $h \to 0^+$, where $v_p$ is the volume of $\mathcal{B}_p(0,1)$ and $g_{ij}$ is the p.d.f. of $V_i - V_j$. Hence the expected contribution to the value of $D(n,p,h)$ by the sequences, $\mathcal{S}_j$, of $M_{p,j}$–dependent random variables $(j = 1, \ldots, p-1)$ is of order of magnitude $nh^p$ as $n \to \infty$ and $h \to 0^+$. Also, the expected contribution to the value of $D(n,p,h)$ by the sequences, $\mathcal{S}_j$, of independent random variables $(j = p, \ldots, n'-i)$ is of order of magnitude $n^2h^p$ under the same limits.

Recall that a Binomial random variable with index $N$ and parameter $\pi$ converges in distribution to a Poisson random variable with parameter $\lambda$ as $N \to \infty$ provided only that $N\pi \to \lambda$ as well. Along the same lines as this, the preceding paragraph suggests that, provided $n^2h^p$ tends to a constant, say $L$, as $n \to \infty$ and $h \to 0^+$, $D(n,p,h)$ will be asymptotically bounded in expectation, and Poisson convergence may be possible. Indicate this multiple limit with the notation $(n, h, n^2h^p) \to (\infty, 0^+, L)$. Specifically, if $0 \le \sup E\{D(n,p,h)\} \le \beta < \infty$, where the supremum is taken over $n$, then $\mathrm{pr}\Big\{D(n,p,h) > d\Big\} \le E\{D(n,p,h)\}/d \le \beta/d$, by Markov's inequality, as $D(n,p,h) \ge 0$ with probability 1. Hence the probability mass of $D(n,p,h)$ is concentrated towards the origin, and this is consistent with $D(n,p,h)$ attaining a Poisson distribution in the limit.

With the obvious modification of the notation of Arratia *et al.* (1989), let $\mathcal{I} = \Big\{(i,j); 1 \le i < j \le n'\Big\}$, $\mathcal{I}_{ij} = \Big\{(k,\ell)\ ;\ X_{ij} \text{ is dependent upon } X_{k\ell}\Big\}$, and so $(i,j) \in$

$\mathcal{I}_{ij}$. Since $X_{ij}$ is dependent on $X_{k\ell}$ exactly when $i-p+1 \le k \le i+p-1$ or $j-p+1 \le \ell \le j+p-1$, $\#(\mathcal{I}_{ij}) \le (2p-1)^2$ and $\#(\mathcal{I}) = n'(n'-1)/2$. Thus

$$b_1 \le \#(\mathcal{I})\,\#(\mathcal{I}_{ij}) \left\{ h^p v_p \max_{(i,j)\in\mathcal{I}} g_{ij}(0) \right\}^2 \le \binom{n'}{2} (2p-1)^2 h^{2p} v_p^2 \max_{(i,j)\in\mathcal{I}} \{g_{ij}(0)\}^2 .$$

Since $g$ is bounded so too is $g_{ij}$, and so $b_1 \to 0$ as $(n, h, n^2h^p) \to (\infty, 0^+, L)$.

Now consider $b_2$. Choose $(i,j) \in \mathcal{I}$ and $(k,\ell) \in \mathcal{I}_{ij}$. Since $i<j$ and $k<\ell$ there is an index $s \in S_1 = \Big\{i,\ldots,i+p-1\Big\} \bigcup \Big\{j,\ldots,j+p-1\Big\}$ such that $s \notin S_2 = \Big\{k,\ldots,k+p-1\Big\} \bigcup \Big\{\ell,\ldots,\ell+p-1\Big\}$. Let

$$\mathcal{X} = \Big\{X_t\,;\, t \in S_1 \bigcup S_2\Big\} \Big\backslash \Big\{X_s\Big\}.$$

Then

$$\begin{aligned}
E(X_{ij}X_{k\ell}) &= \mathrm{pr}\Big\{V_i - V_j \in \mathcal{B}_p(0,h) \;\&\; V_k - V_\ell \in \mathcal{B}_p(0,h)\Big\} \\
&= \mathrm{pr}\Big\{V_k - V_\ell \in \mathcal{B}_p(0,h) \,\Big|\, V_i - V_j \in \mathcal{B}_p(0,h)\Big\}\mathrm{pr}\Big\{V_i - V_j \in \mathcal{B}_p(0,h)\Big\} \\
&\le \mathrm{pr}\Big\{V_k - V_\ell \in \mathcal{B}_p(0,h) \,\Big|\, V_i, V_j\Big\}\mathrm{pr}\Big\{V_i - V_j \in \mathcal{B}_p(0,h)\Big\} \\
&\le \mathrm{pr}\Big\{V_k - V_\ell \in \mathcal{B}_p(0,h) \,\Big|\, \mathcal{X}\Big\}\mathrm{pr}\Big\{V_i - V_j \in \mathcal{B}_p(0,h)\Big\} \\
&= \mathrm{pr}\Big(|X_s - x_0| \le h \,\Big|\, \mathcal{X}\Big)\mathrm{pr}\Big\{V_i - V_j \in \mathcal{B}_p(0,h)\Big\} \\
&= \left\{\int_{-h}^{h} g_X(x - x_0)\,dx\right\} \mathrm{pr}\Big\{V_i - V_j \in \mathcal{B}_p(0,h)\Big\},
\end{aligned}$$

where $g_X$ is the p.d.f. of $X_s$,

$$\begin{aligned}
&\le \left\{\sup_u g_X(u)\right\} \left(\int_{-h}^{h} dx\right) \mathrm{pr}\Big\{V_i - V_j \in \mathcal{B}_p(0,h)\Big\} \\
&\sim 2h \left\{\sup_u g_X(u)\right\} h^p v_p g_{ij}(0),
\end{aligned}$$

as $h \to 0^+$. Thus

$$b_2 \le \binom{n'}{2} \Big\{(2p-1)^2 - 1\Big\}\, 2h \left\{\sup_u g_X(u)\right\} h^p v_p g_{ij}(0) \to 0$$

as $(n, h, n^2h^p) \to (\infty, 0^+, L)$.

By construction of $\mathcal{I}_{ij}$, $b_3 = 0$.

Finally, from equation (3),

$$4 \sum_{(i,j)\in\mathcal{I}} \{E(X_{ij})\}^2 \to 0$$

as $(n, h, n^2h^p) \to (\infty, 0^+, L)$.

Hence we have that $D(n,p,h)$ has a limiting Poisson distribution as described. In practice, the choice of the comparison distance, $h$, will be governed by the relationship $h \propto n^{-2/p}$.

## 3. Testing procedure

### *3.1. Calculation of the mean in the Poisson law*

Under the prescribed rate of convergence, the asymptotic mean of the correlation count is $\frac{1}{2}Lv_p\overline{g}_0$, where $\overline{g}_0$ denotes the average of the densities of each $V_i - V_j$ $(i < j)$.

Since $v_p = \pi^{p/2}/\Gamma\{(p+2)/2\}$ (Courant, 1937, p. 304), it remains to evaluate the density of $V_i - V_j$ at zero, which is a function of the p.d.f. of $Y_i - Y_j$. Let $g^*$ be the p.d.f. of $Y_1 - Y_2$. Then

$$g^*(u_1) = \int_{\mathcal{R}} g(u_1 + u_2)g(u_2)du_2, \qquad g^*(0) = \int_{\mathcal{R}} \{g(u)\}^2\, du.$$

Observe that $V_i$ and $V_j$ $(i < j)$ are stochastically independent whenever $j - i \geq p$. Suppose this to be the case. Then, for infinitesimals $dw_0, \ldots, dw_{p-1}$, the p.d.f., $g_{ij,1}$, of $V_i - V_j$, evaluated at zero, is given by

$$\begin{aligned} g_{ij,1}(0)dw_0 \cdots dw_{p-1} &= \mathrm{pr}\Big(0 < Y_{i+k} - Y_{j+k} \leq dw_k,\ 0 \leq k \leq p-1\Big) \\ &= \prod_{k=1}^{p-1} \mathrm{pr}\Big(0 < Y_{i+k} - Y_{j+k} \leq dw_{k+1}\Big), \end{aligned}$$

since the entries of $V_i - V_j$ are mutually independent, which yields

$$g_{ij,1}(0) = \{g^*(0)\}^p = \left[\int_{\mathcal{R}} \{g(y)\}^2\, dy\right]^p. \tag{4}$$

Consider the case $\lfloor (p+1)/2 \rfloor \leq j - i < p$, where $\lfloor x \rfloor$ denotes the largest integer not exceeding $x$. Let $\mathcal{Y}_{ij} = \Big\{Y_k; Y_k \text{ is an entry in both } V_i \text{ and } V_j, 1 \leq k \leq n\Big\}$. Under the condition $j - i = m$, say, no two elements of $\mathcal{Y}_{ij}$ will occupy a corresponding entry of both $V_i$ and $V_j$. In particular, precisely the first $p - m$ entries of $V_i$ and the last $p - m$ entries of $V_j$ are contained in $\mathcal{Y}_{ij}$. It is important to note this in what follows, as we shall condition on $\mathcal{Y}_{ij}$. It is easily checked that $2(p - m) \leq p$. Again, let $V_i - V_j$

have p.d.f. $g_{ij,2}$. Then, for infinitesimals $dw_0, \ldots, dw_{p-1}$,

$$\begin{aligned} & g_{ij,2}(0) dw_0 \cdots dw_{p-1} \\ & = \operatorname{pr}\Big(0 < Y_{i+k} - Y_{j+k} \le dw_k,\ 0 \le k \le p-1\Big) \\ & = \int \cdots \int \operatorname{pr}\Big(0 < Y_{i+k} - Y_{j+k} \le dw_k,\ 0 \le k \le p-1 \\ & \qquad \Big| y_r < Y_r \le y_r + dy_r,\ Y_r \in \mathcal{Y}_{ij}\Big) \times \left\{ \prod_{\ell=m}^{p-1} g(y_{i+\ell}) dy_{i+\ell} \right\}, \end{aligned}$$

since $\operatorname{pr}\Big(y_r < Y_r \le y_r + dy_r\Big) = g(y_r)dy_r$ for infinitesimal $dy_r$, and where integration is over $y_{i+m} \in \mathcal{R}, \ldots, y_{i+p-1} \in \mathcal{R}$. So, as $dw_k \to 0$ $(k = 0, \ldots, p-1)$, by continuity of $g$,

$$g_{ij,2}(0) = \{g^*(0)\}^{2m-p} \int \cdots \int \left\{ \prod_{k=m}^{p-1} g(y_{i+k}) g(y_{i+k}) \right\} \left\{ \prod_{\ell=m}^{p-1} g(y_{i+\ell}) dy_{i+\ell} \right\} \tag{5}$$

$$\begin{aligned} & = \{g^*(0)\}^{2m-p} \left[ \int_{\mathcal{R}} \{g(y)\}^3 dy \right]^{2p-2m} \\ & = \left[ \int_{\mathcal{R}} \{g(y)\}^2 dy \right]^{2m-p} \left[ \int_{\mathcal{R}} \{g(y)\}^3 dy \right]^{2p-2m}. \end{aligned} \tag{6}$$

We make two remarks. First, the product of squares of $g(y_k)$ in equation (5) arises from the first $(i+p-1)-(i+m)+1 = p-m$ entries of $V_i - V_j$, and the last $p-m$ of same, since, in those entries, we condition on exactly one term of the difference $Y_{i+k} - Y_{j+k}$. The remaining factors in equation (5) contribute to the integral of cubes of $g$ in (6). Secondly, when $m = p$ we obtain the result of the first case, namely when $j - i \ge p$.

Finally, let $1 \le j - i < \lfloor (p+1)/2 \rfloor$ $(p > 1)$, and define $\mathcal{Y}_{ij}$ as in the case $\lfloor (p+1)/2 \rfloor \le j - i < p$. Then the number of entries in $V_i - V_j$ which comprise a difference of two elements of $\mathcal{Y}_{ij}$ is $2(p-m) - p = p - 2m$, since precisely the first $p - m$ entries of $V_i$ and the last $p - m$ entries of $V_j$ are elements of $\mathcal{Y}_{ij}$, and $V_i - V_j$ is of length $p$. So the p.d.f. of $V_i - V_j$ at zero is given by

$$\begin{aligned} & g_{ij,3}(0) dw_0 \cdots dw_{p-1} \\ & = \operatorname{pr}\Big(0 < Y_{i+k} - Y_{j+k} \le dw_k,\ 0 \le k \le p-1\Big) \\ & = \int \cdots \int \operatorname{pr}\Big(0 < Y_{i+k} - Y_{j+k} \le dw_k,\ 0 \le k \le p-1 \\ & \qquad \Big| y_r < Y_r \le y_r + dy_r,\ Y_r \in \mathcal{Y}_{ij}\Big) \times \left\{ \prod_{\ell=m}^{p-1} g(y_{i+\ell}) dy_{i+\ell} \right\} \\ & = \int \cdots \int \operatorname{pr}\Big(0 < Y_{i+k} - y_{j+k} \le dw_k,\ 0 \le k \le m-1\ \& \end{aligned}$$

$$
\begin{aligned}
&0 < y_{i+\ell} - Y_{j+\ell} \le dw_\ell,\ p-m \le \ell \le p-1\ \& \\
&0 < y_{i+r} - y_{j+r} \le dw_r,\ m \le r \le p-m-1\Big) \\
&\quad \times \left\{ \prod_{s=m}^{p-1} g(y_{i+s}) dy_{i+s} \right\},
\end{aligned}
$$

where integration is over $y_{i+m} \in \mathcal{R}, \ldots, y_{i+p-1} \in \mathcal{R}$. As $dw_k \to 0$ $(k = 0, \ldots, m-1, p-m, \ldots, p-1)$, we obtain

$$
\begin{aligned}
& g_{ij,3}(0) dw_m \cdots dw_{p-m-1} \\
&= \int \cdots \int \left\{ \prod_{k=0}^{m-1} g(y_{j+k}) \right\} \left\{ \prod_{\ell=p-m}^{p-1} g(y_{i+\ell}) \right\} \left\{ \prod_{r=m}^{p-m-1} I\left(0 < y_{i+r} - y_{j+r} \le dw_r\right) \right\} \\
&\qquad \times \left\{ \prod_{s=m}^{p-1} g(y_{i+s}) dy_{i+s} \right\} \\
&= \int \cdots \int \left[ \prod_{k=m}^{2m-1} \{g(y_{i+k})\}^2 dy_{i+k} \right] \left\{ \prod_{\ell=p-m}^{p-1} g(y_{i+\ell}) \right\} \left\{ \prod_{s=2m}^{p-1} g(y_{i+s}) dy_{i+s} \right\} \\
&\qquad \times \left\{ \prod_{r=m}^{p-m-1} I\left(0 < y_{i+r} - y_{i+m+r} \le dw_r\right) \right\}, \qquad (7)
\end{aligned}
$$

noting that $2m \le p-1$ when we regroup some of the factors. This leads to two subcases.

The first subcase is when $2m-1 < p-m$, or $p \ge 3m$. This is possible, for example if $p = 4$ and $m = 1$. The presence of the indicator functions in (7) restricts the integration subject to $y_{i+2m} = y_{i+m}, \ldots, y_{i+p-m} = y_{i+p-2m}, \ldots, y_{i+p-1} = y_{i+p-1-m}$, a total of $2p - m$ constraints, as $dw_k \to 0$ $(k = m, \ldots, p-m-1)$. It may be shown that

$$
g_{ij,3}(0) = \left[ \int_{\mathcal{R}} \{g(y)\}^3 dy \right]^{p-3m} \left[ \int_{\mathcal{R}} \{g(y)\}^4 dy \right]^{4m-p} \left[ \int_{\mathcal{R}} \{g(y)\}^2 dy \right]^{p-3m}. \qquad (8)
$$

The second subcase is when $p < 3m$. This is possible, for example if $p = 5$ and $m = 2$. The calculations become quite complicated in this case, although it appears that we recover an expression very similar to that in equation (8).

For simplicity in the numerical study in §4, we shall assume that equation (4) is adequate for all values of $j - i$, since the number of cases where $j - i \le p$ is $o(1)$ asymptotically.

It is preferable either to calculate or to estimate the integrals of various powers of $g$ directly, as in Hall & Marron (1987). However, theory in that and related papers deals only with integrals of squares of density derivatives. Motivated by this particular application, Hall & Wolff (1994a) have developed consistent estimators of integrals of all powers of density derivatives, and we shall use their calculations in the sequel.

*3.2. A sequence of statistical hypotheses*

The result of §2.2, namely that, under the stated conditions, the correlation count is asymptotically distributed according to a Poisson distribution, enables us to construct a test of the hypothesis

$$H_0 : Y_1, \ldots, Y_n \text{ are i.i.d.}$$

against an alternative hypothesis specified by some non-linear model for the series $Y_1, \ldots, Y_n$, as in Brock *et al.* (1986).

As discussed in Cox & Hinkley (1974, §4.8), the general unavailability of uniformly most powerful tests leads to the consideration of local alternatives: loosely, small departures from the null hypothesis.

For power considerations in the present problem, we may consider a particular family of local alternatives, defined by

$$H_q : Y_i = \sum_{j=0}^{q} Z_{i+j} \quad (i = 1, \ldots, n),$$

where $Z_1, \ldots, Z_{n+q}$ are i.i.d.. Moreover, $H_q$ indeed specifies $H_0$ when $q = 0$.

If we assume that $Z_1$ possesses a bounded, absolutely continuous p.d.f., it is not difficult to show that $D(n,p,h)$ will converge in law to a Poisson distribution as $n \to \infty$ and $h \to 0^+$ provided that, simultaneously, $n^2h^p$ converges to a constant. Just as under $H_0$, we need only determine the density of $V_i - V_j$, evaluated at zero, in order to specify the limiting distribution. In principle this can be done in the same fashion as in §3.1, but the expression will be cumbersome and tedious to evaluate. We describe a special case below, and study some more cases in §4.

Suppose that $Z_1, \ldots, Z_{n+m} \sim N(0, \tau^2)$ and are, of course, independent. Let $q \geq 1$. Under $H_q$, $Y_i \sim N(0, q\tau^2)$ and $\text{cov}(Y_i, Y_j) = (j-i)\tau^2 I(1 \leq m < q)$, where $m = j-i$, as before. Thus $V_i - V_j = A\Big(Z_i, \ldots, Z_{j+p-1+q}\Big)'$ where $A$ is a matrix of order $p \times (m+p+q)$ with the row arrangement

$$\underbrace{1, \ldots, 1}_{a}, \underbrace{0, \ldots, 0}_{b}, \underbrace{1, \ldots, 1}_{a}$$

beginning in the $(k,k)$ position $(k = 1, \ldots, p)$, and 0's everywhere else. Here, $a = q$ and $b = m - q$ if $m \geq q$, while $a = q - m$ and $b = m$ if $1 \leq m < q$. Consequently, $V_i - V_j$ has a multivariate Normal distribution with mean 0 and variance matrix $\tau^2 AA'$, and the value of its density at zero is $\{(2\pi\tau^2)^p \det(AA')\}^{-1/2}$. With regard to hypothesis testing, this information now enables us to specify the distribution of the test statistic under any of the local alternative hypotheses, assuming Normality of the components of the moving average representation of $Y_i$.

## 4. Numerical study

Table 1

*Probability of a Type I error (i.i.d. uniform data): exact calculations*

| $p$ | 2 | 3 | 4 | 5 | 6 | 7 | 8 |
|---|---|---|---|---|---|---|---|
| $n = 100$ | 0.044 | 0.042 | 0.042 | 0.046 | 0.056 | 0.040 | 0.028 |
| $n = 250$ | 0.044 | 0.046 | 0.056 | 0.054 | 0.052 | 0.046 | 0.042 |
| $n = 1000$ | 0.046 | 0.060 | 0.062 | 0.058 | 0.054 | 0.042 | 0.032 |

Table 1: The table is headed by the dimension of the phase space. The remaining rows show the simulated probability of rejecting the null hypothesis that the given data are an i.i.d. time series, for independent time series of lengths 100, 250 and 1000. The parameter of the null Poisson distribution was evaluated according to equations (4), (6) and (8). There were 500 independent replications for each sample size.

In all of the simulations, parameters were chosen or estimated subject to the mean of the test statistic being unity. Of course, this implies that the test statistic has a larger coefficient of variation than another with a larger mean. In the following results, this may be the cause of large variability, which is occasionally non–systematic. For example, convergence to unity of the power of the test against particular alternatives, illustrated below, can be better achieved with a larger mean for the test statistic.

Our test statistic is a discrete random variable, and a test with exact significance of 5% is not available. In all of the simulations, we employed appropriate randomised rejection. That is, if $c$ is such that $0 < \mathrm{pr}(D > c) = \theta < 0.05$ and $0 < \mathrm{pr}(D = c) = \theta'$ $(\theta + \theta' > 0.05)$, then reject the null hypothesis of independence if $D > c$, and reject the null hypothesis with probability $(0.05 - \theta)/\theta'$ if $D = c$; if $D < c$ then retain the null hypothesis in preference to the alternative.

Table 1 shows the probability of a Type I error when the given time series comprises i.i.d. uniform random variables on the interval [0,1]. The mean of the null Poisson distribution was calculated exactly, using the correct functional form of the respective densities. The sizes of the various simulated tests are roughly in accordance with the specified theoretical significance level of 5%. However, over the range of phase dimensions we considered, for each series length, the size increased slightly over the range $p = 2, \ldots, 5$ or so, and then fell, a curious result. This is discussed in §5.2.

Table 2 shows the probability of a Type I error when the given independent time series comprises (i) i.i.d. uniform random variables on the interval [0,1], and (ii) i.i.d. standard Normal random variables. Randomised rejection was used, and the parameters of the respective test statistics were estimated non–parametrically. Whilst larger sample sizes improve accuracy, the size of the test increases faster with phase dimension than it does in the case of non–randomised rejection, the results of which are omitted here for brevity. The results are especially pleasing in the Normal case, from which one may conjecture that the test performs well on long–tailed data.

The effect of different kinds of non–linearity on the present test is illustrated in

Table 2

*Probability of a Type I error (i.i.d. uniform and Normal data): non–parametric approximations*

| $p$ | | 2 | 3 | 4 | 5 | 6 | 7 | 8 |
|---|---|---|---|---|---|---|---|---|
| Uniform | $n = 100$ | 0.059 | 0.087 | 0.126 | 0.142 | 0.152 | 0.167 | 0.182 |
| | $n = 250$ | 0.064 | 0.080 | 0.101 | 0.138 | 0.129 | 0.157 | 0.179 |
| | $n = 1000$ | 0.053 | 0.082 | 0.096 | 0.119 | 0.112 | 0.136 | 0.164 |
| Normal | $n = 100$ | 0.056 | 0.075 | 0.093 | 0.117 | 0.133 | 0.146 | 0.209 |
| | $n = 250$ | 0.066 | 0.076 | 0.089 | 0.102 | 0.113 | 0.143 | 0.172 |
| | $n = 1000$ | 0.051 | 0.056 | 0.059 | 0.072 | 0.086 | 0.106 | 0.120 |

Table 2: The table is headed by the dimension of the phase space. The remaining rows show the simulated probability of rejecting the null hypothesis that the given data are an i.i.d. time series, for independent time series of lengths 100, 250 and 1000. The parameters of the null Poisson distribution were evaluated according to the non–parametric method described in the text. There were 1000 independent replications for each sample size.

Table 3, in which a Poisson distribution was fitted to the correlation integral in the presence of time series arising from various parametric models. The models used were

$$Y_i = \alpha Y_{i-1} + \epsilon_i; \tag{9}$$
$$Y_i = \epsilon_i(1 + \alpha Y_{i-1}^2)^{1/2}; \tag{10}$$
$$Y_i = (\alpha + \beta\epsilon_{i-1})Y_{i-1} + \epsilon_i; \tag{11}$$
$$Y_i = \alpha\epsilon_{i-1} + \epsilon_i; \tag{12}$$
$$Y_i = \epsilon_{i-1}(\alpha + \epsilon_i); \text{ and} \tag{13}$$
$$Y_i = \begin{cases} \alpha Y_{i-1} + \epsilon_i & ; \ Y_{i-1} > 1 \\ \beta Y_{i-1} + \epsilon_i & ; \ Y_{i-1} \leq 1 \end{cases}, \tag{14}$$

where, in each case, $\epsilon_i \sim IN(0,1)$. In the simulations, we chose $\beta = 4$ always, and $\alpha = 0.3$, 0.5, $-0.4$, 0.4, 1.0 and 0.3 in model equations (9), (10), (11), (12), (13) and (14), respectively. The respective descriptions of these processes are autoregressive of order 1, AR(1); autoregressive conditional heteroscedastic of order 1, ARCH(1); bilinear of order 1,1,1, BL(1,1,1); moving average of order 1, MA(1); non–linear moving average of order 1, NLMA(1); and threshold autoregressive of order 1, TAR(1). Whilst this table gives some detail about overdispersion with respect to the unit mean Poisson distribution, consideration of variance–to–mean ratios might be more informative.

Figure 1 illustrates the power of the test for two particular alternatives, plotted against the parameter, $\alpha$, where the models governing the alternative hypothesis are a bilinear (BL) process given by $X_t = (\alpha + \beta\epsilon_{t-1})X_{t-1} + \epsilon_t$ $(\alpha^2 + \beta^2 < 1)$ (here, $\beta = 0.4$),

Table 3

*Proportions of exceedances above 3 of $D(n,p,h)$: non–parametric calculations*

| $p$ | 2 | 3 | 4 | 5 | 6 | 7 | 8 |
|---|---|---|---|---|---|---|---|
| AR(1) | 0.030 | 0.046 | 0.063 | 0.095 | 0.101 | 0.104 | 0.158 |
| | 0.017 | 0.025 | 0.052 | 0.046 | 0.083 | 0.098 | 0.164 |
| ARCH(1) | 0.032 | 0.031 | 0.081 | 0.117 | 0.165 | 0.247 | 0.334 |
| | 0.024 | 0.025 | 0.050 | 0.090 | 0.123 | 0.169 | 0.258 |
| BL(1) | 0.034 | 0.048 | 0.073 | 0.080 | 0.117 | 0.115 | 0.128 |
| | 0.032 | 0.031 | 0.053 | 0.074 | 0.089 | 0.097 | 0.125 |
| MA(1) | 0.017 | 0.045 | 0.071 | 0.102 | 0.110 | 0.152 | 0.165 |
| | 0.012 | 0.036 | 0.045 | 0.069 | 0.109 | 0.120 | 0.135 |
| NLMA(1) | 0.031 | 0.051 | 0.096 | 0.112 | 0.170 | 0.163 | 0.163 |
| | 0.033 | 0.048 | 0.062 | 0.090 | 0.110 | 0.149 | 0.163 |
| TAR(1) | 0.019 | 0.042 | 0.051 | 0.076 | 0.101 | 0.140 | 0.163 |
| | 0.034 | 0.028 | 0.047 | 0.057 | 0.060 | 0.110 | 0.131 |

Table 3: The table is headed by the dimension of the phase space. The remaining rows give the proportions of statistics exceeding 3, and correspond, respectively, to model equations (9), (10), (11), (12), (13) and (14) in the text, where the upper row for each model corresponds to series of length 100 and the lower to series of length 250. The nominal significance rate is 1.899%. There were 1000 independent replications for each sample size.

and a non–linear moving average (NLMA) process given by $X_t = \epsilon_{t-1}(\alpha + \epsilon_t)$. These are two models considered in Skaug & Tjøstheim (1993a) and a related preprint. The embedding dimension is 2. The curve labelled $I_0$ refers to the present test; the curves labelled $I_1$ and $I_2$ correspond to Skaug & Tjøstheim's kernel–based statistic; the curve labelled $I_3$ corresponds to their statistic based on the empirical distribution function; and $I_4$ corresponds to a correlation measure, similar to the classical Durbin–Watson statistic, as in Durbin & Watson (1950, 1951, 1971). Broadly speaking, it would appear from these and other simulations that the present test compares well with other methods, except possibly in the case of some linear models where dependence is fairly weak. Whilst its power improves as the embedding dimension increases, the size also increases, a feature observed by Brock *et al.* (1986). Our simulations improved on the reported values of size and power from their numerical study.

## 5. Remarks

### *5.1. Present results*

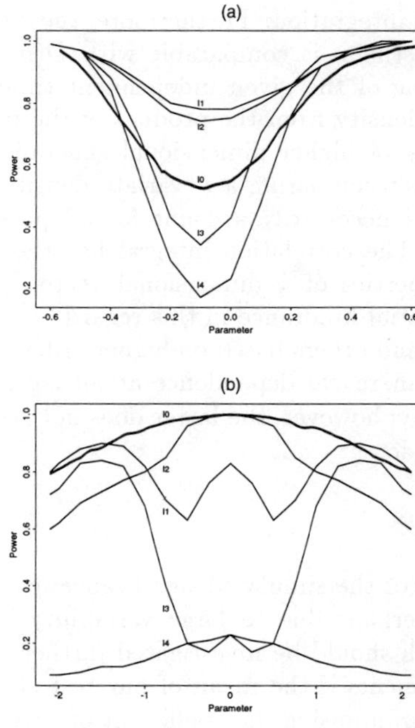


Fig. 1. Graphs of power *versus* the parameter, $\alpha$, of the process governing the alternative hypothesis, namely (a) a bilinear process and (b) a non–linear moving average process, as specified in §4 of the text. The power curve, $I_0$, relates to the present test; curves $I_1, \ldots, I_4$ relate to the non–parametric tests considered by Skaug & Tjøstheim (1993a), as mentioned in the text, and have been transcribed from their paper and a related preprint.

We have shown that the unnormalised correlation integral may have a Poisson distribution in the presence of an arbitrary time series, and we have indicated how the comparison distance should be chosen to enable this.

Our distributional results improve on those of Brock *et al.* (1986) in two ways. First, we specify a suitable comparison distance for the correlation integral. Secondly, we improve slightly on simulated sizes of some tests.

One appealing property of our method is that it does not involve much more numerical computation than in the method of Skaug & Tjøstheim (1993a). They must find estimates of three densities, a bivariate and two marginals, on a grid sufficiently fine to facilitate numerical integration. Furthermore, the nature of the test statistic proposed by Skaug & Tjøstheim is comparable with $D(n, 2, h)$: they examine the two–dimensional embedding of the given independent time series, particularly the departure of its bivariate density from the product of the two marginals. They also perform a numerical study of higher dimensional generalisations of their statistic. However, it is not clear that comparing a $k$–variate density with the product of its $k$ marginals ($k > 2$) must necessarily account for all possible structures of serial dependence up to lag $k$. The correlation integral has the potential to do this, by computing clustering properties of $k$–dimensional vectors; of course, the curse of dimensionality is a substantial hindrance in this regard.

Furthermore, this test and others based on higher order spectra reside in different classes. Whilst spectra can reveal dependence at all lags, the correlation integral examines only finitely many; however, the latter does not require moment conditions whereas spectral methods do.

### *5.2. Further considerations*

The non–monotonicity of the simulated significance levels, as the embedding dimension increased, was perhaps due to large variability in the correlation count, another phenomenon which should be investigated further. It is believed that there would be further improvements if the mean of our test statistic were to be chosen optimally, in the sense of minimising the coefficient of variation of the test statistic or seeking a variance–bias trade–off. More generally, the consistency of this test, as $h \to 0$, has not as yet been studied.

In order to compensate for the increase in the size of the test as the embedding dimension increases, it would be helpful to quantify the effect of the exponential growth of the phase space 'emptiness' on variation in the correlation count, with a view to introducing a correction factor to the comparison distance. Silverman (1986, Ch. 4) discusses related issues in the context of multivariate density estimation.

An unresolved issue is the choice of embedding dimension for our test. Using the geometric analogue of Skaug & Tjøstheim's (1993a,b) methods, we need only consider two dimensions, for which the asymptotic distribution theory appears to apply in the finite sample case. On one hand, using only a two–dimensional embedding would appear to neglect autocorrelations at lags three or more; similarly for other embeddings. Nevertheless, it is the total geometric structure of the embedding of which the correlation integral is a measure. Given that there is no saturation in the power law for the correlation integral, higher–dimensional embeddings of independent data should contain neither any more nor any less structure than in two dimensions. However, it is not clear how to assess which aspects of dependence, other than possibly autocorrelations, are overlooked by an embedding dimension which is too small.

## Acknowledgements

It is a pleasure to record my gratitude to Sir David Cox for supervision of this research, which comprises part of a recently completed doctoral thesis. This work was supported by an Australian Research Council grant to the Centre for Mathematics and Its Applications, Australian National University, a computing equipment grant from the Wingate Foundation, London, and a travel grant from the Nuffield Foundation.

## References

Arratia, R., Goldstein, L. & Gordon, L. 1989 Two moments suffice for Poisson approximations: the Chen–Stein method. *Ann. Prob.* **17**, 9–25.

Arratia, R., Goldstein, L. & Gordon, L. 1990 Poisson approximation and the Chen–Stein method. *Statist. Sci.* **5**, 403–34.

Barbour, A.D., Holst, L. & Janson, S. 1992 *Poisson Approximation.* Clarendon Press: Oxford.

Berliner, L.M. 1992 Statistics, probability and chaos (with Discussion). *Statist. Sci.* **7**, 69–90.

Berman, M. & Eagleson, G.K. 1983 A Poisson limit theorem for incomplete symmetric statistics. *J. Appl. Prob.* **20**, 47–60.

Brock, W.A., Dechert, W.D. & Scheinkman, J.A. 1986 A test for independence based on the correlation integral. Technical report: University of Wisconsin–Madison.

Chen, L.H.Y. 1975 Poisson approximation for dependent trials. *Ann. Prob.* **33**, 69–74.

Courant, R. 1937 *Differential and Integral Calculus*, volume II. Blackie: London.

Cox, D.R. & Hinkley, D.V. 1974 *Theoretical Statistics.* Chapman & Hall: London.

Diggle, P.J. 1983 *Statistical Analysis of Spatial Point Patterns.* Academic Press Inc.: London.

Durbin, J. & Watson, G.S. 1950 Testing for serial correlation in least–squares regression. *Biometrika* **37**, 409–428.

Durbin, J. & Watson, G.S. 1951 Testing for serial correlation in least–squares regression. *Biometrika* **38**, 159–178.

Durbin, J. & Watson, G.S. 1971 Testing for serial correlation in least–squares regression. *Biometrika* **58**, 1–19.

Grassberger, P. & Procaccia, I. 1983 Characterization of strange attractors. *Phys. Rev. Lett.* **50**, 346–349.

Hall, P. & Marron, J.S. 1987 Estimation of integrated squared density derivatives. *Statist. Prob. Lett.* **6**, 109–15.

Hall, P. & Wolff, R.C.L. 1994a Estimators of integrals of powers of density derivatives. *Statist. Prob. Lett.*, in press.

Hall, P. & Wolff, R.C.L. 1994b On the strength of dependence of a time series generated by a chaotic map. Submitted to *J. Time Ser. Anal.*

Hannan, E.J. 1970 *Multiple Time Series.* Wiley: New York.

Hoeffding, W. 1948 A nonparametric test of independence. *Ann. Math. Statist.* **19**, 546–57.

Isham, V. 1993 Some statistical aspects of chaos: a review. In *Chaos and Networks: Statistical and Probabilistic Aspects* Proceedings of a Workshop held at Sandbjerg Manor and the Department of Theoretical Statistics, Aarhus University, 25 April 1992 to 07 May 1992 (O.E. Barndorff–Nielsen, D.R. Cox, J.L. Jensen & W.S. Kendall, eds). Chapman & Hall: London.

Kendall, M. & Ord, J.K. 1990 *Time Series.* Edward Arnold: London.

Silverman, B.W. 1986 *Density Estimation for Statistics and Data Analysis.* Chapman & Hall: London.

Skaug, H.J. & Tjøstheim, D. 1993a A nonparametric test of serial independence based on the empirical distribution function. *Biometrika* **80**, 591–602.

Skaug, H.J. & Tjøstheim, D. 1993b Nonparametric tests of serial independence. In *Developments in Time Series Analysis*, the M.B. Priestley Birthday Volume, Ed. T. Subba Rao. Academic Press: London.

Smith, L.A. 1988 Intrinsic limits in dimension calculations. *Phys. Lett. A* **133**, 283.

Stein, C.M. 1972 A bound for the error in the Normal approximation to the distribution for a sum of dependent random variables. In *Proc. Sixth Berkeley Symp. Math. Statist. Probab.* **2**, 583–602. University of California Press: Berkeley.

Takens, F. 1981 On the numerical determination of the dimension of an attractor. In *Lecture Notes in Mathematics* **898**, (eds D.A. Rand & L.S. Young), 366–381.

Tong, H. 1990 *Nonlinear Time Series: A Dynamical Systems Approach.* Clarendon Press: Oxford.

CHAPTER 6

# CHAOS AND NONLINEAR FORECASTABILITY IN ECONOMICS AND FINANCE [0]

Blake LeBaron
Department of Economics
University of Wisconsin - Madison
1180 Observatory Drive
Madison, WI 53706
(608) 263-2516
blebaron@facstaff.wisc.edu

**Abstract**

Both academic and applied researchers studying financial markets and other economic series have become interested in the topic of chaotic dynamics. The possibility of chaos in financial markets opens important questions for both economic theorists as well as financial market participants. This paper will clarify the empirical evidence for chaos in financial markets and macroeconomic series. It will also compare these two concepts from a financial market perspective contrasting the objectives of the practitioner with those of economic researchers. Finally, the paper will speculate on the impact of chaos and nonlinear modeling on future economic research.

## 1. Introduction

It has now been almost ten years since economists began searching for chaotic dynamics in economic time series. This search has yielded deeper understandings of the dynamics of many different series, and has led to the development of several useful tests for nonlinear structure. However, the direct evidence for deterministic chaos in many economic series remains weak. This paper will survey the existing results and

[0]The author is grateful to the Alfred P. Sloan Foundation and the University of Wisconsin Graduate School for support.

give some intuition about why they were probably not unexpected. It will also argue that chaotic dynamics still needs to be taken seriously for economic systems, but the tools and methods of study will probably differ from those used in the past.

The possibilities of chaos in economic systems brought an enormous amount of initial interest. The concepts of limited forecastability and complex dynamical properties has very strong intuitive appeal for economics. From forecasting movements in foreign exchange and stock markets, to understanding international business cycles, chaos in economics had a broad range of potential applications. This led to an explosion of empirical work searching for possible chaos in all types of economic and financial time series. These studies have found little or no evidence for chaos in any economic time series, but they have turned up a surprising amount of unexplained nonlinear structure in many series. In hindsight most of these results should have been expected to some extent. Researchers looking at macroeconomic and financial series face certain constraints which make the likelihood of directly seeing chaos small. In macroeconomics the problem of short and noisy time series coming from a system whose dynamics and measurement probes may be changing over time impedes the ability to precisely estimate nonlinear processes. In financial markets traders' ability to perceive complex patterns and trade against them reduces the strength of these patterns yielding series which, although not completely random, are probably some of the most difficult to forecast of all real world time series.

This paper will argue that even though the case for economic chaos appears weak the issue is still a very open question for economic research. Many economic questions are concerned with what fraction of observed fluctuations are coming from exogenous shocks versus underlying structures in the system. In macroeconomics this would be related to the amount of business cycle fluctuations attributable to structures that are part of the dynamics of the system versus outside shocks. Similarly, in financial markets we are concerned with the amount of price movements and trading activity coming from the flow of new information into the system, versus the system generating this through a dynamic of trading and learning. Analysis along these lines will require a greater attention to economic theory along with new empirical tests which put a greater emphasis on how noise is processed in an economic environment.

These issues will be covered in detail in the next three sections. First, the results from macroeconomics will be discussed. Second, financial forecasting will be reviewed. The next to last section presents a brief example of some new directions for empirical work. The final section will discuss some of the most interesting directions for future work in economics.

## 2. Macroeconomics

The most likely candidates for nonlinear dynamic structures in economics were macroeconomic time series. These series which exhibit a large amount of structure through business cycle fluctuations seemed like a natural place to look for unseen determinism.

The first tests for chaotic dynamics in macro economic series used several different diagnostic tests. Many began the search with an application of the Grassberger-Procacia(1982) dimension estimation algorithm.[1] Most came to the same conclusion. The series were probably not deterministic chaos, but many showed evidence for interesting nonlinear structure.[2]

Initially, the diagnostics estimated were traditional invariants such as information dimension, or lyapunov exponents. Eventually, some of the studies included a test statistic for independence, the BDS test (Brock et al. 1988). This statistic tests the independence of a time series using the fact that for any independent series, $x_t$,

$$P(d\, m(x_t, x_s) < \epsilon) = P(d\, 1(x_t, x_s) < \epsilon)\, m,$$

$$d\, m(x_t, x_s) = \sup_{j=0,m-1} |x_{t+j} - x_{s+j}|,$$

where $P()$ is the probability of each event, and sup is the supremum norm or maximum. This tests the much more restrictive null hypothesis that the series is independent and identically distributed. It is not a test for chaos. It is useful because it is a well defined, and easy to apply test which has power against any type of structure in a series. This feature can be viewed as both a cost and a benefit. On the one hand it can detect many types of nonlinear dependence that might be missed by other tests. On the other hand, a rejection using this test is not very informative. One extension of this test is to use it as a residual diagnostic. The idea is that a certain linear or nonlinear model specification can be tested by looking at

$$\hat{e}_t = x_t - f(x_{t-1}, \ldots, x_{t-l}),$$

where $f()$ is some model specification. The diagnostic test for independence is then applied to $\hat{e}_t$, the estimated model residuals. There is a well defined asymptotic distribution theory for the test which is worked out in (Brock et al. 1988), and it can also be applied to the residual test in some specifications.[3] There are many other diagnostic tests for nonlinearity available. Several authors mentioned used the test developed by Tsay(1986). Also, the bispectral test developed by Hinich(1982) is

[1]See Grassberger et al.(1991). for a useful survey of good and bad uses of this algorithm, and Jaditz and Sayers(1993) for a good survey of some of these results from macroeconomics. Papers such as Barnett and Chen(1986), Brock and Sayers(1988), Frank and Stengos(1989), Frank, Gencay, and Stengos(1988), Sayers(1989), and Scheinkman and LeBaron(1989), all are applications to various macroeconomic time series.

[2]There were a few disputes in this literature in that one set of authors did make strong claims about finding chaos in a monetary index(Barnett and Chen 1986). However, their findings were disputed in Ramsey et al.(1990) which concluded that their claims about chaos were premature. The recent results in Barnett and Hinich(1993) correct some of the criticisms. The claims for chaos are weakened, but it looks like there is some kind of unexplained nonlinearity in these monetary series.

[3]The approach of using residual diagnostics in nonlinearity testing which was first proposed in Brock(1986) has come under some question recently. Theiler and Eubank(1992) find that in some cases analyzing a time series after linear residuals are taken may reduce power against some chaotic null hypothesis.

applied to several macro economic series in Ashley and Patterson(1989) and Barnett and Hinich(1993). One advantage of this diagnostic is that it has no power against linear stochastic processes. This means that there is no need for estimating residuals before testing. The careful researcher should probably be ready with a toolbox of different tests, knowing a little bit about what the properties of each one is before applying them.

Most of these series suffer from a major limitation of economic data, short, noisy, and possibly nonstationary time series. The number of points available are probably small by any of the data point requirements that have been previously suggested.[4] Most of these studies check the reliability of their results given the small amount of data available by comparing estimates with distributions from simulated stochastic processes. This methodology draws much of its inspiration from bootstrap techniques of Efron(1979).[5] In general this approach has helped keep researchers honest about what they can confidently claim to have found, and it has probably kept down the number of false positive tests of chaos.

A second related branch of research has attempted to directly fit nonlinear specifications to macroeconomic data. This approach has generally been more successful at finding strong evidence for nonlinearities in these series. Papers such as Beaudry and Koop(1993), Hamilton(1989), McQueen and Thorley(1993), Neftci(1984), Potter(1990), and Terasvirta and Anderson(1992) find evidence for nonlinear behavior in aggregate macro time series. They are generally supportive of the conjecture that business cycles behave differently during expansions and contractions, and that there are differences in the impact on future growth from positive and negative shocks today. Two recent papers directly use a nonlinear forecasting framework to evaluate nonlinearities in macro economic time series. Granger et al.(1993) and Jaditz and Sayers(1993). Both papers find that a nonlinear forecasting framework does not add much in terms of out of sample forecast performance. This may suggest some problems in terms of stability/stationarity of the previously documented results. In macroeconomic series the greatest gains have come from fitting nonlinear models which address certain well known features which are not part of a linear framework. This approach often has detected interesting nonlinearities which were missed using diagnostic testing alone.

## 3. Finance

Chaos and its implications for forecasting has been even more hotly debated in financial markets. Financial series provide potentially longer and cleaner series on which to do estimation and out of sample testing. Also, the obvious potential gains to forecasting stock price and foreign exchange rate series has drawn a lot of interest.[6]

---

[4]See Smith(1992) and Ruelle(1990). Most macro series range from about 100-800 data points depending on length and frequency.

[5]Similar techniques have been used in Theiler et al.(1992). See also Efron and Tibshirani(1993) for a more recent treatment of bootstrapping.

[6]One example of this interest is the recent survey in the Economist(1993).

Tests similar to those used for macroeconomic time series have been applied here as well.[7] On the issue of chaos the results often find strong evidence for nonlinear dependence, but no convincing evidence for chaotic dynamics.[8]

The issue of whether a financial series is indeed chaotic may not be of great importance to a financial forecaster who is only interested in adjusting dynamic trading strategies according to apparent predictability in time series. The fact that the previously mentioned diagnostics all found evidence for some kind of nonlinear structure should be a tantalizing indicator for financial forecasters. Just what kinds of structure are in financial time series that these tests are picking up? One of the largest deviations from pure randomness in financial series is volatility persistence. Return movements are very hard to forecast, but magnitudes of the movements are predictable. This has led to a large literature on trying to model this phenomenon.[9] The fact that stock returns exhibit this sort of structure alone is somewhat of a mystery, but the puzzle was further strengthened by LeBaron(1992,1992b) which showed that autocorrelations in stock and foreign exchange market returns were changing depending on an estimate of recent volatility. LeBaron showed that many financial series followed a process that looked like,

$$r_t = \log(p_t) - \log(p_{t-1})$$
$$r_t = f(\sigma\ 2_t) r_{t-1} + \epsilon_t$$
$$\sigma\ 2_t = \sum_{i=1} N r_{t-i}\ 2.$$

It turns out that $f()$ is a decreasing function of conditional variance indicating that local predictability in the series is higher during periods of lower volatility. This phenomenon can be used to achieve some small out of sample improvements in forecasts especially in weekly foreign exchange series.[10]

One final curious feature of financial forecasting which may be related to nonlinearities concerns the analysis of technical trading rules. These rules are heuristic forecasting methods which traders claim give them an edge in forecasting the movements of financial markets. Once generally accepted as being worthless by much

[7]Blank(1992), Brock et al.(1991), De Grauwe et al.(1993), Decoster et al.(1992), Frank and Stengos(1988) Hinich and Patterson(1985), Hsieh(1991), Mayfield and Mizrach(1992), Peters(1991), and Scheinkman and LeBaron(1989).

[8]One exception here is Peters(1991) who finds strong evidence for low dimensional ($d < 3$) chaos. His results are are from low frequency monthly data sets, and are not backed up by any convincing simulations. Without further evidence from simulations these results should be viewed with skepticism.

[9]See Bollerslev et al.(1990) for an excellent survey of this large area which includes the ARCH family of models originated by Engle(1982).

[10]Other papers have documented similar out of sample forecast performance for foreign exchange series. One example is Weigend et al.(1991). Other papers find little or no evidence for forecast improvement (Diebold and Nason(1990).

of the academic community several papers have reopened this debate.[11] The predictability appears to be greatest for foreign exchange markets where the magnitude of trading profitability makes up for reasonable estimates of the costs of trading in these markets. Many of the most successful rules used are related to moving average rules which attempt to follow long range trends. They recommend that a trader buy when the price is above a long range moving average,

$$p_t > \frac{1}{N} \sum_{i=0} N - 1 p_{t-i},$$

and sell when the price is below.[12]

To summarize these results, there is interesting evidence of potential predictability in many of these series. Before one concludes that there is lots of money to be made forecasting financial series several cautions should be made. First, the actual implementation of a forecasting rule for trading may involve unforeseen costs, and prices taken from recorded data sets may not actually be tradeable. Second, taking on some of these dynamic strategies may involve exposure to extensive risks. The large expected returns might be included with a high probability of the strategy losing a considerable amount of money.

This nonlinear forecastability has still ignored the original question about chaos in these series. Given the support for some kind of nonlinear structure the question of chaos still appears very interesting. It is possible that identifying chaos for actual returns series may run up against another barrier. The problem this time is not the lack of data.[13] The problem this time may be related to how much forecastability can be left around in a financial time series. To estimate a lyapunov exponent a researcher needs some idea of forecast degradation over a short horizon. This obviously implies that over the shortest horizon good forecasts were available to analyze how their performance drops after several periods have gone by. Forecasts of this high quality may be unreasonable for financial time series.

As an example of what a small amount of predictability means in a financial market the weekly British Pound/U.S. Dollar foreign exchange series was used. It was sampled every Wednesday from January 1974 through July 1992 at noon New York time. A trader with access to just the correct sign forecast to this series would still have to work hard to claim that the series was chaotic, but I doubt that the trader would care much since the daily return to a strategy of longing or shorting the pound according to the directional forecast would yield a return of about 1% per week, or about 68% compounded over a year. Even if the trader were charged

[11] Papers such as Brock et al.(1992), Curcio and Goodhart(1992), LeBaron(1991), Levich and Thomas(1993), Schulmeister(1987), Sweeney(1986), and Taylor(1993) all present some evidence that there is predictive power contained in some of the rules used by technical traders.

[12] This is the simplest implementation of these types of rules. Real traders use many more complicated variants of this type of strategy, but this is close to the basic idea.

[13] For financial series high quality series are available with millions of data points for high frequency data sets.

a large transaction cost of 0.5% per trade,[14] the strategy would at worst earn about 30% per year.[15] If a positive lyapunov exponent was reliably estimated it would probably imply even greater predictability over the short horizon. It is unlikely that such predictable structure exists for any financial time series.

These are only approximations and conjectures, but they emphasize an important point. For financial series there may be an extremely wide gap between successful nonlinear forecasting, and actual identification of chaotic dynamics in a financial market. The researcher looking for chaos in some of these series probably should cast the search a little wider than just looking a raw price movements. Several possibilities are mentioned in the next section.

## 4. On the Joint Stability of Volume and Volatility

This section gives a quick example reflecting some of the issues mentioned in the previous section. First, it will directly analyze the dynamics of a dual two component time series. Second, it will analyze the dynamics in a framework that considers the stochastic and deterministic aspects of a nonlinear time series. This approach is inspired by Tong(1990), and Yao and Tong(1992), and looks at the local stability properties of the dual series. The idea here is to consider the properties of a nonlinear stochastic process for stock market trading volume and price volatility. Specifically, to look at some of the local stability properties to see if the dynamical system exhibits any regions where trajectories are spreading apart. Even if a global system is stable it may still be of great interest to understand the stability properties of certain subregions. This is especially true when the system is being continually impacted by outside noise.

The series are constructed from daily NYSE trading volume and volatility constructed using the square of a returns series using the CRSP value weighted index from July 1962 through June 1988. Both series are log transformed,

$$s_t = \log(r_t\ 2),$$

and,

$$v_t = \log(V_t),$$

where $r_t$ is the returns series, and $V_t$ is trading volume.[16] Table 1 shows parameter estimates for a simple 1 lag VAR for the volatility and volume processes. The results show the strong persistence in both series, and also some amount of dynamic cross correlations. The strongest of these being a negative relation from lagged volume to future volatility.

A quick glance at the parameters suggests that the response of this system to new shocks is probably stable. Shocks to the system would be quickly dissipated.

---

[14]Actual costs are probably less than 1/5 this level.

[15]This is a lower bound since it is assuming that trader is trading every day.

[16]The trading volume series is detrended using a 100 day moving average.

Table 1: Global Linear Volume Volatility Estimates

$$s_t = a_s + b_s s_{t-1} + c_s v_{t-1}$$
$$v_t = a_v + b_v s_{t-1} + c_v v_{t-1}$$

| Dependent | $a_i$ | $b_i$ | $c_i$ |
|---|---|---|---|
| $s_t$ | 0.000<br>(0.030) | 0.128<br>(0.012) | -0.651<br>(0.128) |
| $v_t$ | 0.000<br>(0.002) | 0.004<br>(0.001) | 0.665<br>(0.009) |

Global simple vector autoregression estimated over entire time series. $s_t$ is volatility $\log(r_t\ 2)$, and $v_t$ is log trading volume (turnover detrended using 100 day moving average).

One measure of the overall speed at which new shocks to the system disappear is the largest eigenvalue of $A\ TA$ where $A$ is the estimated linear system from table 1. Let $\lambda$ be the largest eigenvalue of $A\ TA$. It follows that the deterministic distance one period in the future will be bounded by,

$$||F(x_t + \delta) - F(x_t)|| \leq |\lambda|||\delta|| + o(||\delta||),$$

where $||.||$ is the Euclidean norm $L\ 2$, and $F()$ is the deterministic portion of the dynamics. For this estimated system this number is 0.87. This again shows a strongly stable, and somewhat persistent series, quickly dampening new shocks.

It is possible that the joint dynamics of the two series may depend on the current state of the system. Also, the stability properties of the system may be different in different regions. To explore this, a local linear approximation is used. To get a general interpretation a very simple form of local linear fit will be used. The state space $(s_{t-1}, v_{t-1})$ is split into four subregions using the median value for each series as the splitting level. These four regions will be referred to as high and low volume and volatility respectively. The two series are contemporaneously correlated so the four regions will not contain the same number of points. There will be more points in the low volume - low volatility region, and in the high volume - high volatility region.

The one lag VAR from Table 2 is estimated for each of the four regions in table 2. The results from the first table change dramatically in certain regions. For example, the coefficient of volatility on lagged volume in the low-low region is -1.880, indicating a larger than 1-1 response of volatility on a lagged volume shock. Also, there is some indication of a larger than 1-1 response in the high volatility - low volume region where the coefficient is -1.362.

Another interesting pattern that appears is in the high volume region where the lagged volume to volatility channel changes sign from the low volume region, with parameter estimates of 0.547 and 0.174 in the high and low volume periods respectively. The parameter estimates suggest some instability in the low volume/low volatility region, and this is reflected in a estimate for $\lambda$ of 3.85 in this region. The estimate is also large for the low volume/ high volatility region with an estimate of 2.37. Both

Table 2: Local Linear Volume Volatility Estimates

$$s_t = a_s + b_s s_{t-1} + c_s v_{t-1}$$
$$v_t = a_v + b_v s_{t-1} + c_v v_{t-1}$$

| | | Low Volume | | | High Volume | | |
|---|---|---|---|---|---|---|---|
| | Dependent | $a_i$ | $b_i$ | $c_i$ | $a_i$ | $b_i$ | $c_i$ |
| Low Volatility | $s_t$ | -0.258 | 0.071 | -1.880 | -0.571 | 0.007 | 0.547 |
| | | (0.104) | (0.026) | (0.391) | (0.119) | (0.032) | (0.493) |
| | $v_t$ | -0.026 | 0.002 | 0.557 | 0.020 | -0.001 | 0.507 |
| | | (0.007) | (0.002) | (0.028) | (0.007) | (0.002) | (0.032) |
| High Volatility | $s_t$ | -0.538 | 0.376 | -1.362 | -0.602 | 0.389 | 0.174 |
| | | (0.168) | (0.075) | (0.469) | (0.124) | (0.054) | (0.320) |
| | $v_t$ | -0.038 | 0.018 | 0.630 | -0.025 | 0.021 | 0.674 |
| | | (0.012) | (0.006) | (0.035) | (0.010) | (0.004) | (0.026) |

Local simple vector autoregression estimated on portions of the time series. $s_t$ is volatility $\log(r_t\ 2)$, and $v_t$ is log trading volume (turnover detrended using 100 day moving average).

high volume regions give estimates less than 1 reflecting the stability shown in the local linear models fit in those regions. While the local instabilities in parts of the state space do not make this a chaotic system they do suggest that interesting dynamics may occur at certain times. Since the system is driven by noise it will be continuously swept through the unstable regions at different times in its evolution. It cannot be said that this is indicative of chaos either, but it does display some of the richness of the joint dynamics present in these series. Also, it shows some evidence that there are portions of phase space over which trajectories are locally spreading. Further tests and details are given in LeBaron(1993).

## 5. Toward the Future of Chaos in Economics

Many of the results up till now have been very negative about the presence of chaos in economic time series. This final section turns more upbeat in suggesting several untried and promising paths for the future.

The most obvious extension is to move toward new data sets. The most interesting will probably involve more detailed examination of multivariate series such as the study of volume and volatility shown here. Other work will move to direct analysis of components of economic systems from GNP product accounts to the spatial and geographic information on growing economies. These remain relatively uncharted for nonlinear empirical studies in economics.

Many of the early tests used have been generally atheoretic. There have been few connections from empirical results to underlying economic theories. This can be a plus or a minus at times, but in terms of chaotic dynamics the connections drawn from theories to data have been far too rare. The theorist operating in this area will need to use new techniques, since old standards of estimation and diagnostic testing may fail.[17] A few interesting papers have begun to appear which try to make some connections.[18] It is probably likely that extensive simulations from economic models will be compared with empirical data. However, the features matched may not be those from traditional chaos analysis.

Some recent work has extended the set of available nonlinear tools, and several of these tests have not been applied to economic data sets.[19] Although these tests may prove useful still further research is necessary in attempting to understand these series in relation to some of the important economic questions involved. Are business cycle fluctuations driven more by randomness or inherent structures in the economy? Is trade in financial markets self-generating? These questions still remain unanswered.

Another new branch is to look more closely at large scale interconnected systems as metaphors for economic behavior. This "complex systems" approach is used in Bak et al.(1992) and Brock and LeBaron(1993) in modeling some macroeconomic phenomenon. Neither of these papers says much about chaos per se, but there is a connection. One example of this connection is in the paper by Doyon et al.(1993). In this paper the authors examine the dynamics of a neural network as the overall connectivity is changed. They find a link between nodewise connection strengths and an onset of overall chaotic dynamics in the system. Whether results such as this can impact studies of macroeconomic phenomenon remains to be seen, but they look promising.

In some ways we are only beginning to cut the surface of how to analyze real world data in light of the knowledge about what kinds of dynamics can be generated by nonlinear processes. We have tried to directly move tests from one field to another in an attempt to gain new insights into our home fields. At times this has brought in fresh ideas, and added to the development of some new and interesting diagnostics. However, this transfer should not be the end of the story. We probably should rethink what the important problems were that we were attacking and readjust our searches in these directions. We also need to move closer to understanding and estimating theoretical models for explaining the empirical puzzles that we are interested in.

## References

Ashley, R. A. and Patterson, D. M. 1989 Linear versus nonlinear macroeconomics: A statistical test. *International Economic Review*, **30**, 685-704.

Bak, P., Scheinkman, J., Chen, K. and Woodford, M. 1993 Aggregate fluctuations

[17] Geweke(1993) provides an example of why there may be problems with traditional econometric methods when examining chaotic models.

[18] See for example Chavas and Holt(1991) for their model of hog cycles, and De Grauwe et al.(1993) for their work on foreign exchange markets. Mosekilde and Larson(1988) provide another related attempt to connect a simple chaotic model to some experimental results.

[19] These include Lee, White and Granger(1993), Dechert and Gencay(1992a,b), Nychka et al.(1992), and Sugihara and May(1990).

from independent sectoral shocks: self-organized criticality in a model of production and inventory dynamics. *Ricerche Economiche*, **47**, 3-30.

Barnett, W. and Chen, P. 1986 The aggregation-theoretic monetary aggregates are chaotic and have strange attractors. In *Dynamic Econometric Modelling: Proceedings of the Third Austin Symposium in Economics* (ed. W. Barnett, E. Berndt, H. White), Cambridge University Press.

Barnett, W. A. and Hinich, M. 1993 Has chaos been discovered with economic data? In *Nonlinear Dynamics and Evolutionary Economics* (ed. R. H. Day and P. Chen), pp. 254-265. Oxford University Press, Oxford.

Beaudry, P. and Koop, G. 1993 Do recessions permanently change output? *Journal of Monetary Economics*, **31**, 149-64.

Blank, S. C. 1992 "chaos" in futures markets? a nonlinear dynamical analysis. *Journal of Futures Markets*, **11**, 711-728.

Bollerslev,T., Chou, R. Y., Jayaraman, N. and Kroner, K. F. 1990 ARCH modeling in finance: A review of the theory and empirical evidence. *Journal of Econometrics*, **52(1)**, 5-60.

Brock, W. A. 1986 Distinguishing random and deterministic systems: Abridged version. *Journal of Economic Theory*, **40**, 168-195.

Brock, W. A., Dechert, W. D., Scheinkman, J. A. and LeBaron, B. 1988 A test for independence based on the correlation dimension. Technical report, University of Wisconsin, Madison, WI.

Brock, W. A., Hsieh, D. and LeBaron, B. 1991 *Nonlinear Dynamics, Chaos, and Instability: Statistical Theory and Economic Evidence.* MIT Press, Cambridge, MA.

Brock, W. A., Lakonishok, J. and LeBaron, B. 1992 Simple technical trading rules and the stochastic properties of stock returns. *Journal of Finance*, **47**, 1731-1764.

Brock, W. A. and LeBaron, B. 1993 Using structural modelling in building statistical models of volatility and volume of stock market returns. Technical report, University of Wisconsin - Madison, Madison, Wisconsin.

Brock, W. A. and Sayers, C. L. 1988 Is the business cycle characterized by deterministic chaos? *Journal of Monetary Economics*, **22**, 71-90.

Chavas, J. P. and Holt, M. T. 1991 On nonlinear dynamics: The case of the pork cycle. *American Journal of Agricultural Economics*, **73(3)**, 819-828.

Curcio, R. and Goodhart, C. A. E. 1992 When support/resistance levels are broken, can profits be made? evidence from the foreign exchange market. Technical report, London School of Economics, London, UK.

De Grauwe, P., Dewachter, H. and Embrechts, M. 1993 *Exchange Rate Theory: Chaotic Models of Foreign Exchange Markets.* Blackwell, Oxford.

Dechert, W. D. and Gencay, R. 1992 An algorithm for the n lyapunov exponents of an n-dimensional unknown dynamical system. *Physic D*, **59**, 142-157.

Dechert, W. D. and Gencay, R. 1992 Lyapunov exponents as a nonparametric diagnostic for stability analysis. *Journal of Applied Econometrics*, **7**, S41-S60.

Decoster, G. P., Labys, W. C. and Mitchell, D. W. 1992 Evidence of chaos in commodity futures prices. *Journal of Futures Markets*, **12**, 291-305.

Diebold, F. X. and Nason, J. M. 1990 Nonparametric exchange rate prediction? *Journal of International Economics*, **28**, 315-332.

Doyon, B., Cessac, C., Quoy, M. and Samuellides, M. 1993 Control of the transition to chaos in neural networks with random connectivity. *International Journal of Bifurcation and Chaos*, **3(2)**, 279-291.

Efron, B. 1979 Bootstrap methods: Another look at the jackknife. *The Annals of Statistics*, **7**, 1-26.

Efron, B. and Tibshirani, R. 1993 *An Introduction to the Bootstrap.* Chapman and Hall.

Engle, R. F. 1982 Autoregressive conditional heteroskedasticity with estimates of the variance of united kingdom inflation. *Econometrica*, **50**, 987-1007.

Frank, M. and Stengos, T. 1989 Measuring the strangeness of gold and silver rates of return. *Review of Economic Studies*, **56**, 553-68.

Frank, M. Z., Gencay, R. and Stengos, T. 1988 International chaos? *European Economic Review*, **32**, 1569-1584.

Frank, M. Z. and Stengos, T. 1988 Some evidence concerning macroeconomic chaos. *Journal of Monetary Economics*, **22**, 423-438.

Geweke, J. 1993 Inferences and forecasting for deterministic non-linear time series observed with measurement error. In *Nonlinear Dynamics and Evolutionary Dynamics* (ed. R. H. Day and P. Chen), Oxford University Press.

Granger, C. W. J., Terasvirta, T. and Anderson, H. M. 1993 Modeling nonlinearity over the business cycle. In *Business Cycles, Indicators, and Forecasting* (ed. J. H. Stock and M. W. Watson), pp. 311-326. University of Chicago Press.

Grassberger, P. and Procaccia, I. 1982 Characterization of strange attractors. *Physical Review Letters*, **50**, 346-349.

Grassberger, P., Schreiber, T. and Schaffrath, C. 1991 Nonlinear time sequence analysis. *International Journal of Bifurcations and Chaos*, **1**, 521-547.

Hamilton, J. D. 1989 A new approach to the economic analysis of nonstationary time series and the business cycle. *Econometrica*, **57**, 357-384.

Hinich, M. J. 1982 Testing for Gaussianity and linearity of a stationary time series. *Journal of Time Series Analysis*, **3**, 169-176.

Hinich, M. J. and Patterson, D. M. 1985 Evidence of nonlinearity in daily stock returns. *Journal of Business and Economic Statistics*, **3**, 69-77.

Hsieh, D. 1991 Chaos and nonlinear dynamics: Applications to financial markets. *Journal of Finance*, **46**, 1839-1878.

Jaditz, T. and Sayers, C. 1993 Is chaos generic in economic data? *International journal of bifurcation and chaos*, **3(2)**, 745-755.

Jaditz, T. and Sayers, C. 1993 Using out of sample forecasting performance to evaluate model specification. Technical report, Bureau of Labor Statistics, Washington, DC.

LeBaron, B. 1991 Technical trading rules and regime shifts in foreign exchange. Technical report, University of Wisconsin - Madison, Madison, WI.

LeBaron, B. 1992 Forecast improvements using a volatility index. *Journal of Applied Econometrics*, **7**, S137-S150.

LeBaron, B. 1992 Some relations between volatility and serial correlations in stock market returns. *Journal of Business*, **92**.

LeBaron, B. 1993 The joint dynamics and stability of stock prices and volume. Technical report, The University of Wisconsin - Madison, Madison, Wisconsin.

Lee,, T., White, H. and Granger, C. W. 1993 Testing for neglected nonlinearity in time series models: A comparison of neural network methods and alternative tests.

*Journal of Econometrics*, **56**, 269-290.

Levich, R. M. and Thomas, L. R. 1993 The significance of technical trading-rule profits in the foreign exchange market: A bootstrap approach. *Journal of International Money and Finance*, **12**, 451-474.

Mayfield, E. S. and Mizrach, B. 1992 On determining the dimension of real-time stock price data. *Journal of Business and Economic Statistics*, **10**, 367-374.

McQueen, G and Thorley, S. 1993 Asymmetric business cycle turning points. *Journal of Monetary Economics*, **31**, 341-362.

Mosekilde, E and Larsen, E. R. 1988 Deterministic chaos in the beer production-distribution model. *System Dynamics Review*, **4**, 131-147.

Neftci, S. N. 1984 Are economic time series asymmetric over the business cycle. *Journal of Political Economy*, **92**, 307-328.

Nychka, D., Ellner, S., Gallant, A. R. and McCaffrey, D. 1992 Finding chaos in noisy systems. *Journal of the Royal Statistical Society*, **B52(2)**, 399-426.

Peters, E. 1991 A chaotic attractor for the s+p 500. *Financial Analysts Journal*, pp. 55-81 (March-April

Potter, S. 1990 *Nonlinear Time Series and Economic Fluctuations.* PhD thesis, University of Wisconsin, Madison, WI.

Ramsey, J. B., Rothman, P. and Sayers, C. L. 1990 The statistical properties of dimension calculations using small data sets: Some economic applications. *International Economic Review*, **31**, 991-1020.

Ridley, M. 1993 Frontiers of finance. *The Economist.*

Ruelle, D. 1990 Deterministic chaos: The science and fiction. *Proceedings of the Royal Society Series A*, **427**, 241-248.

Sayers, C. 1989 Chaos and the business cycle. In S. Krasner, editor, *The Ubiquity of Chaos.* American Association for the Advancement of Science Publications, Washington, D.C.

Scheinkman, J. and LeBaron, B. 1989 Nonlinear dynamics and gnp data. In W. A. Barnett, J. Geweke, and K. Shell, editors, *Economic Complexity: Chaos, Sunspots, Bubbles, and Nonlinearity.* Cambridge University Press, Cambridge, UK.

Scheinkman, J. A. and LeBaron, B. 1989 Nonlinear dynamics and stock returns. *Journal of Business*, **62**, 311-38.

Schulmeister, S. 1987 An essay on exchange rate dynamics. Technical report, Wissenschaftszentrum Berlin fur Sozialforschung, Berlin, Germany.

Smith, R. L. 1992 Estimating dimension in noisy chaotic time series. *Journal of the Royal Statistical Society B*, **54**, 329-351.

Sweeney, R. J. 1986 Beating the foreign exchange market. *Journal of Finance*, **41**, 163-182.

Taylor, S. J. 1992 Rewards available to currency futures speculators: Compensation for risk or evidence of inefficient pricing? *Economic Record*, **68**, 105-116.

Terasvirta, T. and Anderson, H. M. 1992 Characterizing nonlinearities in business cycles using smooth transition autoregressive models. *Journal of Applied Econometrics*, **3**.

Theiler, J. and Eubank, S. 1993 Don't bleach chaotic data. *Chaos*, **3**, 771-782.
Theiler, J., Eubank, S., Longtin, A., Galdrikian, B. and Farmer, J. D. 1992 Testing for nonlinearity in time series: the method of surogate data. *Physica D*, **58**, 77-94.

Tong, H. 1990 *Non-linear Time Series: A Dynamical Systems Approach.* Oxford University Press, Oxford, UK.

Tsay, R. S. 1986 Nonlinearity tests for time series. *Biometrika*, **73**, 461-466.

Weigend, A. S., Huberman, B. A. and Rumelhart, D. E. 1991 Predicting sunspots and exchange rates with connectionist networks. In *Proceedings of the 1990 NATO Workshop on Nonlinear Modeling and Forecasting* (ed. S. Eubank and M. Casdagli), Redwood City, CA. Addison-Wesley.

Yao, Q. and Tong, H. 1994 Quantifying the influence of initial values on nonlinear prediction. *Journal of the Royal Statistical Society B*, **56**, 301-325.

CHAPTER 7

# PARADIGM CHANGE IN PREDICTION

A. S. Weigend
Department of Computer Science and Institute of Cognitive Science
University of Colorado Boulder, Colorado 80309-0430, U.S.A.*

**Abstract**

This paper describes the fundamental change in paradigm in time series prediction that has occurred in the last decade. The examples are drawn from the *Santa Fe Time Series Prediction and Analysis Competition.* In 1979, the results of an earlier competition were published by the Royal Society. In that competition, and also in the follow up five years later, relatively large numbers of time series were provided (111 and 1001, respectively), taken from business (forecasting sales), economics (predicting recovery from the recession), finance, and the social sciences. However, all of the series used were very short, generally less than 100 values long. Most of the algorithms entered were fully automated, and most of the discussion centred around linear models.

In contrast, the Santa Fe Competition focused on only six data-sets, ranging from 1000 to 100 000 points. All of the successful entries were fundamentally nonlinear and, even though significantly more computer power was used to analyse the larger data-sets with more complex models, the application of the techniques required more careful manual control than in the past. There was a general failure of simplistic 'black-box' approaches. In all successful entries, exploratory data analysis preceded the application of the algorithm. The Santa Fe Competition showed examples of nonlinear results going far beyond what is possible within the canon of linear systems analysis, but also showed that there are unprecedented opportunities for the analysis to go astray.

*http://www.cs.colorado.edu/home/andreas/public_html/home.htme

## 1. Introduction: Before the Santa Fe Competition

The desire to predict the future and understand the past drives the search for laws that explain the behaviour of observed phenomena; examples range from the irregularity in a heartbeat to the volatility of a currency exchange rate. If there are known underlying deterministic equations, in principle they can be solved to forecast the outcome of an experiment based on knowledge of the initial conditions. To make a forecast if the equations are not known, one must find both the rules governing system evolution and the actual state of the system. In this article we will focus on phenomena for which underlying equations are not given; the rules that govern the evolution must be inferred from regularities in the past. For example, the motion of a pendulum or the rhythm of the seasons carry within them the potential for predicting their future behaviour from knowledge of their oscillations without requiring insight into the underlying mechanism.

Time series analysis has three goals: forecasting, modelling, and characterization. The aim of *forecasting* (also called *predicting*) is to accurately predict the short-term evolution of the system; the goal of *modelling* is to find a description that accurately captures features of the long-term behaviour of the system. These are not necessarily identical: finding governing equations with proper long-term properties may not be the most reliable way to determine parameters for good short-term forecasts, and a model that is useful for short-term forecasts may have incorrect long-term properties. The third goal, system *characterization*, attempts with little or no *a priori* knowledge to determine fundamental properties, such as the number of degrees of freedom of a system or the amount of randomness. This overlaps with forecasting but can differ: the complexity of a model useful for forecasting may not be related to the actual complexity of the system.

Before the 1920s, forecasting was done by simply extrapolating the series through a global fit in the time domain. The beginning of 'modern' time series prediction might be set at 1927 when Yule invented the autoregressive technique to predict the annual number of sunspots. His model predicted the next value as a weighted sum of previous observations of the series. In order to obtain 'interesting' behaviour from such a linear system, outside intervention in the form of external shocks must be assumed. For the half-century following Yule, the reigning paradigm remained that of linear models driven by noise.

However, there are simple cases for which this paradigm is inadequate. For example, a simple iterated map, such as the logistic equation (Ulam & von Neumann 1946), can generate a broadband power spectrum that cannot be obtained by a linear approximation. The realization that apparently complicated time series can be generated by very simple equations (May 1976) pointed to the need for a more general theoretical framework for time series analysis and prediction.

Historically, an important step beyond linear models for prediction was taken in 1980 by Tong & Lim. After more than five decades of approximating a system with one globally linear function – i.e. one hyperplane – they suggested the use of two hyperplanes. This threshold autoregressive model (TAR) is globally nonlinear: it consists of choosing one of two local linear autoregressive models based on the value of the system's state. A general TAR model allows for any indicator variable beyond a single past value of the endogenous variable. However, it remains difficult to build strongly nonlinear function (such as a parabola or the logistic map) out of a small number of linear parts.

Two crucial developments occurred in the last decade; both were enabled by the general availability of powerful computers that permitted much longer time series to be recorded, more complex algorithms to be applied to them, and the data and the results of these algorithms to be interactively visualized. The first development, state-space reconstruction by time-delay embedding, drew on ideas from differential topology and dynamical systems to provide a technique for recognizing when a time series has been generated by deterministic governing equations and, if so, for understanding the geometrical structure underlying the observed behaviour. The second development was the emergence of the field of machine learning, typified by neural networks, that can adaptively explore a large space of potential models. With the shift in artificial intelligence from rule-based methods towards data-driven methods, the field was ready to apply itself to time series, and time series, now recorded with orders of magnitude more data points than were available previously, were ready to be analysed with machine-learning techniques requiring relatively large data-sets.

However, the evaluation of the new methods has been difficult. Adequate benchmarks were lacking, and most of the literature has been fragmentary and anecdotal. To foster a comparison of the methods process, we organized the *Santa Fe Time Series Prediction and Analysis Competition.* With the assistance of an advisory board from the relevant disciplines,† we collected a large library of candidate data-sets, select a representative small subset, specified the competition tasks, and ran the competition during the fall of 1991. The goal was not to pick 'winners' and 'losers', but rather to provide a structure for researchers from the interested disciplines to compare quantitatively the results of their analysis of a group of data-sets selected to span the range of studied problems. In more detail, we selected the following six data-sets for the competition (three of them, A, D, and E, are shown in figure 1):

(A) A clean physics laboratory experiment.

(B) Physiological data from a patient with sleep apnea.

(C) High-frequency currency exchange rate data.

(D) A numerically generated series designed for this competition.

(E) Astrophysical data from a variable star.

(F) J. S. Bach's final (unfinished) fugue from *The Art of the Fugue.*

†The advisors were Leon Glass (biology), Clive Granger (economics), Bill Press (astrophysics and numerical analysis), Maurice Priestley (statistics), Itamar Procaccia (dynamical systems), T. Subba Rao (statistics), and Harry Swinney (experimental physics).

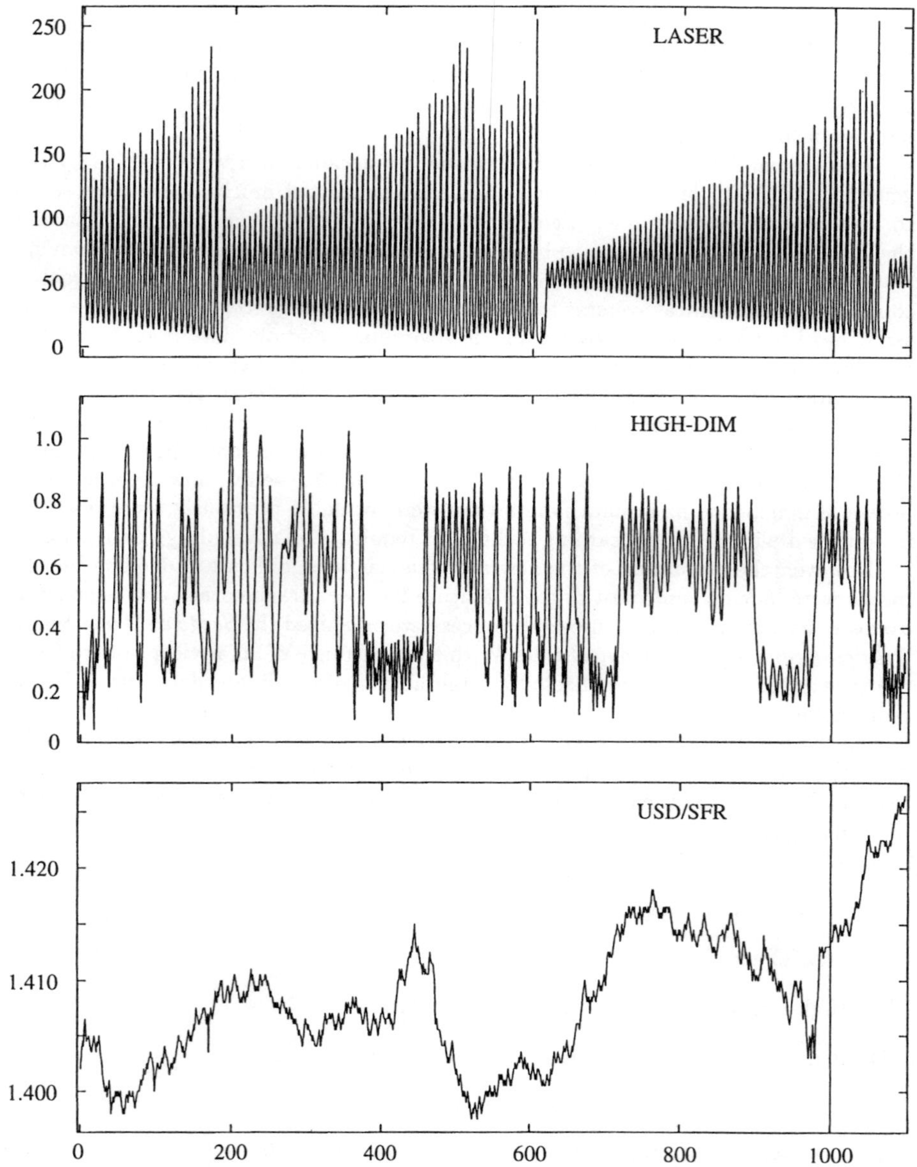


Fig. 1. Graphs of the laser series, the high-dimensional computer-generated series and the tickwise exchange rate data, plotted on the same timescale. The 1000 points for the Swiss Franc data correspond to 21–25 March 1991.

Table 1: Attributes of the sets with prediction tasks

| | laser | computer-generated | exchange rates | music |
|---|---|---|---|---|
| origin | physics laboratory | computer | a Swiss bank | human |
| characterization | low-dim det chaos | high-dimensional | stochastic | fugue |
| noise | clean | some | noisy | |
| stationarity | stationary | small drift | several timescales | |
| length | 1000 | 100 000 | 30 000 | 3808 |
| predict: | 100 values & error bars | 500 values | 60 values (total) | *ad lib* |

In this paper, we focus on the sets with prediction tasks. Some of their attributes are summarized in table 1. (The remaining sets were only used in the analysis category, not in the prediction category.)

To explore the results of the competition, a NATO Advanced Research Workshop was held after the close of the competition in the spring of 1992; workshop participants included members of the competition advisory board, representatives of the groups that had collected the data, successful participants, and interested observers. The results of the competition and the workshop are summarized in the overview article by Gershenfeld & Weigend (1994).

The reminder of this paper is organized as follows: in § 2, we start with the distinction between strong and weak models and place successful competition entries into this framework. In the § 3, we give the specific results of the competition. Section 4 closes with an outlook on the future.

## 2. Paradigms in the Santa Fe Competition

### *2.1. Strong and weak models*

Strong models have strong assumptions. They are usually expressed in a few equations with a few parameters, and can often explain a plethora of phenomena. In weak models, on the other hand, there are only a few domain-specific assumptions. To compensate for the lack of explicit knowledge, weak models usually contain many more parameters (which can make a clear interpretation difficult). It can be helpful to conceptualize models in the two-dimensional space spanned by the axes data-poor↔data-rich and theory-poor↔theory-rich. Due to the dramatic expansion of the capability for automatic data acquisition and processing, it is increasingly feasible to venture into the theory-poor and data-rich domain.

The idea of weak modelling (data-rich and theory-poor) is by no means new – an autoregressive moving average (ARMA) model is a good example. What is new is the emergence of weak models (such as neural networks) that combine broad generality with insight into how to manage their complexity. For such models with broad approximation abilities and few specific assumptions, the distinction between memorization and generalization becomes important. To clearly separate memorization from generalization, the true continuation of the competition data was kept secret until the deadline, ensuring that the continuation data could not be used by the participants for tasks such as parameter estimation or model selection. Successful forecasts of the withheld *test set* (also called *out-of-sample predictions*) from the provided *training set* (also called *fitting set*) were produced by two general classes of techniques: those based on state-space reconstruction (which make use of explicit understanding of the relationship between the internal degrees of freedom of a deterministic system and an observable of the system's state to build a model of the rules governing the measured behaviour of the system), and connectionist modelling (which uses potentially rich models along with learning algorithms to develop an implicit model of the system). We will see that neither is uniquely preferable. The domains of applicability are not the same, and the choice of which to use depends on the goals of the analysis (such as an understandable description versus accurate short-term forecasts).

### *2.2. Geometrical reconstruction in state space*

Yule's (1927) original idea for forecasting was that future predictions can be obtained by superimposing immediately preceding values,

$$x_t = \sum_{i=1}^{d} a_i x_{t-i} \ (+\text{`noise'}), \tag{1}$$

where $x_t$ denotes the values of the time series at time $t$, and $a_i$ the weighting of the value $i$ time steps back. $d$ stands for 'dimension' and corresponds to the order of such an autoregressive model. This superposition of past values, also called a tapped delay line, is used routinely in the context of signal processing and time series analysis, suggesting that it is more than just a typographical convenience.

In fact, there is a deep connection between time-lagged vectors and underlying dynamics. This connection was was proposed in 1980 by Ruelle (personal communication), Packard *et al.* (1980), and Takens (1981; he published the first proof), and later strengthened by Sauer *et al.* (1991) and by Casdagli *et al.* (1991). Delay vectors of sufficient length are not just a representation of the state of a linear system – it turns out that delay vectors can recover the full geometrical structure of a nonlinear system. These results address the general problem of inferring the behaviour of the intrinsic degrees of freedom of a system when a function of the state of the system is measured. If the governing equations and the functional form of the observable are known in advance, then a Kalman filter is the optimal linear estimator of the state of the system. We, however, focus on the case where there is little or no *a priori* information available about the origin of the time series.

Time-delay embedding differs from traditional experimental measurements in three fundamental respects:

1. It provides detailed information about the behaviour of degrees of freedom other than the one that is directly measured.

2. It rests on probabilistic assumptions and – although it has been routinely and reliably used in practice – it is not guaranteed to be valid for any system.

3. It allows precise questions only about quantities that are invariant under such a transformation, since the reconstructed dynamics have been modified by an unknown smooth change of coordinates.

This last restriction may be unfamiliar, but it is surprisingly unimportant: embedded data can be used for forecasting a time series and for characterizing the essential features of the dynamics that produced it. Time-delay embedding, while appearing similar to traditional state-space models with lagged vectors, makes a crucial link between behaviour in the reconstructed state space and the internal degrees of freedom. Sauer (1994) very successfully applied these ideas of a geometrical reconstruction of the low-dimensional manifold to the competition data, yielding the best longer-term predictions for the laser data where the simple geometrical structure was readily apparent. We now turn to the problem of what can be done if we are unable to understand the system in such explicit terms. The main idea will be to learn to emulate the behaviour of the system.

*2.3. Learning: neural networks*

In the competition, the majority of contributions, and also the best predictions for each set used neural networks, also called connectionist models. They provide a convenient language game for nonlinear modelling. They are typically used in pattern recognition, where a collection of features (such as an image) is presented to the network, and the task is to assign the input image to one or more classes. Another typical use for neural networks is (nonlinear) regression, where the task is to find a smooth interpolation between points. In both these cases, all the relevant information is presented simultaneously. In contrast, time series prediction involves processing of patterns that evolve over time – the appropriate response at a particular point in time depends not only on the current value of the observable but also on the past. Time series prediction has had an appeal for neural networkers from the very beginning of the field. In 1964, Hu weather forecasting. In the post-backpropagation era, Lapedes & Farber (1987) trained their (nonlinear) network to emulate the relationship between output (the next point in the series) and inputs (its predecessors) for computer-generated time series, and Weigend *et al.* (1990, 1992) addressed the issue of finding networks of appropriate complexity for predicting observed (real-world) time series.

State-space embedding 'solves' the forecasting problem for a low-dimensional deterministic system. If there is understandable structure in the embedding space, it can be detected and modelled; the open questions have to do with finding good representations of the surface and with estimating the reliability of the forecast. This approach of reconstructing the geometry of the manifold will fail if the system is high-dimensional, has stochastic inputs, or is non-stationary, because in these cases, there is no longer a simple surface to model.

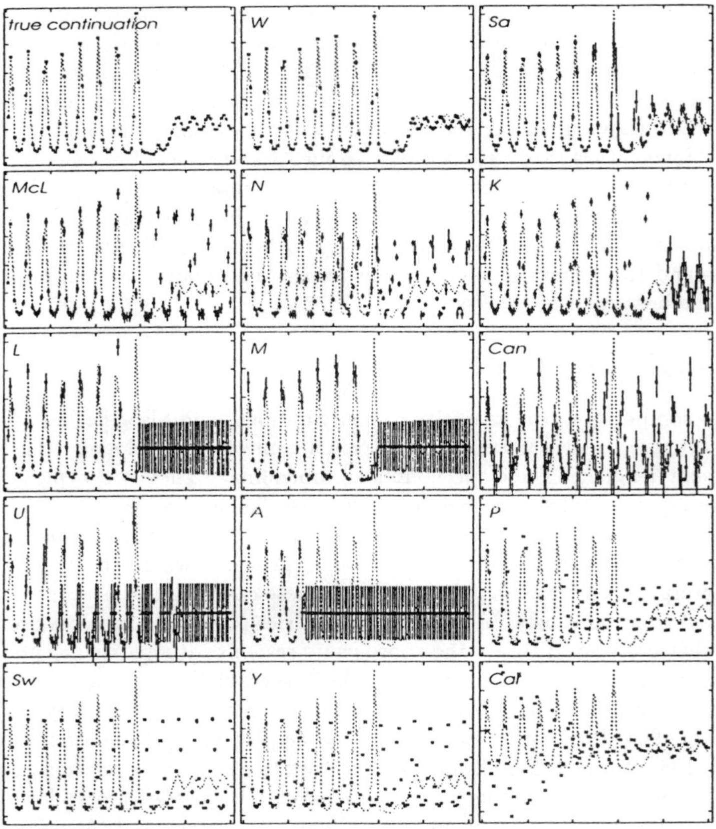


Fig. 2. Continuations of the laser data. The letters correspond to the code of the entrant. Grey lines indicate the true continuation, $\times$ the predicted values, and vertical bars the predicted uncertainty.

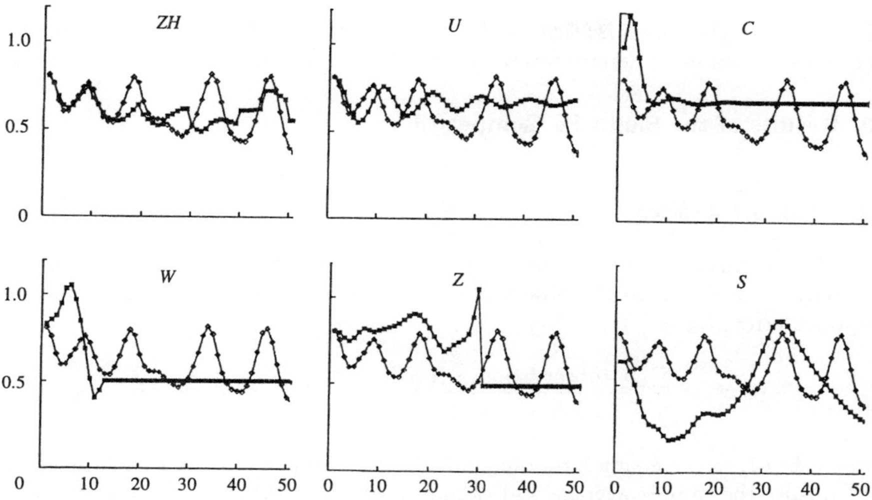


Fig. 3. Predicted continuations of the computer generated data for the first 50 steps of the prediction period.

Neural networks do not build an explicit description of a surface. On the one hand, this makes it harder to interpret them even for simple forecasting tasks. On the other hand, they promise to be applicable (and mis-applicable) to situations where simpler, more explicit approaches fail: much of their promise comes from the hope that they can learn to emulate unanticipated regularities in a complex signal. In the competition, the best as well as many of the worst forecasts of the laser data were obtained with neural networks.

A network that is to predict the future must know about the past. The simplest approach is to provide time-delayed samples to its input layer. A network without (nonlinear) hidden units is equivalent to one linear filter (i.e. an autoregressive model, equation (1)). However, a network with nonlinear hidden units can be viewed as combining a number of 'squashed' filters. Mozer (1994) classifies discusses architectures for connectionist time series prediction, and the general theoretical framework for connectionist modelling is given by Rumelhart *et al.* (1994).

The best short-term predictions were obtained by Eric Wan, at the time a graduate student at Stanford (Wan 1994). At first sight, his choice of a network with 1105 parameters to fit 1000 data points seems absurd. How is it possible not to overfit if there are more parameters than data points? The key is knowing when to stop. At the outset of training, the parameters have random values, and so changing any one coefficient has little impact on the quality of the predictions. As training progresses and the fit improves, the effective number of parameters grows (Weigend & Rumelhart 1991; Weigend 1994). The overall error in predicting points out of the training set will initially decrease as the network learns to do something, but then will begin to increase once the network learns to do too much; the location of the minimum of the cross-validation error determines when the effective network complexity is right.

## 3. Results of the Santa Fe Competition

### *3.1. Data-set A: laser*

The submissions were evaluated in two ways. First, using the predicted values $\widehat{x}_k$ only (in addition to the observed values $x_k$), we compute the normalized mean squared error, $E_{\mathrm{NMS}}$:

$$E_{\mathrm{NMS}}(N) = \frac{\sum_{k\in\mathcal{T}}(\text{observation}_k - \text{prediction}_k)^2}{\sum_{k\in\mathcal{T}}(\text{observation}_k - \text{mean}_{\mathcal{T}})^2} \approx \frac{1}{\widehat{\sigma}_{\mathcal{T}}^2}\frac{1}{N}\sum_{k\in\mathcal{T}}(x_k - \widehat{x}_k)^2 , \qquad (2)$$

where $k = 1, \ldots, N$ enumerates the points in the withheld test set $\mathcal{T}$, and $\text{mean}_{\mathcal{T}}$ and $\widehat{\sigma}_{\mathcal{T}}^2$ denote the sample average and sample variance of the observed values (targets) in $\mathcal{T}$. A value of $E_{\mathrm{NMS}} = 1$ corresponds to simply predicting the average.

Second, the submitted error bars $\widehat{\sigma}_k$ enter the computation of the likelihood of the observed data, given the predicted values and the predicted error bars, based on an assumption of independent gaussian errors. Although this assumption may not be justified, it provides a simple form that captures some desirable features of error weighting. Since the original real-valued data is quantized to integer values, the probability of seeing a given point $x_k$ is found by integrating the gaussian distribution over a unit interval (corresponding to the rounding error of 1 bit of the analogue-to-digital converter) centred on $x_k$:

$$p(x_k|\widehat{x}_k, \widehat{\sigma}_k) = \frac{1}{\sqrt{2\pi\widehat{\sigma}_k^2}}\int_{x_k-0.5}^{x_k+0.5} \exp\left(\frac{-(\xi - \widehat{x}_k)^2}{2\widehat{\sigma}_k^2}\right) \mathrm{d}\xi. \qquad (3)$$

If the predicted error is large, then the computed probability will be relatively small, independent of the value of the predicted point. If the predicted error is small and the predicted point is close to the observed value, then the probability will be large, but if the predicted error is small and the prediction is not close to the observed value, then the probability will be very small. The potential reward, as well as the risk, is greater for a confident prediction (small error bars). Under the assumption of independent errors, the likelihood of the whole test for the the observed data given the submitted model is then

$$p(\mathrm{D}|\mathrm{M}) = \prod_{k=1}^{N} p(x_k|\widehat{x}_k, \widehat{\sigma}_k). \qquad (4)$$

Finally, we take the logarithm of this probability of the data given the model (this turns products into sums and also avoids numerical problems), and then scale the

Table 2: Entries received before the deadline for the prediction of the laser

| | method | | computer | time | $E_{\text{NMS}}$ | $-\log(\text{lik.})$ |
|---|---|---|---|---|---|---|
| W | conn | 1-12-12-1; lag 25,5,5 | SPARC 2 | 12 h | 0.028 | 3.5 |
| Sa | loc lin | low-pass embd, 8 dim, 4nn | DEC 3100 | 20 min | 0.080 | 4.8 |
| McL | conn | feedforward, 200-100-1 | CRAY Y-MP | 3 h | 0.77 | 5.5 |
| N | conn | feedforward, 50-20-1 | SPARC 1 | 3 weeks | 1.0 | 6.1 |
| K | visual | look for similar stretches | SG Iris | 10 s | 1.5 | 6.2 |
| L | visual | | | | 0.45 | 6.2 |
| M | conn | feedforward, 50-350-50-50 | 386 PC | 5 days | 0.38 | 6.4 |
| Can | conn | recurrent, 4-4c-1 | VAX 8530 | 1 h | 1.4 | 7.2 |
| U | tree | k-d tree; AIC | VAX 6420 | 20 min | 0.62 | 7.3 |
| A | loc lin | 21 dim, 30 nearest neighb's | SPARC 2 | 1 min | 0.71 | 10.0 |
| P | loc lin | 3 dim time delay | Sun | 10 min | 1.3 | — |
| Sw | conn | feedforward | SPARC 2 | 20 h | 1.5 | — |
| Y | conn | feedforward, weight-decay | SPARC 1 | 30 min | 1.5 | — |
| Car | linear | Wiener filter, width 100 | MIPS 3230 | 30 min | 1.9 | — |

$E_{\text{NMS}}$ (short for $E_{\text{NMS}}(100)$) gives the normalized sum squared error (or relative mean square error), $-\log(\text{lik.})$ gives the negative logarithm of the likelihood of the data given the predicted values and predicted errors, averaged over the prediction set.

result by the size of the data-set, $N$. This defines the *negative average log likelihood*:

$$-\frac{1}{N}\sum_{k=1}^{N}\log p(x_k|\hat{x}_k,\hat{\sigma}_k). \tag{5}$$

We give these two statistics for the submitted entries for data-set A along with a brief summary of the entries in table 2 and show all predictions received for data-set A in figure 2. The symbols ($\times$) correspond to the predicted values, the vertical bars to the submitted error bars. The true continuation points are connected by a grey line (to guide the eye).

Table 2 lists the methods applied to the laser data, the computers and CPU time required, as well as the relative mean square error the negative average log-likelihood. The submitted continuations of the laser are shown in figure 2; the letter 'W' refers to the entry by Wan (1994), who used a 'time delay neural network'.

None of the competition entries estimated the error bars in a principled way. Progress has been made: Nix & Weigend (1994) describe a method to generate both the prediction and the confidence intervals as a function of the location in input space ('local error bars'). They use a maximum likelihood framework to train a network with two output units, one giving the predicted mean, the other one the error bar. A method to estimate the probability distribution of the next value – a rather data-hungry enterprise – is described by Srivastava & Weigend (1994). They use a method called 'fractional binning' to estimate the conditional density of the forecast.

*3.2. Data-set D: computer-generated data*

Table 3: Entries received before the deadline for the computer-generated data

| | method | | computer | time | $E_{\mathrm{NMS}}$ (15) | $E_{\mathrm{NMS}}$ (30) | $E_{\mathrm{NMS}}$ (50) |
|---|---|---|---|---|---|---|---|
| ZH | conn | ...-30-30-1 and 30-100-5 | CM-2 (16k) | 8 days | 0.086 | 0.57 | 0.87 |
| U | tree | k-d tree; AIC | VAX 6420 | 30 min | 1.3 | 1.4 | 1.4 |
| C | conn | recurrent, 4-4c-1 | VAX 8530 | n/a | 6.4 | 3.2 | 2.2 |
| W | conn | 1-30-30-1; lags 20,5,5 | SPARC 2 | 1 day | 7.1 | 3.4 | 2.4 |
| Z | linear | 36 AR(8), last 4k points | SPARC | 10 min | 4.8 | 5.0 | 3.2 |
| S | conn | feedforward | SPARC 2 | 20 h | 17.0 | 9.5 | 5.5 |

To provide a relatively long series of known high-dimensional dynamics (between the extremes of data-set A and data-set C) with weak non-stationarity, we generated 100 000 points by numerically integrating the equations of motion for a damped, driven particle

$$\mathrm{d}^2\boldsymbol{x}/\mathrm{d}t^2 + \gamma \mathrm{d}\boldsymbol{x}/\mathrm{d}t + \nabla V(\boldsymbol{x}) = \boldsymbol{F}(t) \tag{6}$$

in an asymmetrical four-dimensional four-well potential

$$V(\boldsymbol{x}) = A_4\left(x_1^2 + x_2^2 + x_3^2 + x_4^2\right)^2 - A_2\,|x_1 x_2| - A_1 x_1 \tag{7}$$

with periodic forcing

$$\boldsymbol{F}(t) = F\sin(\omega t)\widehat{\boldsymbol{x}}_3. \tag{8}$$

$\mathbf{x} = (x_1, x_2, x_3, x_4)$ denotes the location of the particle. This system has nine degrees of freedom (four position, four velocity, and one forcing time). The equations were integrated with 64-bit real numbers using a fixed-step fourth-order Runge–Kutta algorithm (to eliminate possible coupling to an active stepper). The potential has four wells that are tilted by the parameter $A_1$. This parameter slowly drifted during the generation of the data according to a small biased random walk from 0.02 to 0.06. As a scalar observable we chose

$$\sqrt{(x_1 - 0.3)^2 + (x_2 - 0.3)^2 + x_3^2 + x_4^2}.$$

This observable projects the four wells onto three distinguishable states. Data-set D was generated with $A_4 = 1$, $A_2 = 1$, $A_1$ drifting from 0.02 to 0.06, $\gamma = 0.01$, $F = 0.135$, and with $\omega = 0.6$ (for these parameters, in the absence of the weak forcing by the drift, this system is not chaotic).

The contributions for the computer generated data that were received before the deadline are listed in table 3; they are evaluated by normalized mean squared error, $E_{\mathrm{NMS}}$, averaged over the first 15, 30, and 50 predictions. Figure 3 shows the predictions received for this data-set over the range of the first 50 points of the prediction range. The code 'ZH' refers to the entry by Zhang & Hutchinson (1994), who also use a neural network to generate their predictions.

## 4. The Future

We have given some examples of what appears to be a steady progress of insight over ignorance in analysing time series. Is there a limit to this development? Can we hope for the discovery of a universal forecasting algorithm that will predict everything about all time series? The answer is emphatically 'no!' Even for completely deterministic systems, there are strong bounds on what can be known. The search for a universal time series algorithm is related to Hilbert's vision of reducing all of mathematics to a set of axioms and a decision procedure to test the truth of assertions based on the axioms (*Entscheidungsproblem*); this culminating dream of mathematical research was dramatically dashed by Gödel (1931) and then by Turing. The most familiar result of Turing is the undecidability of the halting problem: it is not possible to decide in advance whether a given computer program will eventually halt (Turing 1936). But since Turing machines can be implemented with dynamical systems (Moore 1991), a universal algorithm that can directly forecast the value of a time series at any future time would need to contain a solution to the halting problem, because it would be able to predict whether a program will eventually halt by examining the program's output. Therefore, there cannot be a universal forecasting algorithm.

However, the progress in the last decade in analysing time series has been remarkable and is well witnessed by the 650 pages of the summary volume of the Santa Fe Competition (Weigend & Gershenfeld 1994). Where once time series analysis was shaped by linear systems theory, it is now possible to recognize when an apparently complicated time series has been produced by a low-dimensional nonlinear system, to characterize its essential properties, and to build a model that can be used for forecasting. At the opposite extreme, there is now a much richer framework for designing algorithms that can learn the regularities of time series that do not have a simple origin. This progress has been inextricably tied to the arrival of the routine availability of significant computational resources (making it possible to collect large time series, apply complex algorithms, and interactively visualize the results), and it may be expected to continue as the hardware improves.

General access to computer networks has enabled the widespread distribution and collection of common material, a necessary part of the logistics of the competition, thereby identifying interesting results where they once might have been overlooked (such as from students, and from researchers in countries that have only recently been connected to international computer networks). This synchronous style of research is complementary to the more common asynchronous mode of relatively independent publication, and may be expected to become a familiar mechanism for large-scale scientific progress. We hope that, in the short term, the data and results from this competition can continue to serve as basic reference benchmarks for new techniques. We also hope that, in the long term, they will be replaced by more worthy successors.

## Acknowledgments

I deeply thank Neil Gershenfeld: without him this project would not have happened. I am also grateful for the support of Stanford University, Chulalongkorn University, Xerox PARC, the University of Colorado at Boulder, and for the Research Initiation Grant of the National Science Foundation, NSF RIA ECS-9309786.

## References

Casdagli, M., Eubank, S., Farmer, J. D. & Gibson, J. 1991 State space reconstruction in the presence of noise. *Physica* D **51**, 52–98.

Gershenfeld, N. A. & Weigend, A. S. 1994 The future of time series: learning and understanding. In *Time series prediction: forecasting the future and understanding the past* (ed. A. S. Weigend & N. A. Gershenfeld), pp. 1–70. Reading, MA: Addison-Wesley.

Gödel, K. 1931 Über formal unentscheibare Sätze der *Principia Mathematica* und verwandter Systeme. I. *Monatshefte Math. Phys.* **38**, 173–198. (An English translation of this paper is found in *On formally undecidable propositions* by K. Gödel, New York, Basic Books, 1962).

Hu, M. J. C. 1964 Application of the adaline system to weather forecasting. E.E. degree thesis. *Tech. Rep.* 6775-1, Stanford Electronic Laboratories, Stanford, CA, U.S.A.

Lapedes, A. & Farber, R. 1987 Nonlinear signal processing using neural networks. *Tech. Rep.* no. LA-UR-87-2662, Los Alamos National Laboratory, Los Alamos, NM, U.S.A.

Makridakis, S. & Hibon, M. 1979 Accuracy of forecasting: an empirical investigation. *Jl R. statist. Soc.* A **142**, 97–145. With discussion.

Makridakis, S., Andersen, A., Carbone, R., Fildes, R., Hibon, M., Lewandowski, R., Newton, J., Parzen, E. & Winkler, R. 1984 *The forecasting accuracy of major time series methods.* New York: Wiley.

May, R. M. 1976 Simple mathematical models with very complicated dynamics. *Nature, Lond.* **261**, 459.

Moore, C. 1991 Generalized shifts: unpredictability and undecidability in dynamical systems. *Nonlinearity* **4**, 199–230.

Mozer, M. C. 1994 Neural net architectures for temporal sequence processing. In *Time series prediction: forecasting the future and understanding the past* (ed. A. S. Weigend & N. A. Gershenfeld), pp. 243–264. Reading, MA: Addison-Wesley.

Nix, D. A. & Weigend, A. S. 1994 Estimating the mean and variance of the target probability distribution. In *Proc. IEEE Int. Conf. on Neural Networks (IEEE-ICNN), Orlando, Florida, June 1994*, pp. 55–60. IEEE Press.

Packard, N. H., Crutchfeld, J. P., Farmer, J. D. & Shaw, R. S. 1980 Geometry from a time series. *Phys. Rev. Lett.* **45**, 712–716.

Rumelhart, D. E., Durbin, R., Golden, R. & Chauvin, Y. 1994 Backpropagation: the basic theory. In *Backpropagation: theory, architectures and applications* (ed. Y. Chauvin & D. E. Rumelhart), Hillsdale, NJ: Lawrence Erlbaum.

Sauer, T. 1994 Time series prediction using delay coordinate embedding. In *Time series prediction: forecasting the future and understanding the past* (ed. A. S. Weigend & N. A. Gershenfeld), pp. 175–193. Reading, MA: Addison-Wesley.

Sauer, T., Yorke, J. A. & Casdagli, M. 1991 Embedology. *J. statist. Phys.* **65**, 579–616.

Srivastava, A. N. & Weigend, A. S. 1994 Computing the probability density in connectionist regression. In *Proc. Int. Conf. on Artificial Neural Networks (ICANN), Sorrento, Italy, May 1994*, pp. 685–688. Springer-Verlag.

Takens, F. 1981 Detecting strange attractors in turbulence. In *Dynamical systems and turbulence* (ed. D. A. Rand & L.-S. Young), *Lecture Notes in Mathematics*, vol. 898, pp. 336–381. Berlin: Springer-Verlag.

Tong, H. & Lim, K. S. 1980 Threshold autoregression, limit cycles and cyclical data. *Jl R. statist. Soc.* B **42**, 245–292.

Turing, A. M. 1936 On computable numbers, with an application to the *Entscheidungsproblem*. *Proc. Lond. math. Soc.* **42**, 230–265.

Ulam, S. & von Neumann, J. 1947 *Bull. Am. math. Soc.* **53**, 1120.

Wan, E. A. 1994 Times series prediction using a connectionist network with internal delay lines. In *Time series prediction: forecasting the future and understanding the past* (ed. A. S. Weigend & N. A. Gershenfeld), pp. 195–217. Reading, MA: Addison-Wesley.

Weigend, A. S. 1994 On overfitting and the effective number of hidden units. In *Proc. 1993 Connectionist Models Summer School* (ed. M. C. Mozer, P. Smolensky, D. S. Touretzky, J. L. Elman & A. S. Weigend), pp. 335–342. Hillsdale, NJ: Erlbaum Associates.

Weigend, A. S. & Gershenfeld, N. A. (eds) 1994 Time series prediction: forecasting the future and understanding the past. *Proc. Santa Fe Institute Studies in the Sciences of Complexity*, vol. XV. Reading, MA: Addison-Wesley.

Weigend, A. S. & Rumelhart, D. E. 1991 Generalization through minimal networks with application to forecasting. In *INTERFACE '91 – 23rd Symp. on the Interface: Computing Science and Statistics* (ed. E. M. Keramidas), pp. 362–370. Interface Foundation of North America.

Weigend, A. S., Huberman, B. A. & Rumelhart, D. E. 1990 Predicting the future: a connectionist approach. *Int. J. neur. Sys.* **1**, 193–209.

Weigend, A. S., Huberman, B. A. & Rumelhart, D. E. 1992 Predicting sunspots and exchange rates with connectionist networks. In *Nonlinear modeling and forecasting* (ed. M. Casdagli & S. Eubank), pp. 395–432. Redwood City, CA: Addison-Wesley.

Yule, G. 1927 On a method of investigating periodicity in disturbed series with special reference to Wolfer's sunspot numbers. *Phil. Trans. R. Soc. Lond.* A **226**, 267–298.

Zhang, X. & Hutchinson, J. 1994 Simple algorithms on fast machines: practical issues in nonlinear time series prediction. In *Time series prediction: forecasting the future and understanding the past* (ed. A. S. Weigend & N. A. Gershenfeld), pp. 219–241. Reading, MA: Addison-Wesley.

CHAPTER 8

# PREDICTING NONUNIFORM CHAOTIC ATTRACTORS IN AN ENZYME REACTION

L.F. Olsen[1], K.R. Valeur[1], T. Geest[1], C.W. Tidd[2] and W.M. Schaffer[1,2]
[1]Physical Biochemistry Group, Institute of Biochemistry, Odense University, Campusvej 55, DK-5230 Odense M, Denmark.
[2]Department of Ecology and Evolutionary Biology, The University of Arizona, Tucson, Arizona 85721, U.S.A.

**Abstract**

Nonlinear forecasting was used to predict the time evolution of fluctuating concentrations of dissolved oxygen in the peroxidase-oxidase reaction. This reaction entails the oxidation of NADH with molecular oxygen as the electron acceptor. Depending upon the experimental conditions, either regular or highly irregular oscillations obtain. Previous work suggests that the latter fluctuations are almost certainly chaotic. In either case, the dynamics contain multiple time scales, which fact results in an uneven distribution of points in the phase space. Such "nonuniformity", as it is called, is a rock on which conventional methods for analyzing chaotic time series often founder. The results of the present study are as follows: 1. Short-term forecasting with local linear predictors yields results that are consistent with a hypothesis of low-dimensional chaos. 2. Most of the evidence for nonlinear determinism disappears upon the addition of small amounts of observational error. 3. It is essentially impossible to make predictions over time intervals longer than the average period of oscillation for time series subject to continuous and frequent sampling. 4. Far more effective forecasting is possible for points on Poincaré sections. 5. An alternative means for improving forecasting efficacy using the continuous data is to include a second variable (NADH concentration) in the analysis. Since nonuniformity is common in biological time series, we conclude that the application of nonlinear forecasting to univariate time series requires care both in implementation and interpretation.

## 1. Introduction

In the past decade, numerous methods have been proposed to distinguish low-dimensional chaos from the output of stochastic processes. Of these, two of the most widely utilized entail estimating the system's fractal dimension (Grassberger & Procaccia, 1983) and its maximum Lyapunov exponent (Wolf *et al.*, 1985). Specifically, non-integer dimensions and positive exponents, so-called "fingerprints" of chaos, have been taken as evidence of nonlinear determinism (Abraham *et al.*, 1989; Tsonis, 1992). In part, the popularity of this approach reflects the fact that dimension and Lyapunov exponents feature prominently in the theory of chaotic dynamical systems. Moreover, the algorithms in question are easily implemented on the computer. Unfortunately, there is a growing body of evidence (Osborne & Provenzale, 1989; Ruelle, 1990; Ellner, 1991) to suggest that these and other methods designed to estimate dynamical invariants can be 'fooled' by stochastic time series in which there is demonstrably no chaotic behavior at all.

One approach to resolving this dilemma is the "method of surrogate data" whereby analyses of experimental data are compared with corresponding calculations carried out on stochastic data sets reproducing the linear properties of the original (Theiler *et al.*, 1992). An alternative methodology, and the subject of the present report, is due to Casdagli (1992). Here, techniques from nonlinear forecasting (see below) are utilized to evaluate stochastic alternatives to nonlinear determinism. Rejection of the null hypotheses is taken as evidence favoring, though not necessarily proving, that the data are, in fact, chaotic.

Nonlinear forecasting (Farmer & Sidorowich, 1987) is based on the observation that even for chaotic systems, nearby points in the phase space travel together over the short term and, in some instances, for extended time periods. Accordingly, it is possible to approximate the "rules" by which a chaotic system evolves, even though one has no idea as to the actual equations governing the motion. Sugihara & May (1990) were among the first to apply nonlinear forecasting to biological time series. They argued that chaotic data exhibit a characteristic "prediction profile", a consequence of sensitivity to initial conditions (Ruelle, 1979), whereby the initial decline in predictability follows a negative exponential (see also, Farmer & Sidorowich, 1987; Wales, 1991). This pattern, they maintained was relatively insensitive to short data sets and noise, and could be used to distinguish chaos from regular dynamics in the presence of observational error.

In fact, certain types of stochastic behavior (Lefebvre *et al.*, 1993) exhibit similar declines in predictability. However, the scaling behavior of predictive power with prediction time of such data is different from that of chaotic data (Sugihara & May, 1990; Tsonis & Elsner, 1992). Additionally, Casdagli (1992) has proposed a variant on the basic method which he suggests can be used to distinguish low-dimensional chaos from stochastic dynamics. Essentially, one uses local linear maps, induced from the reconstructed flow, to forecast the time evolution of the system. By varying what is meant by "local", one can determine whether the dynamics are more in accordance with chaos or stochastic autoregression. That is, for a chaotic system, increasing the size of the neighborhoods for which the linear maps are computed should lead to declining predictability — *i.e.*, the nonlinear nature of the flow means that linear approximations to the vector field change over the phase space. Conversely, for linear stochastic systems, the rules of motion do not change over the phase space. In this case, increasing neighborhood size results in more points being used to induce the maps. Hence, the error resulting from finite data sets will be reduced and forecasting efficacy will improve or remain constant.

Most of the algorithms used to estimate fractal dimensions and the like were developed and tested using data from dynamical systems in which points are more or less evenly distributed in the phase space. However, in biology, we often encounter highly unequal distributions. Following Nicolis *et al.* (1983), we shall refer to systems exhibiting such distributions as "nonuniform". Biological examples of nonuniform dynamics include the irruptive behavior of ecological populations including insects and small mammals (May, 1978), fluctuations in incidence rates of childhood diseases (Schaffer *et al.*, 1988; Tidd *et al.*, 1993), cardiac activity as indexed by the electrocardiogram (Babloyantz & Destexhe, 1988), and the relaxation oscillators observed in certain enzyme-catalyzed reactions (Geest *et al.*, 1993). Even in the absence of noise, nonuniformity can play havoc with nonlinear methods. For example, in the case of irruptive time series, the standard techniques underestimate both the fractal dimension and the maximum Lyapunov exponent (Schaffer *et al.*, 1988; Tidd *et al.*, 1993). Conversely, when applied to relaxation oscillators, these methods underestimate the dimension and overestimate the exponent (Geest *et al.*, 1993).

Nonlinear forecasting schemes also have problems with nonuniform dynamics. In particular, predictability is often much lower than one would anticipate from other estimates of the system's sensitivity to initial conditions (Tidd *et al.*, 1993; Geest *et al.*, 1993). Nonetheless nonlinear forecasting can provide useful information about nonuniform time series which cannot be obtained from attempts to estimate fractal dimensions and Lyapunov characteristic exponents. For example, forecasting may help to discriminate between different deterministic or stochastic models of the data (Tidd *et al.*, 1993; Geest *et al.*, 1993).

In the present paper, we use nonlinear forecasting to study nonuniform data from a biochemical system, the peroxidase-oxidase reaction. The results support previously reported evidence (Olsen & Degn, 1977; Olsen, 1983; Geest *et al.*, 1992, 1993) for low-dimensional chaos in this system. However, if the experimental data are perturbed by the addition of even small amounts of noise, much of the evidence for nonlinear determinism disappears. A further consequence of nonuniformity in the experimental data is that predictions based on embedded time series of $O_2$ concentration are essentially impossible for prediction times longer than an average period of oscillation. Predictions over longer time scales can be achieved if a second measured variable is used as a reference variable or if predictions are made using return maps rather than continuous time series.

## 2. The experimental system

The peroxidase-oxidase (PO) reaction is the oxidation of reduced nicotinamide adenine dinucleotide (NADH) with molecular oxygen as the electron acceptor:

$$O_2 + 2NADH + 2H^+ \rightarrow 2H_2O + 2NAD^+$$

When the reaction takes place at low pH with continuous supplies of both NADH and $O_2$, and in the presence of 2,4-dichlorophenol and methylene blue, the concentrations of $O_2$ and NADH oscillate (Nakamura *et al.*, 1969; Olsen & Degn, 1978). In addition to periodic behavior, both quasiperiodicity (Samples *et al.*, 1992; Hauck & Schneider, 1993) and chaos (Olsen & Degn, 1977; Geest *et al.*, 1992, 1993; Rys & Wang, 1992; Förster *et al.*, 1994, Hauck & Schneider, 1994) have been observed. In addition, there is experimental evidence for a period doubling route to chaos followed by stable period-3 oscillations as the concentration of dichlorophenol is increased (Geest *et al.*, 1992; Steinmetz *et al.*, 1993). Since the evidence for chaos in this system is unequivocal, the data constitute an ideal test case for determining the efficacy and reliability of published methods for detecting nonlinearity and chaos in experimental time series. An additional advantage of the PO system is that the experimental results are largely reproducible by a simple mathematical model consisting of four nonlinear differential equations (Olsen, 1983).

## 3. Materials and methods

Horseradish peroxidase (RZ 3.0) and NADH were purchased from Boehringer, Mannheim. 2,4-dichlorophenol and methylene blue (analytical grade) were obtained from Merck.

Experiments were conducted in a 20 × 20 mm quartz cuvette fitted with a stirrer. The cuvette was mounted in an Aminco DW 2000 dual wavelength spectrophotometer or in a Shimadzu UV-1201 single beam spectrophotometer. NADH was measured as the absorbance change at 360 nm or as the difference in absorbance at 360 nm and 380 nm. The oxygen concentration was measured with a Clark electrode inserted into the side of the cuvette. The data from the spectrophotometer and the oxygen electrode were both sampled at 1 s intervals by a personal computer through an interface board and stored on disk for later analyses. There were two experimental protocols:

Protocol I is described by Geest *et al.* (1992, 1993). Here, a 0.08–0.2 M solution of NADH is pumped at a rate of 20–40 μl/h into a 7 ml reaction mixture containing 0.1 M sodium acetate, pH 5.1, 30–40 μM 2,4-dichlorophenol, 0.1 μM methylene blue and 0.5–1 μM peroxidase. The oxygen was transported into the reaction mixture by diffusion from a gas head space above the liquid containing a mixture of oxygen and nitrogen with an $O_2$ content of 1.42% (v/v). The oxygen transfer constant was 0.0037 $s^{-1}$.

Protocol II involves a continuous flow stirred tank reactor (CSTR) with an effective liquid volume of 6.9 ml. The flow rate was controlled by a Harvard Apparatus model 22 syringe pump. Reactants were fed into the CSTR as two solutions in gas-tight syringes (Hamilton). One syringe contained 0.2 M sodium acetate, pH 5.1, 0.5 μM peroxidase, 0.2 μM methylene blue and 100 μM 2,4-dichlorophenol and the other contained 0.4 mM NADH in distilled water. The solutions were equilibrated with pure nitrogen before use. The dilution rate for the inflow of reactants was held constant at $4.4 \times 10^4$ $s^{-1}$. In these experiments the oxygen content of the gas phase was 0.63% (v/v) and the oxygen transfer constant was 0.0048 $s^{-1}$. The temperature was 28°C in all experiments.

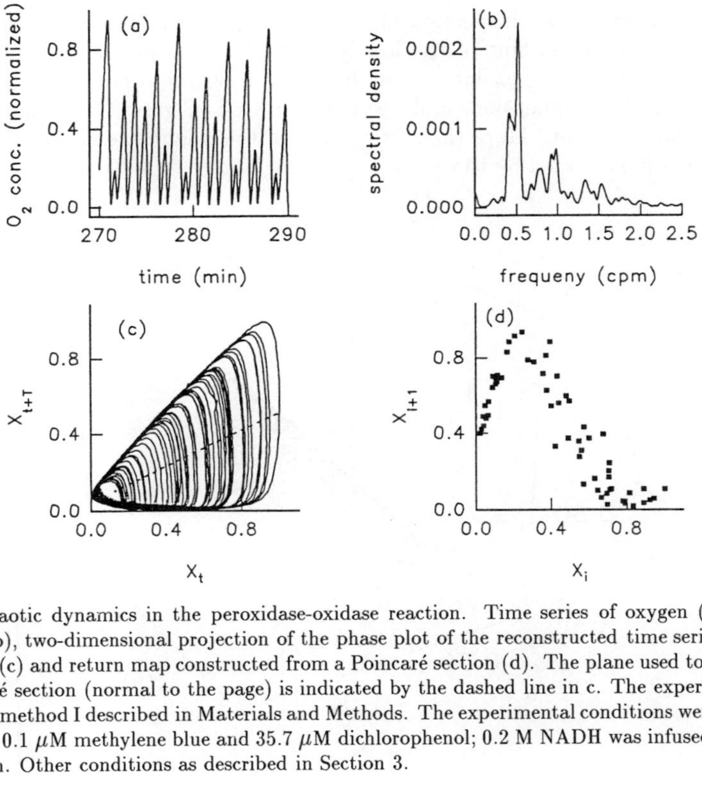


Fig. 1. Chaotic dynamics in the peroxidase-oxidase reaction. Time series of oxygen (a), power spectrum (b), two-dimensional projection of the phase plot of the reconstructed time series using a delay of 6 s (c) and return map constructed from a Poincaré section (d). The plane used to construct the Poincaré section (normal to the page) is indicated by the dashed line in c. The experiment was made using method I described in Materials and Methods. The experimental conditions were: 0.7 $\mu$M peroxidase, 0.1 $\mu$M methylene blue and 35.7 $\mu$M dichlorophenol; 0.2 M NADH was infused at a rate of 22.5 $\mu$l/h. Other conditions as described in Section 3.

Four apparently chaotic experimental time series ($O_2$ and NADH concentrations) were the subjects of analysis. Of the four data sets, three were obtained via Protocol I described above, while the fourth was obtained via Protocol II. In each case, the data were subdivided and the ability of the first part, the "atlas", to predict the time evolution of the second assessed. Predictions were generated by zero$^{th}$, first and second order local predictors following the approach of Farmer & Sidorowich, 1987 (see also, Schaffer & Tidd, 1990). Predictability was indexed by the square of the standard correlation coefficient of predicted *vs.* observed concentrations and also by the normalized root mean square error.

## 4. Results

In Figure 1, we give an example of chaotic fluctuations of $[O_2]$ in the experimental system. Here we display a representative time series, a power spectrum, a 2-dimensional projection of a reconstructed (Takens, 1981) phase portrait, and a return map. We emphasize that qualitatively similar dynamics were observed for both experimental protocols. Inspection of the time series and projected phase portrait in Figure 1 gives little hint of the highly nonuniform nature of the data. This is because we have connected data points by lines. If on the other hand, the data are embedded in three dimensions and the lines omitted (Figure 2), the nonuniformity becomes more evident. More than two thirds of the points are concentrated in the narrow bundle indicated by the arrow.

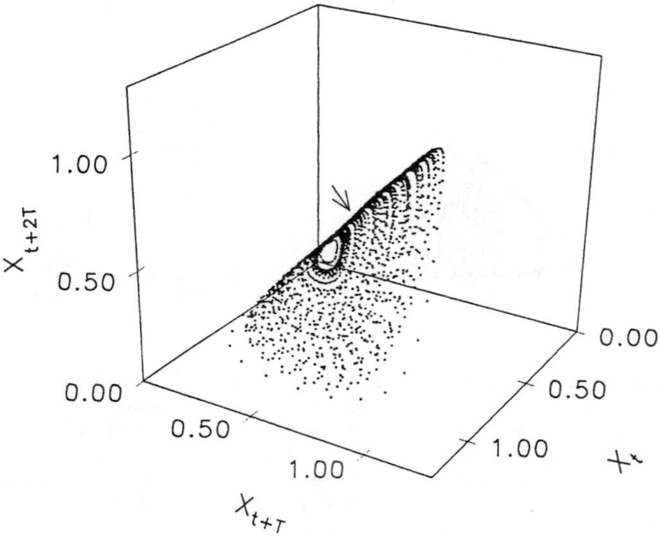


Fig. 2. Nonuniform dynamics of oxygen in the peroxidase-oxidase reaction. 3780 oxygen measurements were reconstructed in three dimensions using a time delay of 6 s. Experimental conditions as in Figure 1.

One approach to identifying nonlinearity in experimental time series involves nonlinear forecasting via local linear predictors (Casdagli, 1992). Here, the data are embedded in an $m$-dimensional space by application of Takens' method of delays. For each embedding dimension, $m$, one varies the number, $k$, of neighboring (in the phase space) atlas points used to forecast the time evolution of the target point in question. For each $k$, we calculate the normalized root-mean-square error, $E_m(k)$, averaged over all the target points and looks for the $k$-value, $k_{\min}$, which gives the lowest $E_m(k)$. Casdagli argues that if $k_{\min}$ is small, then the data is best described by a deterministic, nonlinear model. Conversely, intermediate values of $k_{\min}$ suggest that the best descriptor of the data is nonlinear stochastic models, while high values of $k_{\min}$ correspond to linear stochastic models.

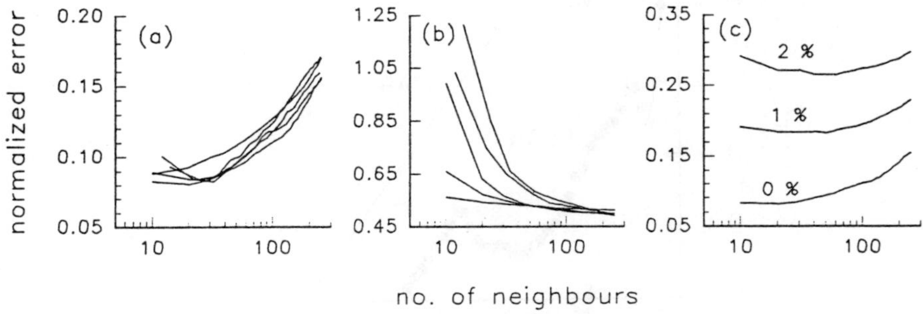


Fig. 3. Forecasting errors for measurements of oxygen from the PO reaction and corresponding surrogate data generated by randomizing the phases of the discrete Fourier transform of the original data and inverting the randomized spectrum. a, original data: b, surrogate data; c, original data with 0, 1 and 2% observational noise. 3780 data were embedded in dimensions 2 to 6 (in c the embedding dimension was 3) using a delay of 6 s. The first 3000 points were used to predict the remaining 780 points. A local linear prediction scheme was used and the number of nearest neighbors was varied. Prediction time was 6 s. Experimental conditions as in Figure 1.

Figure 3a shows a plot of $E_m(k)$ *vs.* $k$ for various embedding dimensions. Overall, predictability declines as the number of points (and hence the size of the neighbourhood) used to generate the predictions increases. According to Casdagli, this suggests that the data are best described by a deterministic, nonlinear model. Further evidence for this conclusion is presented in Figure 3b. Here we plot $E_m(k)$ for a set of 'surrogate data' which reproduce its power spectrum but with the phases randomized. As discussed by Theiler *et al.* (1992), the surrogate data are equivalent to output from a linear stochastic process. In this case, forecasting efficacy declines as we increase the number of points used to induce predictions. This, of course, is what Casdagli tells us should happen, and it indicates in a dramatic way that the experimental time series contains functionally important nonlinearities. In short, application of Casdagli's method provides yet another bit of evidence supporting a chaotic interpretation of irregular fluctuations in the reaction. Sadly, the efficacy of this approach turns out to be sensitive to the level of observational error which in many systems is far greater than that encountered in the case of the PO reaction. Specifically, if we corrupt the experimental data with small (one to two percent) amounts of noise, $k_{\min}$ is shifted toward higher values or vanishes completely (Figure 3c). At 2% noise predictability is lost to the extent that, according to Casdagli, one can only conclude that the time series is nonlinear, but not that it is low-dimensional, i.e. chaotic. In other words, a small increase in observational error can weaken or entirely eliminate the evidence for low-dimensional chaos as indicated by the present technique. Elsewhere, we argue that this sensitivity is a consequence of the data's nonuniform character.

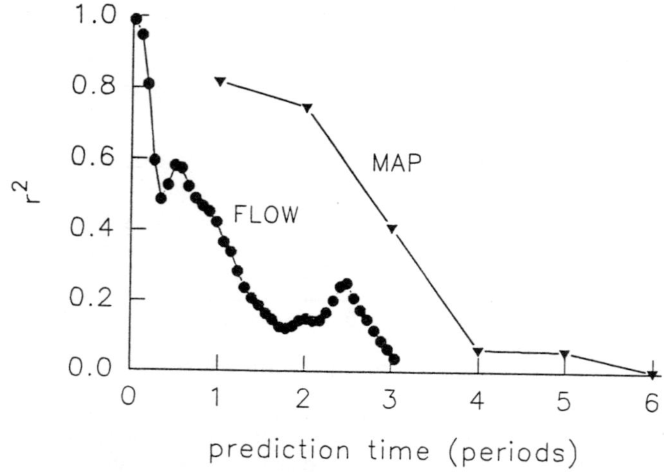


Fig. 4. Prediction profile for continuous oxygen data and return maps. The squared correlation coefficient is plotted against prediction time. Prediction time is normalized to the average period of oscillation which was 63 s. The continuous data consisted of 3780 points. The first 2000 points were used to predict the remaining 1780 points. The embedding dimension was 6 and the embedding delay was 8 s. The return map consisted of 60 points. The first 30 points were used to predict the remaining 30 points. Embedding dimension was 3 and the delay was 1. A zero$^{\text{th}}$ order prediction scheme was used for both data sets. Experimental conditions as in Figure 1.

Because of sensitivity to initial conditions, it is expected (*e.g.*, Farmer & Sidorowich, 1987; Sugihara & May, 1990; Wales, 1991, Tsonis & Elsner, 1992) that predictability for chaotic data should decline with increasing length of the prediction interval. Moreover, there exist scaling statistics which, at least in principle, relate predictability profiles to the rate at which neighboring trajectories diverge in the phase space.

Figure 4 shows two prediction profiles, $r^2$ *vs.* the prediction interval, $T_p$, which were obtained from the data shown in Figure 1. The curve labelled "flow" is for the continuously sampled data, whereas the curve labelled "map" is for points on the return map. Note that each point on the map corresponds to a single orbital excursion and that $T_p$ is expressed relative to the average period of oscillation. Both plots show a decline in $r^2$ with increasing $T_p$ which, as noted above, is to be expected. What is not expected, is the fact that the predictive power of the return map is much greater than that of the continuously sampled time series. Additionally, the scaling statistics of the two profiles are different, to the point that the profile for the continuously sampled data is inconsistent with an interpretation of low dimensional chaos (Wales, 1991; Tsonis & Elsner, 1992).

Why are return maps more predictable than the continuous data? To answer this question we made a number of forecasts of points from different regions of the reconstructed attractor. Two such forecasts are shown in Figure 5. Here we compare the predictability of a point on the Poincaré section used to construct the return map with that of a second point in the compressed region of the attractor. In the latter case predicted and observed values diverge rapidly. Conversely trajectories based at points on the section are much more predictable. This suggests that the limited predictive power of the continuous data is due to the preponderance of points in the compressed region, which has low predictability, and hence to the nonuniformity.

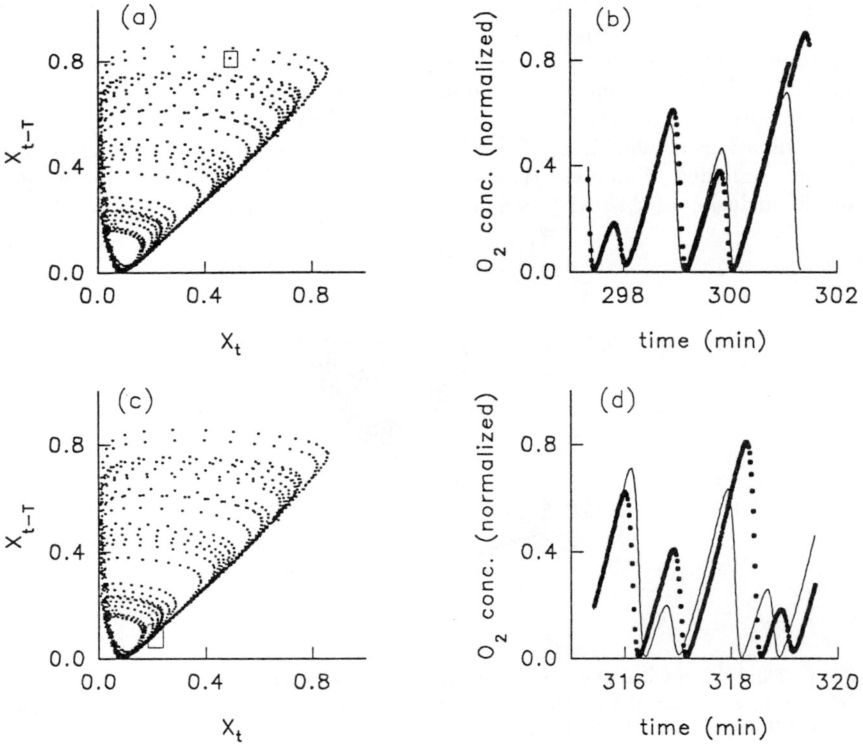


Fig. 5. Nonlinear forecasting of different regions of the reconstructed phase plot. In a) and c) the boxes indicate the points to be predicted. In b) and d) the predicted (filled circles) and observed values (solid lines) are shown for a prediction time of 240 s. The data were embedded in 3 dimensions using a delay of 10 s. The first 2000 points were used to predict the remaining 1780 points. A zero$^{th}$ order prediction scheme was used. Experimental conditions as in Figure 1.

Similar results (not shown) were obtained using data produced by a mathematical model (Olsen, 1983) of the PO reaction, *i.e.*, the predictive power of points in the compressed region of the attractor is essentially zero for values of $T_p$ in excess of the average period of oscillation. Why is this compressed region so unpredictable? To understand this we studied Poincaré sections of this region for the model data. We found that these sections evidence a complex mixing of trajectories with the consequence that nearby points on the section do not necessarily correspond to nearby trajectories in state space at a later time. Obviously, what is needed is a way of "unfolding" these regions such that points which are initially close together do not wind up too far apart. In principle, one can do this by increasing the embedding dimension. However, for experimental systems subject to dynamical and observational noise, this introduces yet another source of error — *i.e.*, the "dimensions" are lagged values of the observed variable and the longer the lag the greater the error. Alternatively, one can include a second variable in the analysis. Figure 6 shows as phase plot obtained by supplementing the $[O_2]$ time series with simultaneous measurements of [NADH], *i.e.* the oxygen data were embedded in two dimensions and the NADH data used as a third variable. We note that this phase plot differs from that obtained from the $O_2$ determinations alone (Figure 2) in that the compressed region of the latter has now become unfolded. Consequently the long term predictive power is greatly enhanced by the inclusion of this extra variable as illustrated in Figure 7.

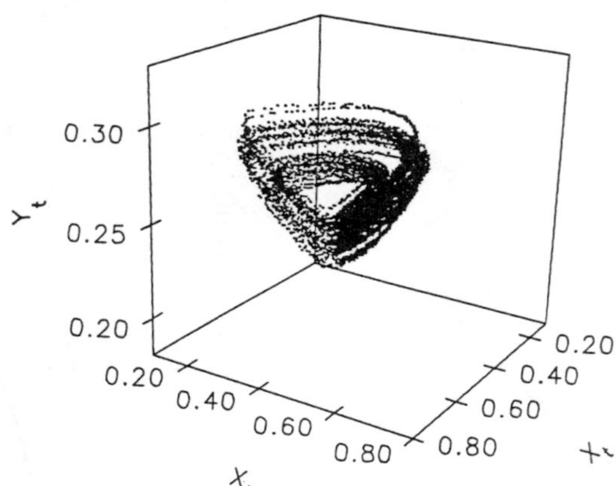


Fig. 6. Three-dimensional phase plot obtained by embedding oxygen data in two dimensions (time delay 20 s) and using the NADH data as the third variable. The experiment was made using method II described in Materials and Methods.

## 5. Discussion

The foregoing analysis may be viewed as an exercise in the study of nonuniform data. Having investigated the effects of noise on the prediction statistics for such data, we conclude that evidence for nonlinear determinism is easily obscured by small

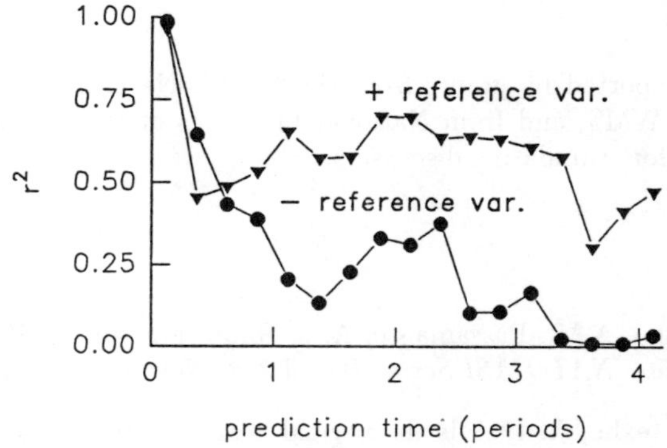


Fig. 7. Prediction profile for continuous oxygen data and continuous oxygen *plus* NADH data. Prediction time is normalized to the average period of oscillation which was 200 s. The data set contained 10000 measurements of oxygen and NADH. First the oxygen data were embedded in three dimensions using an embedding delay of 20 s. The first 7000 points were used to predict the remaining 3000 points and the squared correlation coefficient was plotted against prediction time. Next the oxygen data were embedded in two dimensions and the NADH data were used as the third variable as illustrated in Figure 6. Again the first 7000 points were used to predict the last 3000 points and the squared correlation coefficient was plotted against prediction time. Experimental conditions as in Figure 6.

amounts of measurement error. Moreover, prediction profiles of nonuniform data reveal a scaling behavior inconsistent with the hypothesis of low-dimensional chaos, *i.e.* predictability declines too rapidly. In short, nonlinear forecasting applied to nonuniform chaotic data can easily lead to the *wrong* conclusion, *i.e.*, that the time series evidence stochasticity as opposed to chaos. As noted above, nonuniform data in biological systems are a commonplace and are often contaminated with far more noise than the PO data which are obtained in a carefully controlled laboratory situation. A much studied example of nonuniform biological data are monthly notifications of measles in large First World cities. Following the application of nonlinear forecasting to such time series, Ellner (1992) and Casdagli (1992) concluded that the evidence for low-dimensional chaos in the data is at best ambiguous. Ellner, in particular, pointed to the fact that log transforming the data resulted in an increase in long term predictive power from which he concluded that the shape of the prediction profile is indecisive. In fact, for irruptive time series, the effect of such a transformation is to reduce the degree of nonuniformity, which is also the effect of restricting one's analysis to return maps computed from Poincaré sections of the sort shown in Figure 1. Accordingly, we would argue that Ellner's results are consistent with the thrust of the present paper which is that nonuniformity is a source of Type II error whereby one incorrectly accepts the null hypothesis of stochastic dynamics. For the particular case of measles, this point of view is further supported by the results reported by Tidd *et al.* (1993).

**Acknowledgements**

This work was supported by grants from the Danish Natural Science Research Council to LFO and WMS, and from National Institutes of Health to WMS. We thank Bruce Kendall for stimulating discussions.

**References**

Abraham, N.B., Albano, A.M., Passamante, A. & Rapp, P.E. (eds.) 1989 *Measures of complexity and chaos. NATO ASI Series B*, vol 208. New York: Plenum Press.

Babloyantz, A. & Destexhe, A. 1988 Is the normal heart a periodic oscillator? *Biol. Cybern.* **58**, 203–211.

Casdagli, M. 1992 Chaos and deterministic versus stochastic nonlinear modelling. *Jl. R. Statist. Soc.* B **54**, 303–328.

Ellner, S. 1991 Detecting low-dimensional chaos in population dynamics data: A critical review. In *Chaos and insect ecology* (ed. J.A. Logan & F.P. Hain), pp. 63–90. Blacksburg, Virginia: University Press of Virginia.

Farmer, J.D. & Sidorowich, J.J. 1987 Predicting chaotic time series. *Phys. Rev. Lett.* **59**, 845–848.

Förster, A., Hauck, T. & Schneider, F.W. 1994 Chaotic resonance of a focus and entrainment of a limit cycle in a periodically driven peroxidase–oxidase reaction in a CSTR. *J. Phys. Chem.* **98**, 184–189.

Geest, T., Steinmetz, C.G., Larter, R. & Olsen, L.F. 1992 Period-doubling bifurcations and chaos in an enzyme reaction. *J. Phys. Chem.* **96**, 5678–5680.

Geest, T., Olsen, L.F., Steinmetz, C.G., Larter, R. & Schaffer, W.M. 1993 Nonlinear analyses of periodic and chaotic time series from the peroxidase–oxidase reaction. *J. Phys. Chem.* **97**, 8431–8441.

Grassberger, P. & Procaccia, I. 1983 Measuring the strangeness of strange attractors. *Physica* D **9**, 189–208.

Hauck, T. & Schneider, F.W. 1993 Mixed-mode and quasiperiodic oscillations in the peroxidase–oxidase reaction. *J. Phys. Chem.* **97**, 391–397.

Hauck, T. & Schneider, F.W. 1994 Chaos in a Farey sequence through period doubling in the peroxidase–oxidase reaction. *J. Phys. Chem.* **98**, 2072–2077.

Lefebvre, J.H., Goodings, D.A., Kamath, M.V. & Fallen, E.L. 1993 Predictability of normal heart rhythms and deterministic chaos. *Chaos* **3**, 267–276.

May, R.M. 1978 Fluctuations in abundance of tropical insects. *Nature, Lond.* **278**, 505–507.

Nakamura, S., Yokota, K. & Yamazaki, I. 1969 Sustained oscillations in a lactoperoxidase, NADH and $O_2$ system. *Nature, Lond.* **222**, 794.

Nicolis, J.S., Meyer-Kress, G. & Haubs, G. 1983 Non-uniform chaotic dynamics with implications to information processing. *Z. Naturforsch.* A **38**, 1157–1169.

Olsen, L.F. 1983 An enzyme reaction with a strange attractor. *Phys. Lett.* A **94**, 454–457.

Olsen, L.F. & Degn, H. 1977 Chaos in an enzyme reaction. *Nature, Lond.* **267**, 177–178.

Olsen, L.F. & Degn, H. 1978 Oscillatory kinetics of the peroxidase–oxidase reaction in an open system. Experimental and theoretical studies. *Biochim. Biophys. Acta* **523**, 321–334.

Osborne, A.R. & Provenzale, A. 1989 Finite correlation dimension for stochastic systems with power law spectra. *Physica* D **35**, 357–381.

Ruelle, D. 1979 Sensitive dependence on initial condition and turbulent behavior of dynamical systems. *Ann. N.Y. Acad. Sci.* **316**, 408–416.

Ruelle, D. 1990 Deterministic chaos: the science and the fiction. *Proc. R. Soc. Lond.* A **427**, 241–248.

Rys, P. & Wang, J. 1992 Hydrogen/deuterium isotope effects in the oscillating peroxidase–oxidase reaction. *Biochem. Biophys. Res. Commun.* **186**, 612–616.

Samples, M.S., Hung, Y.-F. & Ross, J. 1992 Further experimental studies of the horseradish peroxidase–oxidase reaction. *J. Phys. Chem.* **96**, 7338–7342.

Schaffer, W.M., Olsen, L.F., Truty, G.L., Fulmer, S.L. & Graser, D.J. 1988 Periodic and chaotic dynamics in childhood infections. In *From chemical to biological organization* (ed. M. Markus, S.C. Müller & G. Nicolis), pp. 331–347. Berlin: Springer-Verlag

Schaffer, W.M. & Tidd, C.W. 1990 *NLF: nonlinear forecasting for chaotic dynamical systems*. Tucson, Arizona: Dynamical Systems, Inc.

Steinmetz, C.G., Geest, T. & Larter, R. 1993 Universality in the peroxidase–oxidase reaction: Period doublings, chaos, period three, and unstable limit cycles. *J. Phys. Chem.* **97**, 5649–5653.

Sugihara, G., Grenfell, B. & May, R.M. 1990 Distinguishing error from chaos in ecological time series. *Phil. Trans. R. Soc. Lond.* B **330**, 235–251.

Sugihara, G. & May, R.M. 1990 Non-linear forecasting as a way of distinguishing chaos from measurement error in time series. *Nature, Lond.* **344**, 734–741.

Takens, F. 1981 Detecting strange attractors in turbulence. In *Dynamical systems and turbulence* (ed. D. Rand & L.S. Young), pp. 366–381. Berlin: Springer-Verlag.

Theiler, J., Galdrikian, B., Longtin, A., Eubank, S. & Farmer, J.D. 1992 Using surrogate data to detect nonlinearity in time series. In *Nonlinear modelling and forecasting* (ed. M. Casdagli & S. Eubank), pp. 163–188. Redwood City, California: Addison-Wesley.

Tidd, C.W., Olsen, L.F. & Schaffer, W.M. 1993 The case for chaos in childhood epidemics. II. Predicting historical epidemics from mathematical models. *Proc. R. Soc. Lond.* B **254**, 257–273.

Tsonis, A.A. 1992 *Chaos: from theory to applications*, New York: Plenum Press.

Tsonis, A.A. & Elsner, J.B. 1992 Nonlinear prediction as a way of distinguishing chaos from random fractal sequences. *Nature, Lond.* **358**, 217–220.

Wales, D.J. 1991 Calculating the rate of loss of information from chaotic time series by forecasting. *Nature, Lond.* **350**, 485–488.

Wolf, A., Swift, J.B., Swinney, H.L. & Vastano, J.A. 1985 Determining Lyapunov exponents from a time series. *Physica* D **16**, 285–317.

CHAPTER 9

# CHAOS IN GEOPHYSICAL FLUIDS

Raymond Hide
Department of Physics
(Atmospheric, Oceanic and Planetary Physics)
Clarendon Laboratory
Parks Road
Oxford OX1 3PU, England, U.K.

**Abstract**

Irregular buoyancy-driven flows occur in the atmospheres and fluid interiors of the Earth and other planets, and of the Sun and other stars, where they influence and often control the transfer of heat. Their presence is manifest in or implied by a wide variety of observed phenomena, including external magnetic fields generated by self-exciting magnetohydrodynamic (MHD) dynamo action. Based on the laws of classical mechanics, thermodynamics and, in the case of electrically-conducting fluids, electrodynamics, the governing mathematical equations are well known, but they are generally intractable owing to their essential nonlinearity. Computers play a key role in modern theoretical research in geophysical and astrophysical fluid dynamics, where ideas based on chaos theory are being applied in the analysis of models and the assessment of predictability. The aim of this paper is to provide a largely qualitative survey for non-specialists. The survey comprises two parts, namely a general introduction (Part I) followed by a discussion of two representative areas of research (Part II). Each of these areas is concerned with phenomena attributable to symmetry-breaking bifurcations caused by gyroscopic (Coriolis) forces, namely (a) large-scale waves and eddies in the atmospheres of the Earth, Jupiter and other planets (where, exceptionally, laboratory experiments have been influential), and (b) MHD dynamos. Various combinations of Faraday disk dynamos have been studied numerically as low-dimensional nonlinear electromechanical analogues of MHD dynamos, particularly in efforts to elucidate the complex time series of geomagnetic polarity reversals over geological time. The ability of the intensively-studied Rikitake coupled disk dynamo system to behave chaotically appears to be a consequence of the neglect of mechanical friction, the inclusion of which renders the system structurally unstable.

## Part I: General introduction

### 1. Buoyancy-driven flows

Flows in the atmospheres and interiors of the Earth and other planets and of the Sun and other stars are driven by buoyancy forces due to the action of gravity on spatial variations of density. These density variations are associated with temperature variations produced and maintained by differential heating and cooling, and they are modified by variations in pressure and chemical composition. Through their ability to transport ("advect" or "convect") heat in both horizontal and vertical directions such fluid flows influence and often control the overall heat balance and evolution of the systems within which they occur.

Buoyancy-driven flows derive their kinetic energy from gravitational potential energy, with heavier fluid tending to sink and lighter fluid to rise. Concomitant advection of heat in the vertical is always in the upward direction (unless the thermal coefficient of cubical expansion of the fluid is negative). Upward advection of heat by large-scale waves and jet streams in the lower reaches of the Earth's atmosphere–the troposphere–helps to maintain the vertical temperature gradient there at a value somewhat less than the adiabatic value. This in turn influences the scales and other characteristics of tropospheric flows, thereby providing a dynamically-important feedback mechanism. Horizontal advective heat transfer by this flow–and also by the currents driven in the underlying oceans by atmospheric surface winds and by buoyancy forces–keeps the equator-to-pole temperature difference at the Earth's surface at a fraction of the radiative equilibrium value that would otherwise obtain in order to balance differential solar heating.

The two general conditions for the occurrence of stable hydrostatic equilibrium–with no fluid flow and buoyancy forces everywhere balanced by pressure gradients alone (see (2.1) below) – are quite strict. The first is that density gradients nowhere possess a horizontal component, for otherwise fluid elements would experience gravitational torques, which cannot be balanced by pressure gradients. The second is that the vertical density gradient is either (a) "bottom heavy" (i.e. the "potential density" nowhere increases upwards (implying that the actual density everywhere decreases upwards at a rate less than the so-called "adiabatic gradient", which vanishes when the fluid is incompressible), or (b) "top- heavy" but of insufficient magnitude for buoyancy forces to be able to promote instability against inhibiting effects due to viscosity, thermal conduction and radiation, and also to gyroscopic (Coriolis) forces when the whole system is in general rotation, and to Lorentz forces when the fluid can conduct electricity and magnetic fields are present. When the first of these conditions is satisfied but the second is not, convection of the Rayleigh-Bénard type occurs, such as the flow that gives rise to the "granular" appearance of the solar photosphere. Another example of Rayleigh - Bénard convection is the very slow flow – centimetres per year – in the highly viscous mantle of the Earth that geophysicists invoke to account for heat transfer there and also for various tectonic and other processes manifest in geophysical and geological data. When the second condition is satisfied but first is not, as in extensive regions of the atmospheres of the Earth and other planets, where the vertical density gradient is "bottom heavy" but there are impressed horizontal density gradients, flow must always occur no matter how small the magnitude of these gradients. Many factors determine the form and speed of the flow, and in the important case when Coriolis forces exert a dominant influence, the process of "sloping convection" can occur, in which the typical trajectory of a fluid element is inclined at an angle to the horizontal less than the slope of the surfaces of equal potential density. Through their differential action on the axial and non-axial components of the flow velocity vector, Coriolis forces tend to promote sloping convection, which is the dominant large-scale dynamical process in the extra-tropical regions of the Earth's atmosphere.

Irregularity characterizes the spatio-temporal behaviour of geophysical (and astrophysical) fluids, as evinced by both direct and, more commonly, indirect observations (see below). Attempts to interpret the observations in terms of underlying fluid - dynamical processes or to predict future behaviour require in the first instance the separation of manifestations in the data of effects due to random forcing (noise) of various kinds from any that might be attributable to deterministic chaos. The "predictability horizon" of a system exhibiting irregular behaviour depends *inter alia* upon the extent to which the underlying processes are dominated by deterministic chaos rather than by noise. In principle it should be possible to identify deterministic chaos in observational time series, by using sophisticated modern methods (see e.g. Drazin and King 1992, Ghil *et al.* 1991 Moon 1987, Mullin 1993, Ott 1993, Smith 1988, 1992, 1994, Thompson and Stewart 1986) developed for adducing evidence of low-dimensional attractors in the data. This can often be done in investigations of well-controlled laboratory systems (see e.g. Guckenheimer and Buzyna 1983, Haken 1981, Mullin 1993, Read 1993, Read *et al.* 1992), but it is more difficult or even impossible in the study of observational time series from geophysical and astrophysical fluids and other natural systems, owing to the inadequate length of available time series and to measurement errors. So other kinds of evidence for chaos must be sought, such as that afforded by constructing models of the prototype with varying degrees of simplification (see Section 2 below) and analyzing their behaviour. The clear identification in the prototype of regimes of behaviour determined by bifurcations at critical values of key parameters as revealed by the models would be evidence for chaos.

Optimism – albeit limited – shown by workers concerned with basic problems of predictability of the terrestrial atmosphere (see e.g. Ghil *et al.* 1991, Lorenz 1963, 1967, 1980, 1993, Mason *et al.* 1986, Monin 1972, Mullin 1993, Nicolis and Nicolis 1987, Palmer *et al.* 1994, Thompson, 1957, 1988, Webster and Keller 1975, White 1990) derives to some extent from the practical experience of weather forecasters and climatologists. Also influential according to the literature have been laboratory experiments such as those carried out on sloping convection in a rotating fluid annulus subject to a steady axisymmetric temperature gradient (Hide 1953). In these experiments key parameters were identified and several flow regimes of varying degrees of spatial and temporal complexity were first delineated, ranging from steady axisymmetric flow through periodic and quasi-periodic non-axisymmetric flows to highly irregular "geostrophic turbulence". The quasi-periodic wave-like flows discovered in these experiments and termed "vacillation" have attracted some attention, but they have not yet been reproduced in satisfactory detail in numerical models (see e.g. Hignett *et al.* 1985, Lorenz 1963 (b), Quinet 1973, Thompson, 1988, White 1988). Elsewhere in this volume, Palmer *et al.* (1994) explain how modern ideas in the theory of deterministic chaos are guiding meteorologists armed with very powerful computers and highly sophisticated numerical models of the troposphere and the overlying stratosphere in their efforts to monitor global-scale atmospheric flow and forecast its future behaviour. Underlying such exercises is the implicit recognition of the possibility "intransitivity" and "multiple equilibria" (in modern parlance) (see Hide 1953) as well as other manifestations of essentially nonlinear and possibly chaotic behaviour such as hysteresis (see Fultz *et al.* 1959) in atmospheric flows. (For further references to laboratory studies and related work, see e.g. Buzyna *et al.* 1984, Ghil and Childress 1987, Hart 1986, Hide 1977, Hide and Mason 1975, Lewis 1992, Read 1988, Read *et al.* 1992, Smith 1994 and articles in Corby 1969, Hopfinger 1992, King and Mobbs 1991, Mullin 1993 and Roberts and Soward 1978.)

Owing to public demands for weather and climate forecasts, meteorologists enjoy abundant observations of wind velocity, pressure, temperature and other variables of direct dynamical relevance, collected frequently and systematically at many levels over the whole globe, and made available rapidly in "user friendly" form, making meteorologists the envy of workers in other areas of geophysical and astrophysical fluid dynamics. And for studying temporal fluctuations in one particularly important and revealing average property of the global atmospheric circulation, namely the total angular momentum, accurate surrogate data can be deduced from geodetic observations of fluctuations in the Earth's rotation vector (see Hide 1984, Hide and Dickey 1991, Rosen 1993). But in other areas of geophysical fluid dynamics such as the study of the atmospheres of other planets, observations are usually indirect in nature and of more restricted coverage in space and time, as they also are in studies of the oceans and of the Earth's atmosphere on the long time scales of interest to climatologists (see e.g. Berger *et al.* 1989, Ingersoll 1990, Ingersoll and Lyons 1993, James 1994, Neelin *et al.* 1994, Philander 1992, Saltzman and Verbitsky 1993, Sreenath 1993, Willebrand and Anderson 1993).

In the study of magnetohydrodynamic (MHD) flow in the Earth's liquid metallic core, where the main geomagnetic field is produced by self-exciting MHD dynamo action, the main observational data comes from magnetic measurements made at or near the surface of the Earth (see e.g. Jacobs 1987–91, 1994, Melchior 1986, also Part II below). Magnetic observations also provide the basis for studies of MHD flows in the electrically-conducting deep interiors of other planets such as Jupiter, Saturn, Uranus and Neptune and in the outer layers of the Sun (see e.g. Ness 1994,

Proctor and Gilbert 1994, Sonett *et al.* 1991, Stevenson 1983.) Additional data in the case of the Sun comprise a long time series of annual sunspot numbers, which exhibits nearly periodic oscillations with a dominant period of about eleven years. In the interpretation of these observations in terms of basic fluid-dynamical processes, highly simplified models analyzed by methods suggested by chaos theory are being applied with some success (e.g. Platt *et al.* 1993, Ruzmaikin *et al.* 1994, Weiss 1990, 1993, 1994).

Many essentially nonlinear phenomena are encountered in astronomy and geophysics (see e.g. Turcotte 1993), including earthquakes, the prediction of which is the topic of another paper in this session (Matsuzaki 1994). In the Part II of this article on chaos in geophysical fluids, two topics are selected for more detailed discussion, namely the study of (a) large-scale motions in planetary atmospheres, where laboratory studies have played a useful role, and (b) planetary magnetism, with emphasis on polarity reversals of the main geomagnetic field where theoretical work has to rely on mathematical modelling alone. The article reflects my own personal scientific interests and point of view. In preparing it I have relied heavily on my research library. This includes many reprints of papers from a wide range of journals, which colleagues have kindly sent to me over a period of more than forty years. A full bibliography citing all relevant original references would run to the editorially-unacceptable length of more than two hundred items, so I have included a selection of books, monographs and review articles where these original references can be found. Also it is impossible to reproduce here the many pictures and diagrams used to illustrate the lecture upon which this article is based.

## 2. Basic equations and modelling

Applying Newton's laws of motion to a fluid element of unit volume which at time $t$ is located at the general point whose vector position is $\mathbf{r}$ in the chosen inertial frame of reference gives the equation

$$\rho\, D\mathbf{u}/Dt = -\nabla p - \rho\nabla V + \mathbf{F}\ . \tag{2.1}$$

Here $\rho$ denotes the mass density of the fluid at $P$, $\mathbf{u}$ is the Eulerian flow velocity, and the operator

$$D/Dt \equiv \partial/\partial t + \mathbf{u}\ .\nabla \tag{2.2}$$

is the "substantial" time derivative following the moving fluid element. $\nabla p$ is the gradient of the pressure $p$, $-\nabla V$ is the acceleration due to gravity, and $\mathbf{F}$ represents all the other forces acting on the fluid element, including viscosity. In the case of electrically - conducting fluids, with which the subject of MHD is concerned, $\mathbf{F}$ also includes the Lorentz force $\mathbf{j}\times\mathbf{B}$ where $\mathbf{j}$ is the electrical current density at $P$ and $\mathbf{B}$ the magnetic field. Continuity of matter is expressed by the equation

$$D\rho/Dt + \rho\nabla.\mathbf{u} = 0. \tag{2.3}$$

All these quantities, $\rho$, $\mathbf{u}$, $p$, $V$, $\mathbf{j}$ and $\mathbf{B}$ are in general functions of $\mathbf{r}$ and $t$. Equations (2.1) and (2.3) effectively comprise four scalar partial differential equations in twelve scalar unknowns, namely $\rho$, $p$ and $V$ and the individual components of the three-dimensional vectors $\mathbf{u}$, $\mathbf{B}$ and $\mathbf{j}$, so additional equations are needed. When $\nabla V$ is variable the law of gravity is used to relate $\rho$ and $V$; and when $\rho$ is variable an equation of state relating $\rho$ at $P$ to the pressure $p$, entropy $S$, composition $K$ and temperature $T$ etc., is needed, together with transport equations for $S$, $K$, and $T$. These include the equation

$$(\partial/\partial t + \mathbf{u}.\nabla)(\rho c T) = J \tag{2.4}$$

expressing the heat balance of a fluid element of unit volume at $P$ if $c$ is the specific heat. Here the term $\mathbf{u}.\nabla(\rho c T)$ represents the contribution of advection to the heat balance and $J$ represents all other contributions, including conduction and radiation.

Finally, in the case of MHD flows in electrically-conducting fluids, such as those found in planetary interiors and in stars, we need the equations of electrodynamics applied to a moving medium. These relate $\mathbf{u}$, $\mathbf{j}$ and $\mathbf{B}$ implicit in (2.1) and bring in further variables such as the electric field vector $\mathbf{E}$ and the electric charge density $\theta$ at $P$. In the case of non-relativistic flows, $\mathbf{B}$ and $\mathbf{j}$ are related by Ampére's law $\nabla \times (\mathbf{B}/\mu) = \mathbf{j}$ (where $\mu$ is the magnetic permeability) and the other equations of electrodynamics (expressing the laws of Gauss, Faraday, etc.), which lead to the following equation involving the time rate of change of $\mathbf{B}$:

$$\partial\mathbf{B}/\partial t - \nabla(\mathbf{u} \times \mathbf{B}) = -\nabla \times (\sigma^{-1}\nabla \times (\mu^{-1}\mathbf{B})) + \nabla \times \mathbf{Z}. \tag{2.5}$$

(cf. Moffatt 1978, Parker 1979). Here $\sigma$ and $\mathbf{Z}$ are defined by a generalized Ohm's law applied to a moving medium:

$$\mathbf{j} = \sigma[\mathbf{E} + \mathbf{u} \times \mathbf{B} + \mathbf{Z}]. \tag{2.6}$$

The full set of nonlinear partial differential equation (PDE's) thus obtained is complete. In addition to $c$, $\mu$ and $\sigma$ (see (2.3) and 2.5)), the equations include other coefficients needed to specify the physical properties of the fluid at a general point $P$ at time $t$ such as the coefficient of viscosity, thermal coefficient of cubical expansion, ratio of principal specific heats, coefficients of thermal conductivity, chemical diffusion, dielectric constant, etc. The mechanical, thermal and electromagnetic boundary conditions under which these equations must be solved in specific cases are also well known, at least in principle. They usually amount to the requirement that all dependent variables and their derivatives as well as certain fluxes be continuous everywhere. It follows that theoretical fluid dynamics is not handicapped primarily by incomplete knowledge of the basic equations to be solved! It is the intractability of these equations owing to their essential nonlinearity that causes the main difficulties. Nonlinearity can arise in important cases from the boundary conditions or from spatial and temporal variability in the various parameters such as viscosity in the governing equations. More commonly however it is advection that it is responsible, as expressed by the terms $(\mathbf{u}.\nabla)\mathbf{u}$, $(\mathbf{u}.\nabla)\rho$, $(\mathbf{u}.\nabla)(\rho c T)$ and $(\mathbf{u}.\nabla)\mathbf{B}$ implicit or otherwise on the left-hand sides of (2.1), (2.3), (2.4) and (2.5) respectively. In spite of these difficulties, thanks to powerful modern computers and a wide range of new techniques in laboratory studies, where computers can be used in the control of apparatus and the analysis of data, the subject of fluid dynamics continues to develop as a lively branch of classical physics with many applications in engineering, geophysics and astronomy.

All mathematical modelling based on the above equations involves the simplification of both the equations and the boundary conditions governing the prototype being studied, and there is an art in constructing such models. In all cases the most sophisticated models such as those now used in numerical weather prediction involve minimal simplifications, still resulting in P.D.E's to be solved for all dynamical variables. At the other extreme are very highly simplified "low-dimensional" models governed by ordinary differential equations (O.D.E.s) usually expressing the time evolution of mean quantities such as spatially averaged velocity, temperature, magnetic field, etc. (see e.g. Lorenz 1963a, Moore and Spiegel 1966). Such models might be trustworthy when they are derived by rational arguments from the full P.D.E's, but the most highly simplified models used in the study of the behaviour of nonlinear systems can be irrational ones, for they are often based to some extent on intuition, giving errors which are hard to assess. Analyses of such models provide results which might be helpful with diagnostic studies of more complicated systems, but they always carry a "health warning". Nevertheless, in the right hands findings based on low-dimensional models can be brought to bear on truly quantitative studies of the prototype, just as metaphysical ideas, though non-scientific, can influence truly scientific work! The most favourable situation arises when the investigator has at his disposal the results not only from numerical integrations of a range of mathematical models but also from relevant laboratory studies. Recent books (e.g. Cvitanović 1984, Haken 1981, Mullin 1993) discuss examples of thermally-driven Rayleigh - Bénard convection and sloping convection (see Section 1 above and Part II below) and mechanically-driven motions such as Taylor-Couette flow. Useful numerical integrations were not available before high-speed computers were introduced and guidance then had to be sought solely from laboratory experiments and from limited and sometimes misleading findings based on linearized models.

There is also an art in the exploitation of results from mathematical or laboratory models in research on much more complicated and uncontrollable natural systems. In his celebrated monograph on the general circulation of the Earth's atmosphere, Lorenz (1967) devotes a whole chapter to the discussion of the results of laboratory experiments on sloping convection in cylindrical apparatus (see also Lorenz 1993). To paraphrase his concluding remarks, the laboratory experiments tell us more about planetary atmospheres in general than about the Earth's atmosphere in particular. They indicate the variety of flow patterns that can occur and the conditions favourable to each of these. Perhaps the most important contribution of laboratory experiments to the theory of the general circulation of the terrestrial atmosphere has been the separation of essential considerations from the minor and irrelevant. Condensation of water vapour, for example, may yet play an essential role in the Tropics but in temperate latitudes it appears to be no more than a modifying influence, since systems occurring in the atmosphere, including even cyclones and fronts, are found in the laboratory system, where there is no analogue of the condensation process. Similar remarks apply to the topographic features of the Earth and the so-called "beta" effect (which represents the latitudinal variation of the Coriolis parameter associated with the near-spherical shape of the Earth) now appears to play a lesser rôle than had once been supposed by theoreticians. Certainly a numerical weather forecast would fail if the beta-effect were disregarded, but the beta-effect does not seem to be required for the development of typical atmospheric systems. The experiments emphasise the necessity for truly quantitative considerations. At the very least these must be sufficient to place the Earth's atmosphere in one of the régimes discovered in the experiments.

Most of the experiments to which he refers were carried out in the early 1950's (for references see e.g. Corby 1969, Hide and Mason 1975), but there have been many subsequent related studies (see e.g Drazin and King 1992, Guckenheimer and Buzyna 1983, Hide *et al.* 1994, Hignett *et al.* 1985, Hopfinger 1991, King and Mobbs 1991, Lorenz 1963, Nicolis and Nicolis 1987, Ott 1993, Quinet 1974, Read 1993, Read *et al.* 1992, Smith 1988, 1992, 1994, White 1988, 1990.) Modern work on sloping convection is based on the powerful combination of laboratory experiments and numerical models, and it is able to exploit some of the new methods of signal processing and time-series analysis to which chaos theory has given rise (see Section 1 above). In such work, where in the search for "low-dimensional attractors" in the output of transducers recording the behaviour of a system the acquisition of long time-series of high signal-to-noise ratio data is crucial, numerical models and laboratory models clearly have a key rôle to play. As already mentioned in Section 1, observational time series from natural geophysical fluid systems are usually much too noisy and limited in duration for chaos theory to be directly applicable in their interpretation.

Just as in meteorology it is of interest to know what the atmospheric general circulation would be like in the absence of various complicating effects such as the presence of continents and oceans, water vapour in the atmosphere, time-varying thermal forcing, etc. in the boundary conditions, a central question in geomagnetism concerns the extent to which it might be necessary to invoke complex thermal, mechanical and electrical boundary conditions imposed on the Earth's liquid metallic fluid core by very slow flow occurring in the highly viscous overlying mantle in order to account for observations such as the highly-variable frequency of polarity reversals of the main geomagnetic field. Should polarity reversals be regarded as being "forced" by irregular and fluctuating boundary conditions, or as an manifestation of "free" instabilities that would occur under simple and fixed boundary conditions? Modern research indicates that details of the pattern of the geomagnetic field at the surface of the Earth are manifestations of both "forced" and "free" processes. Edmond Halley raised money from the Crown at the end of the seventeenth century for the purpose of mapping this pattern (see Chapman 1941) with a view to predicting future changes, before the invention of the marine chronometer effectively solved the main practical problem which Halley had in mind, namely the accurate determination of geographical longitude at sea.

The very existence of the durable Great Red Spot in the atmosphere of Jupiter –

more than three centuries after its discovery by Robert Hooke in 1665 – and of other long-lived dynamical features in the atmospheres of the major planets must have implications for theories of atmospheric predictability. Why are there more of these markings in the southern hemisphere of Jupiter than in the northern hemisphere? What is the rôle of these eddies in the heat balance of the atmosphere? To what extent should the eddies be regarded as "free" or "forced", and what (if any) is their relationship to the field of transient eddies within which they are embedded? Can we infer the vertical structure of Jupiter's atmosphere (or even just its depth) from the flow at the observable upper level? These and many other related questions pose key problems in the essentially nonlinear dynamics of geophysical fluids.

## 3. Symmetry breaking, heat transfer by geostrophic and magnetostrophic flows, and elastoid oscillations

It is inconvenient to use (2.1) when dealing with fluid flow in planets and stars, which usually rotate rapidly relative to an inertial frame. Owing to the rotation of the solid Earth with a period of 24 hours, points on the equator move in space at a speed some 40 times the typical speed (10 $\mathrm{ms}^{-1}$) of atmospheric winds relative to the Earth's surface. The corresponding factor for ocean currents is even higher, about 4,000, and for the Earth's liquid metallic core it is higher still, about $10^6$. For motions in the atmosphere of Jupiter, a planet with a diameter ten times that of the Earth and a rotation period as short as ten hours, the factor is more than 40. And even for the Sun, with a rotation period as long as a month but a diameter ten times that of Jupiter, linear speeds of rotation greatly exceed typical relative speeds of large-scale atmospheric flow. So we must refer our dynamical equations to a more convenient reference frame.

If we choose a frame that rotates steadily with angular velocity $\mathbf{\Omega}$ relative to an inertial frame, (2.1) has to be modified by adding the so-called Coriolis force $2\rho\mathbf{\Omega}\times\mathbf{u}$ to the left-hand side and replacing $-\nabla V$ by $\mathbf{g}$, which now includes centripetal effects. When $\mathbf{\Omega}$ is so large in magnitude that $\mid 2\rho\mathbf{\Omega}\times\mathbf{u}\mid$ greatly exceeds the terms $\mid \rho\partial\mathbf{u}/\partial t \mid$ and $\mid \rho(\mathbf{u}.\nabla)\mathbf{u} \mid$ on typical time and space scales of the whole system, to a first approximation we can write

$$2\rho\mathbf{\Omega}\times\mathbf{u}\approx\mathbf{\Gamma}+\mathbf{F} \tag{3.1}$$

where

$$\mathbf{\Gamma}\equiv-\nabla p+\mathbf{g}\rho. \tag{3.2}$$

When $\mid \mathbf{F} \mid \ll \mid 2\rho\mathbf{\Omega}\times\mathbf{u} \mid$, (3.1) reduces to

$$2\rho\mathbf{\Omega}\times\mathbf{u}\approx\mathbf{\Gamma}, \tag{3.3}$$

which expresses approximate "geostrophic" balance between Coriolis forces and dynamic pressure gradients in the horizontal. To a good approximation, large-scale flows occupying extensive regions in the oceans and in the atmospheres of the Earth and other planets (with the exception of slowly-rotating Venus) are either found or expected to be in geostrophic balance nearly everywhere. Exceptional regions are near the equator, where the vertical component of $\mathbf{\Omega}$ vanishes, and within narrow "fronts" and jet streams, where the nonlinear advective term $\rho(\mathbf{u}.\nabla)\mathbf{u}$ is typically comparable with $2\rho\mathbf{\Omega}\times\mathbf{u}$ in magnitude. Indeed, the existence of such "detached shear layers" in rapidly - rotating fluids is implied by (3.3), which, being of lower order than the full equation, is diagnostic in character rather than prognostic, and therefore incapable of giving solutions that satisfy all the necessary boundary conditions (see Hide 1977). When the Lorentz force $\mathbf{j}\times\mathbf{B}$ makes the dominant contribution to $\mathbf{F}$ in (3.1) and is not negligible in magnitude in comparison with the Coriolis term, the equation becomes:

$$2\rho\mathbf{\Omega}\times\mathbf{u}\approx\mathbf{\Gamma}+\mathbf{j}\times\mathbf{B} \tag{3.4}$$

This diagnostic equation expresses "magnetostrophic" balance of forces acting on individual fluid elements. Such balance nearly everywhere may be the main property that characterizes large-scale MHD flows in the electrically-conducting deep interiors of the planets, and also in the Sun and other stars.

Equation (3.3) is the basis of the meteorologist's "Buys-Ballot law" – stand with your back to the wind in the northern (southern) hemisphere and the low pressure is found to be on the left (right)! It follows from (3.3) that if the pattern of fluid flow $\mathbf{u}$ and the associated fields of $p$ and $\rho$ are symmetric about the axis of rotation, then the component of $\mathbf{u}$ perpendicular to the rotation axis is everywhere zero. This implies that any flow that is both axisymmetric and geostrophic cannot advect heat (or any other quantity) in directions that are perpendicular to the rotation axis. Such flows would therefore make no contribution to the overall heat balance of systems subject to axisymmetric applied heating and cooling. This result suggests, and experience based on many detailed studies of rapidly rotating laboratory systems and natural systems confirms (see Hide 1977), that large-scale non-axisymmetric flow capable of advecting heat perpendicularly to the rotation axis would develop in "generic" (i.e. typical) cases. We thus see a connection between the basic function of thermally-driven flows – namely heat transfer – and symmetry breaking which, from a theoretical point of view, can be associated with bifurcations produced by the action of Coriolis forces when $|\mathbf{\Omega}|$ is large enough.

This symmetry breaking associated with geostrophy and heat transfer in generic systems provides a good starting point in the discussion of the role of rotation in the production of magnetic fields by MHD dynamo action in astronomical bodies such as the Earth and other planets (see Hide 1982). By a theorem due to Cowling and others (for references see Proctor and Gilbert 1994) no magnetic field with an axis of symmetry can be maintained by fluid motions against ohmic dissipation, so that a strictly axisymmetric pattern of fluid motion $\mathbf{u}$ and magnetic field $\mathbf{B}$ would be incapable of dynamo action. It is only possible to find non-decaying solutions to (2.5) (with $\mathbf{Z}=0$) when the configuration of $\mathbf{B}$ has no axis of symmetry.

A magnetic field produced by self-exciting dynamo action can be regarded as having originated as an instability involving the amplification of a much weaker adventitious field through inductive action produced by fluid motion. Equilibration occurs when on average the rate at which buoyancy forces do work on the system is balanced by ohmic and other types of energy dissipation. When the magnetic field is so weak that Lorentz forces are negligible, quasi-geostrophic non-axisymmetric flow is expected to occur. Dynamo action then amplifies the magnetic field, and various lines of evidence indicate that the field strength would typically build up until $\mathbf{j} \times \mathbf{B}$ is comparable in magnitude with $2\rho\boldsymbol{\Omega} \times \mathbf{u}$ (see (3.3) and (3.4)), a process by which geostrophic balance gives way to magnetostrophic balance.

We now turn to an important general property of geophysical fluids which bears directly on their response to any kind of forcing, namely their ability to support wave motions and oscillations in which material particle displacements possess components parallel to the wave fronts (see e.g. Acheson and Hide 1973, Gill 1984, Hide and Stewartson 1972, Lighthill 1978). Ordinarily, the essential mechanical difference between a fluid and a solid is the inability of the former to resist an applied shear stress, rendering it unable to transmit energy and information by shear waves. But "elastoid" shear oscillations are possible in geophysical fluids owing to the action of Coriolis forces, Lorentz forces and, when the density distribution is "bottom heavy", buoyancy (Archimedes!) forces, associated respectively with general rotation, the presence of magnetic fields, and gravity. Elastoid oscillations are often generated by internal instabilities, and their properties, which include anisotropy and dispersion, are influenced by mutual interactions and by interactions with background flows and bounding surfaces. Nonlinear effects do not always increase the degree of disorder in a system, for in the remarkable soliton phenomenon nonlinear advection exactly cancels effects due to linear wave dispersion.

The elastoid oscillations encountered in studies of the oceans and of the atmospheres of the Earth and other planets include the so-called Rossby waves and Kelvin waves. Complex interactions between waves involving Coriolis and buoyancy forces and generated by internal instabilities are also seen in the non-axisymmetric flow régimes investigated in laboratory work on sloping convection. Even richer in variety are the elastoid oscillations encountered in studies of the electrically - conducting fluid interiors of planets and stars, where Lorentz forces arise. Of particular importance in magnetohydrodynamics is a special class of slow quasi-magnetostrophic oscillations in which Lorentz and Coriolis forces act in opposite directions. In the liquid metallic core of the Earth, the time scales of such oscillations could be comparable with those characteristic of the geomagnetic secular variation (decades to centuries), observations of which have been studied, often with a view to prediction, by geophysicists since the times of Halley and other early investigators (see Section 1 above).

This concludes the general introduction to chaos in geophysical fluids (see Hide 1994). In Part II below, two representative areas of research are discussed, both concerned with phenomena attributable to symmetry-breaking bifurcations caused by gyroscopic (Coriolis) forces. These are (a) waves and eddies in the atmospheres of the Earth, Jupiter and Saturn (where, exceptionally laboratory experiments have been influential), and (b) MHD dynamos, where effective laboratory experiments are not technically feasible.

## Part II: Two representative areas of research

## 4. Sloping convection as a paradigm for large-scale waves and eddies in planetary atmospheres

Having presented a general introduction to research on chaotic flows in geophysical fluids we now turn to the first of two representative areas chosen for more detailed discussion (see summary above), namely large-scale waves and eddies in the atmospheres of the Earth, Jupiter and other planets. An important aim of research on motions in planetary atmospheres is to identify dynamical processes capable of accounting for the observations. To this end are needed *inter alia* investigations of model planetary atmospheres in which all the key parameters such as chemical composition, differential heating and cooling, size and rotation rate of the planet, mechanical and thermal conditions at bounding surfaces, etc. are varied systematically, and the findings compared with available observations of the actual planets. Comparatively few atmospheric scientists are engaged in such activities (see e.g. Ingersoll 1990, James 1994, Williams 1979), but there are signs that in future research such models and even better ones will be used to identify régimes of behaviour determined by bifurcations at critical values of basic parameters. This will be an important step in predictability studies where evidence of deterministic chaos must be sought (see Section 2 of Part I). In areas where they can be trusted, numerical models have obvious advantages in diagnostics studies over laboratory experiments, and the powerful combination of laboratory and numerical studies has contributed significantly to geophysical fluid dynamics during the past two decades and holds out great promise for future work as computers become even more effective (see e.g. Hignett *et al.*, 1985, White 1988). But it is worth recording that in spite of these advances, some of the main findings of laboratory experiments (see Part I) have not yet been reproduced satisfactorily in numerical models.

In the original annulus experiments (see Part I), where heat entered the systems by conduction *via* one of the side walls and left *via* the other side wall, non-axisymmetric upper level flows take the form of a jet-stream undulating in a wave-like pattern touching each side walls at $m$ points, where $m$ is the number of waves in the azimuthal direction. Theoretical considerations of the nature of this pattern led to a systematic attempt to change it by altering the thermal boundary conditions, in a series of experiments requiring the use of a fluid annulus subject to internal heating (for references see Hide and Mason 1975, Lewis 1992). With internal heating there are three important cases to consider, namely when all the heat is extracted *via* the inner wall, *via* the outer wall, and *via* both walls at roughly equal rates. In the first and second of these cases, the pattern consists of $m$ waves extending as before across the annulus plus $m$ eddies situated near the insulated wall (where no heat is extracted). In the third case, the waves are absent and the $m$ eddies extend across the annulus. In all cases the eddies circulate in an anticyclonic sense (at the top surface). In accordance with the ideas being tested, cyclonic eddies are thermodynamically impossible in this case of internal heating and side-wall cooling, since they would have to be associated with a temperature gradient that would result in conductive heat flow *away* from the imposed sink of heat. By the same arguments in the case of *internal* cooling and side-wall heating all motions would be reversed: in particular the upper level eddies would then circulate in a *cyclonic* sense.

The waves and eddies are seen as playing a crucial rôle in two processes, namely the conversion of potential energy into kinetic energy and the horizontal advection of heat. They must also advect heat vertically upwards (see Section 1 above), a process which enables the system to control the horizontal scales of motion. The proposal that large oval atmospheric eddies on Jupiter and Saturn including Jupiter's anticyclonic Great Red Spot (GRS) and White ovals and the cyclonic "barges", are all manifestations of sloping convection has led to a variety of new investigations (for references see Hide *et al.*, 1994), all carried out with the aim of testing the proposal and extending our knowledge of sloping convection. It might seem surprising that dynamically-similar phenomena can be produced on length scales differing by such large factors, up to about $10^8$, but quite general considerations of the processes that produce sloping convection can account for this similarity (see Hide 1977, Hide *et al.*, 1994). Central to these considerations is first the recognition that the flow in the main body of the fluid is nearly geostrophic almost everywhere. This requires that the so-called Ekman number $\varepsilon_1$ and Rossby number $\varepsilon_2$ satisfy

$$\varepsilon_1 \ll 1 \quad \text{and} \quad \varepsilon_2 \ll 1 \tag{4.1}$$

Here

$$\varepsilon_1 \equiv \nu/2\Omega \min\,(L^2, D^2) \text{ and } \varepsilon_2 \equiv U/L\Omega \tag{4.2}$$

where $\nu$ is a measure of the effective coefficient of kinematic viscosity, $U$ is a typical flow speed, $\Omega \equiv |\, \mathbf{\Omega} \,|$ and $L$ and $D$ are characteristic horizontal and vertical length-scales respectively. The other main consideration concerns the so-called Brunt-Väisälä frequency $N$ with which a vertically-displaced fluid element would oscillate in a stably-stratified (i.e. bottom heavy) atmosphere in the absence of forces other than buoyancy forces. The essential balance of terms in the governing equations implies that

$$2\Omega L/ND \sim 1. \tag{4.3}$$

In the case of the Earth's atmosphere we can take $L \sim 3 \times 10^6 m$, $2\Omega \sim 10^{-4} s^{-1}$, $N \sim 2 \times 10^{-1} s^{-1}$ (the corresponding period $2\pi/N \sim 300$s) and $D \sim 10^4$m, we have $2\Omega L/ND \sim 1.5$. For typical laboratory experiments with quasi-geostrophic flows, $L \sim 10^{-1}$m, $2\Omega \leq 10 s^{-1}$, $0.1 s^{-1} \leq N \leq 1 s^{-1}$ and $3 \times 10^{-2}\text{m} \leq D \leq 3 \times 10^{-1}$m, the corresponding range of $2\Omega L/ND$ being $0 < 2\Omega L/ND \leq 300$, which includes values characteristic of all planetary atmospheres! For the South Tropical Zone of Jupiter's atmosphere, where the Great Red Spot is situated, we have $L \sim 10^7$m and $2\Omega \sim 10^{-4} s^{-1}$, so that $ND \sim 10^3 \text{ms}^{-1}$ if $2\Omega L/ND \sim 1$. This is not an outlandish value of $ND$ but the extent to which it is consistent with observations is hard to judge, for little is known about the vertical structure of the Great Red Spot. In devising a strategy for research on the dynamics of planetary atmospheres other than the Earth's it has been necessary to keep in mind that so far as planetary science is concerned the most useful conclusions are those bearing on the vertical structure of atmospheres and the nature of underlying regions, for most observations refer only to the outermost regions of the atmosphere. So it is appropriate to consider a wide range of dynamically-plausible processes and possible vertical structures. One could argue that the most important outcome of research in this area has been the attention given by fluid dynamicists to rapidly rotating fluids, and that the planetary features themselves might continue to remain enigmatic for many years to come. Useful determinations of $N$ comprise one of the goals of the space-craft *Galileo* now on its way to Jupiter. Perhaps the collision of the comet *Shoemaker-Levy* with Jupiter starting in July 1994 (see e.g. Harrington *et al.*, 1994) will cause a perceptible disturbance from which further information about $N$ can be deduced!

According to the "sloping convection paradigm" the principal phenomena in planetary atmospheres that would seem to have their counterparts in the above-mentioned laboratory studies and associated numerical models are the irregular large-scale wave motions observed in the Earth's atmosphere, the more regular large-scale wave motions in the atmosphere of Mars (the global circulation of which is occasionally interrupted by extensive dust storms, see e.g., Ingersoll and Lyons 1993), and the long-lived oval eddies occurring on scales ranging from $10^3$–$10^4$ km in the atmospheres of Jupiter and of other major planets, as manifest in the complex patterns seen (by the reflected light of the Sun) at the visible cloud surface. Here is not the place to discuss the laboratory and numerical experiments in detail and to enumerate further proposed tests and various lines of existing evidence (such as high infra-red emission, implying descending motions, in a "collar" surrounding the Great Red Spot) in favour of the hypothesis. The hypothesis implies that the phenomena to which it applies are driven directly by buoyancy forces and contribute significantly to advective heat transfer. But mention should be made that in the laboratory studies the highest local Reynolds number $UL/\nu$ associated with fully-developed sloping convection is not large enough for shear instabilities to arise. Such instabilities would however occur on the vast scale of planetary atmospheres, and on the present hypothesis they may partially account for the highly irregular "turbulence" in which the Great Red Spot and other long-lived eddies are embedded. This view of the relationship between the turbulence and the long-lived eddies differs fundamentally from that held by other investigators in this lively area of research, who see the long-lived eddies as mechanically-driven features deriving their kinetic energy from the smaller-scale turbulent eddies through a nonlinear anticascade process (for references see Hide 1980, Hide *et al.*, 1994, Ingersoll 1990). It is possible that the zonal jets at the cloud levels in the atmospheres of Jupiter and Saturn owe their existence to deep-seated Rayleigh-Bénard convection (see e.g. Condie and Rhines 1994).

## 5. Low-dimensional analogues of self-exciting magnetohydrodynamic dynamos and geomagnetic polarity reversals

We conclude this paper with a brief account of an important process that occurs in many geophysical and astrophysical fluid systems but (unlike sloping convection) cannot be reproduced and studied systematically on the small scale of the laboratory, namely the self-exciting MHD dynamo process. The reason for this can be seen by inspecting equation (2.5). The ability of fluid flow to amplify magnetic fields through the process of motional induction is expressed by the second term on the left-hand side of the equation. In magnitude this inductive term is typically of the order of

$$R_m \equiv U\hat{L}\mu\sigma \tag{5.1}$$

(the "magnetic Reynolds number) times the first term on the right-hand side of equation (2.5), which represents ohmic dissipation. Owing to the presence of the characteristic length scale $\hat{L}$ in the definition of $R_m$, when dealing with planets and stars it is found that quite slow motions in fluids of moderate electrical conductivity $\sigma$ can give large values of $R_m$, which are impossible to achieve even with the highest practical values of $U$ and $\sigma$ attainable in the laboratory. So research on self-exciting dynamos has to be based on mathematical studies of the governing equations.

It is now generally accepted that the main magnetic fields of the Earth and other planets (Jupiter, Saturn, Uranus, Neptune and possibly Mercury) are generated by the self-exciting MHD dynamo mechanism operating in the electrically- conducting – but not necessarily metallic – fluid regions of their interiors. In the early 1950's, when the Earth was still the only planet known to possess a general magnetic field, geophysicists came round to the view that the magnitude of the field and its complex temporal variability implied that the dynamo mechanism must be at work in the Earth's liquid metallic core. It is known that slow relative motions of a rapidly - rotating fluid of low viscosity are strongly influenced by the action of Coriolis forces, which render the flow patterns highly anisotropic. So it was natural to assume that the near alignment between the Earth's magnetic and rotation axes must be a manifestation of such anisotropy, for the pattern of electric currents generated by dynamo action in the core will be related – albeit in a complicated way – to the pattern of fluid motions there. This general notion appeared to gain strength with the findings (starting in 1955 with ground-based observations of decametric and decimetric radiation from Jupiter and followed by NASA's highly successful *Pioneer* and *Voyager* missions to the outer solar system) that the magnetic axes of Jupiter and Saturn are at present closely aligned with their rotation axes. So dynamo theoreticians are being kept on their toes by the subsequent discovery from further *Voyager* observations that the planets Uranus and Neptune have highly eccentric magnetic fields (see Ness 1994)!

By re-examining in the light of these new findings the likely rôle of Coriolis forces in the dynamo process, it is possible to formulate a strategy (see Hide 1991) for future work towards the interpretation of the magnetic fields of all the planets which would not in the first instance involve straying too far from the current paradigm. But new mathematical methods will be needed that allow for the treatment of strong departures from axial symmetry in the instantaneous magnetic field. These methods are now in sight as geophysicists equipped with very powerful computers begin the task of developing realistic numerical models. Systematic research with these models will include investigations of effects due to departures from axial symmetry in the boundary conditions, with a view in the first instance to interpreting certain important details of the geomagnetic field. But it would be wise in the study of the eccentric magnetic fields of Uranus and Neptune to see what progress can be made by examining "free" behaviour, without in the first instance invoking "forced" behaviour caused by non-axisymmetric boundary conditions.

The most striking property of the geomagnetic field is that the dipole has reversed its polarity many times over geological history. In certain respects this is not surprising, because for every solution $(\mathbf{B}(\mathbf{r},t),\ \mathbf{u}(\mathbf{r},t))$ of the MHD dynamo equations, $(-\mathbf{B}(\mathbf{r},t),\ \mathbf{u}(\mathbf{r},t))$, is also a solution. At about $7 \times 10^5$ years, the average interval between geomagnetic polarity reversals is very much longer, by a factor of about 100, than the duration of a reversal event, when the polarity changes sign. But the time series of reversals is highly irregular (see Jacobs 1994): there have been intervals as long as $3 \times 10^7$ years when the polarity apparently remained unchanged! It is likely that this general behaviour reflects to some extent *changing* boundary conditions at the surface of the core produced *inter alia* by mantle convection. It is also possible however that the time series of reversals includes manifestations of "deterministic chaos" typical of the behaviour of nonlinear systems operating under *fixed* boundary conditions.

It will not be possible to resolve such questions satisfactorily until full dynamo models have been investigated by solving the highly nonlinear partial differential equations of MHD under realistic boundary conditions (see Soward 1983). So far as polarity reversals are concerned, relevant studies to date have been largely concerned with the determination of the generic temporal behaviour of low-dimensional analogues. These satisfy nonlinear ordinary differential equations, which are readily integrated using quite modest computers. The first such analogue was proposed by Bullard (1955), who analyzed the behaviour of the simplest imaginable system, namely a Faraday disk dynamo consisting of a single disk and coil arrangement driven by a steady applied couple $G$. He was able to show that the current $I(t)$ in the system remained unidirectional and oscillated periodically but non-harmonically with amplitude and period which depend on the starting conditions. Malkus (1992) recognized that by introducing a resistive shunt between the disk and the coil into the Bullard system (a modification adumbrated by Bullard (1955), see also Allan (1962)) it was possible to produce equations with a degree of complexity known from previous work on nonlinear systems (e.g. Lorenz 1963, Moore and Spiegel 1966, see also Moon 1987, Thompson and Stewart 1986) to give rise to chaos, an expectation confirmed by Robbins (1977).

Much earlier, Rikitake (1958) introduced a system which has attracted much attention from mathematicians interested in the details of its chaotic behaviour, and also from geophysicists concerned with the interpretation of the statistical properties of the highly irregular time series of geomagnetic polarity reversals (for references see Jacobs 1994, Melchior 1986, Moffatt 1978, Parker 1979, Rikitake 1966, Turcotte 1993). Rikitake secured the same degree of complexity as in the shunted single disk dynamo by coupling two identical Bullard-type dynamos together. Dissipation due to ohmic heating is included in the Rikitake system, but as we shall see in what follows, an implicit assumption in previous work with that system, namely that dissipation due to mechanical friction may safely be neglected, appears to be unjustified. We conclude our discussion by showing that mechanical friction renders the dynamical behaviour of the Rikitake system "structurally unstable" (for definition see e.g. Thompson and Stewart 1986) and incapable of chaotic behaviour (see below).

The equations governing the Rikitake system are as follows (see e.g. Ghil and Childress 1987, Jacobs 1994, Moffatt 1978):

$$L dI_1/dt + RI_1 = M\Omega_1 I_2, \tag{5.2}$$

$$L dI_2/dt + RI_2 = M\Omega_2 I_1, \tag{5.3}$$

$$A d\Omega_1/dt = G - MI_1 I_2 \tag{5.4}$$

and

$$A d\Omega_2/dt = G - MI_2 I_1. \tag{5.5}$$

Here $(I_1(t),\, \Omega_1(t))$ are the current and angular speed of rotation of the disk in the first dynamo and $(I_2(t),\, \Omega_2(t))$ the corresponding quantities in the second dynamos; $L$ and $R$ respectively are the self-inductance and resistance of each dynamo, $2\pi/M$ is the mutual inductance of each coil and disk arrangement, and $A$ is the moment of inertia of each disk. By measuring time $t$ in units $(AL/GM)^{\frac{1}{2}}$, current in units $(G/M)^{\frac{1}{2}}$ and angular speed of rotation in units $(GL/AM)^{\frac{1}{2}}$, and writing

$$I_i = (G/M)^{\frac{1}{2}} X_i, \quad \Omega_i = (GL/AM)^{\frac{1}{2}} Y_i, \quad i = 1, 2, \tag{5.6}$$

equations (5.2) to (5.5) become

$$\dot{X}_1 + \gamma X_1 = Y_1 X_2, \tag{5.7}$$

$$\dot{X}_2 + \gamma X_2 = Y_2 X_1 \tag{5.8}$$

and

$$\dot{Y}_1 = \dot{Y}_2 = 1 - X_1 X_2, \tag{5.9}$$

where the dot denotes differentiation with respect to dimensionless time and

$$\gamma \equiv (AR^2/GLM)^{\frac{1}{2}} \ . \tag{5.10}$$

It follows from (5.9) that

$$Y_1 - Y_2 \equiv \Delta Y \tag{5.11}$$

remains constant.

Ito (1980) reported one of the most extensive of several detailed published theoretical studies (see Jacobs 1994, Moffatt 1978) based on these equations. The system has two equilibrium points $\alpha$ and $\beta$ with coordinates $(\pm\delta,\ \pm\delta^{-1},\ \gamma\delta^2)$ in the $(X_1,\ X_2,\ Y_1)$ phase space, where $\delta$ satisfies

$$\Delta Y = \gamma(\delta^2 - \delta^{-2}) . \tag{5.12}$$

To paraphrase a published summary of Ito's main findings, both equilibrium points $\alpha$ and $\beta$ are unstable foci, and around them there is an attracting plane which traps all orbits starting from any point except those on the $Y_1$ axis. An orbit circles around $\alpha$ or $\beta$ in this plane, irregularly travelling from an orbit around one point to one around the other. This corresponds to a reversal of the magnetic field. A phase diagram in the parameter space $(\gamma,\ \delta)$ shows various regions of periodic regime and a chaotic regime, with the transition from the former to the latter characterized by a succession of period-doubling bifurcations. Near the centre of the chaotic regime however there is an area of parameter space where reversals seldom occur and the dynamics are less disordered. The Markov entropy of the Lorenz map for the system has a sharp minimum in this parameter region, which Ito termed the "minimum entropy regime", concluding that the smallness and non-uniformity of the frequency of reversals as shown by the palaeomagnetic data suggest that the geodynamo is in such a state of minimum entropy.

This and related theoretical studies (see Jacobs 1994) cover wide ranges of $\gamma$ and $\delta$. In the work of Ito, for example, $0 < \gamma < 10$ and $1 < \delta < 6$. However the inclusion of mechanical friction in the Rikitake system appears to place restrictions on the possible values of $\delta$ that can be achieved. This is readily seen by adding the term $-k\Omega_1$ to the right-hand side of (5.4) and $-k\Omega_2$ to the right-hand side of (5.5), supposing for simplicity that $k$ is constant, corresponding to linear mechanical friction. By subtracting one of the resultant equations from the other we find

$$A d(\Omega_2 - \Omega_1)/dt = -k(\Omega_2 - \Omega_1) \tag{5.13}$$

which has solutions

$$[\Omega_2(t) - \Omega_1(t)] = [\Omega_2(t=0) - \Omega_1(t=0) \exp\,[-kt/A] . \tag{5.14}$$

It follows that $\Omega_2 - \Omega_1$ is no longer constant unless $k = 0$ or $t \to \infty$. Mechanical friction thus reduces $\Omega_2 - \Omega_1$ to zero, giving $\Delta Y = 0$; it damps out initial transients until $\delta$ takes the value unity (see (5.12)), at which chaotic behaviour cannot occur.

This result is one of several offshoots of an ongoing study, by the writer and Dr. D.J. Acheson, of single Faraday disk dynamos with additional electrical and mechanical elements (another being the discovery of simple novel reversing systems for which during the long interval of time between reversal events $I(t)$ remains virtually constant but $\Omega(t)$ does not). From this work, which was undertaken in the first instance with another geophysical (i.e. not the geodynamo) problem in mind, it became evident that, owing to structural instability the characteristic nonlinear relaxation oscillations of the Bullard dynamo give way under the influence of mechanical friction to steady solutions which (apart from the direction of the current) are independent of the starting conditions, with the current equal to $\pm[(G - kR/M)/M]^{\frac{1}{2}}$ or zero according as $G > kR/M$ or $G \leq kR/M$, the corresponding values of the angular speed of rotation being $R/M$ and $G/K$ respectively. It should not be difficult in future research to take a variety of structurally-stable disk dynamo systems capable of chaotic behaviour and investigate the statistical and other properties of that behaviour in detail, for comparison with observed time series of geomagnetic polarity reversals.

**Acknowledgements**

Kindly commenting on the foregoing results, Professor N.O. Weiss has emphasized that in an early paper on the subject written (without the benefit of modern chaos theory) by Allan (1962), the equations for coupled disk dynamos with $k \neq 0$ are written down but not analyzed. The best known and most influential part of that paper (see e.g. Jacobs 1994) is Allan's detailed analysis of the case when $k = 0$, in which the chaotic behaviour of the Rikitake system was first demonstrated.

The author is grateful to the California Institute of Technology for the award of a Fairchild Distinguished Scholarship, during the tenure of which (October to December 1993) this invited review article was started.

**References**

Acheson, D.J. and Hide, R. 1973 *Repts. Prog. Phys.*, **36**, 159–221.

Allan, D.W. 1962 *Proc. Camb. Philos. Soc.*, **58**, 671–693.

Berger, A., Duplessy, J.-Cl. and Schneider, S.H. (eds.) 1989 *Climate change and geosciences.* Dordrecht: Reidel Publishing Company.

Buchler, J.R, Perdang, J.M. and Spiegel, E.A. (eds.) 1985 *Chaos in astrophysics.* Dordrecht: Reidel Publishing Company.

Bullard, E.C. 1955 *Proc. Camb. Philos Soc.*, **51**, 744–760.

Buzyna, G., Pfeffer, R.L. and Kung, R. 1984 *J. Fluid Mech.*, **145**, 337–403.

Chapman, S. 1941 *Occasional Notes Roy. Astron. Soc.*, **1** part 9, 122–134.

Condie, S.A. and Rhines, P.B. 1994 *Nature*, **367**, 711–713.

Corby, G. (ed.) 1969 *The global circulation of the atmosphere.* London: Royal Meteorological Society.

Cvitanović, P. (ed.) 1984 *Universality in chaos; a reprint selection.* Bristol: Adams

Hilger.

Drazin, P.G. and King, G.P. (eds.) 1992 *Interpretation of time series from nonlinear systems.* The Netherlands: North Holland.

Fultz, D., Long, R.R., Owens, G.V., Bohan, W., Kaylor, R. and Weil, J. 1959 *Meteorol. Monographs Amer. Meteorol. Soc.*, **4**, 1–104.

Ghil, M and Childress, S. 1987 *Topics in geophysical fluid dynamics: Atmospheric dynamics, dynamo theory and climate dynamics.* New York, Springer Verlag.

Ghil, M., Kimoto, M. and Neelin, J.D. 1991 *Rev. Geophys. Supplement*, 46–55.

Gill, A.E. 1982 *Atmosphere-ocean dynamics.* London: Academic Press.

Guckenheimer, J. and Buzyna, G. 1983 *Phys. Rev. Lett.*, **51**, 1438–1441.

Haken, H. (ed) 1981 *Chaos and order in nature.* New York: Springer Verlag.

Harrington, J., Le Beau, Jnr., R.P., Backes, K.A. and Dowling, T.E., 1994 *Nature*, **368**, 525–527.

Hart, J. 1986 *Physica D*, **20**, 350–362.

Hide, R. 1953 pp. 101–117 in *Fluid models in geophysics.* (ed. R.R. Long), Washington D.C.: U.S. Government Printing Office. (Also *Quart. J. Roy. Meteorol. Soc.*, **79**, 161.)

Hide, R. 1977 *Quart. J. Roy. Meteorol Soc.*, **103**, 1–28.

Hide, R. 1980 *The Observatory*, **100**, 182–193.

Hide, R. 1982 *Philos. Trans. Roy. Soc. London*, **A 306**, 223–234.

Hide, R. 1984 *Philos. Trans Roy. Soc. London*, **A 313**, 107–121.

Hide, R. 1991 *Geophys. Astrophys. Fluid Dyn.*, **62**, 183–189.

Hide, R. 1994 *Philos. Trans. Roy. Soc. London*, **A, 348**, (in the press).

Hide, R. and Dickey, J.O. 1991 *Science*, **253**, 629–637.

Hide, R., Lewis, S.R. and Read, P.L. 1994 *Chaos*, **4**, 135-162. (in the press).

Hide, R. and Mason, P.J. 1975 *Advances in Physics*, **24**, 47–100.

Hide, R. and Stewartson, K. 1972 *Rev. Geophys. and Space Phys.*, **10**, 579–598.

Hignett, P., White, A.A., Carter, R.D, Jackson, W.D.N and Small, R. M 1985 *Quart Roy. Meteorol. Soc.*, **111**, 131–154.

Hopfinger, E.J. (ed.) 1991 *Rotating fluids in geophysical and industrial situations.* New York: Springer Verlag.

Ingersoll, A.P. 1990 *Science*, **248**, 308–315.

Ingersoll, A.P. and Lyons, J.R. 1993 *J. Geophys. Res.*, **98**, 10,951–10,961.

Ito, K. 1980 *Earth and Planet. Sci. Lett.*, **51**, 451–456.

Jacobs, J.A. (ed.) 1987–1991 *Geomagnetism* (4 volumes). New York: Academic Press.

Jacobs, J.A. 1994 *Reversals of the Earth's magnetic field.* Cambridge University Press.

James, I.N. 1994 *Introduction to circulating atmospheres.* Cambridge University Press.

King J.C. and Mobbs, S.D. (eds). 1991 *Waves and turbulence in rotating stratified fluids.* Oxford University Press.

Lewis, S.R. 1992 *Geophys. Astrophys Fluid Dyn.*, **65**, 31–55.

Lighthill, J. 1978 *Waves in fluids.* Cambridge University Press.

Lorenz, E.N. 1963(a) *J. Atmos. Sci.*, **20**, 130–141.

Lorenz, E.N. 1963(b) *J. Atmos. Sci.*, **20**, 448–464.

Lorenz, E.N. 1967 *The nature and theory of the general circulation of the Earth's atmosphere.* Geneva: World Meteorological Organization.

Lorenz, E.N. 1980 *Ann. New York Acad. Sci.*, **357**, 282–291.

Lorenz, E.N. 1993 *The essence of chaos.* London: U.C.L. Press Limited.

Malkus, W.V.R. 1972 *E.O.S. Trans. Amer. Geophys. Un.*, **53**, 617.

Mason, J., Mathias, P. and Westcott, J.H. (eds.) 1986 *Predictability in science and society.* London: The Royal Society.

Matsuzaki, M. 1994 *Philos Trans. Roy. Soc London*, **A 443**, (in the press

Melchior, P. 1986 *The physics of the Earth's core.* Oxford: Pergamon Press.

Moffatt, H.K. 1978 *Magnetic field generation in electrically-conducting fluids.* Cambridge University Press.

Monin, A.S. 1972 *Weather forecasting as a problem in physics.* Cambridge, Massachusetts: M.I.T. Press.

Moon, F.C. 1987 *Chaotic vibrations.* New York: John Wiley.

Moore, D.W. and Spiegel, E.A. 1966 *Astrophys. J.*, **143**, 871–887.

Mullin, T. (ed.) 1993 *The nature of chaos.* Oxford: Clarendon Press.

Neelin, J.D., Latif, M. and Jin, F.-F. 1994 *Ann Rev. Fluid Mech*, **26**, 617–659.

Ness, N.F. 1994 *Philos. Trans Roy . Soc. London*, **A**, 349. (in the press

Nicolis, C. and Nicolis, G. (eds.) 1987 *Irreversible phenomena and dynamical systems analysis in geosciences.* Dordrecht: D. Reidel Publishing Company.

Ott, E. 1993 *Chaos in dynamical systems.* Cambridge University Press.

Palmer, T.N., Buizza, R., Molteni, F., Chen, Y-Q. and Corti, S. 1994 *Philos Trans.Roy. Soc. London*, **A 443**, (in the press)

Parker, E.N. 1979 *Cosmical magnetic fields: their origin and activity.* Oxford: Clarendon Press.

Philander, G. 1990 *El Niño, la Niña and Southern Oscillation.* New York: Academic Press.

Platt, N., Spiegel, E.A. and Tresser, C. 1993 *Geophys. Astrophys. Fluid Dyn.*, **73**, 147–161.

Proctor, M.R.E. and Gilbert, A.D. (eds.) 1994 *Lectures on solar and planetary dynamos.* Cambridge University Press.

Quinet, A. 1974 *Advances in Physics*, **17**, 106–186.

Read, P.L. 1993 *Physica D*, **69**, 353–365.

Read, P.L., Bell, M.J., Johnson, D.W. and Small, R.M. 1992 *J. Fluid Mech.*, **238**, 599–632.

Rikitake, T. 1958 *Proc. Camb Philos. Soc.*, **54**, 89–105.

Rikitake, T. 1966 *Electromagnetism and the Earth's interior.* New York: Elsevier Publishing Company.

Robbins K.A. 1977 *Math. Proc. Cambridge Philos. Soc.*, **82**, 309–325.

Roberts, P.H. and Soward, A.M. (eds.) 1978 *Rotating fluids in geophysics.* London: Academic Press.

Rosen, R.D. 1993 *Surveys in geophysics*, **14**, 1–29.

Ruzmaikin, A., Feynman, J. and Robinson, R. 1994 *Solar physics*, **149**, 395–403.

Saltzmann, B. and Verbitsky, M.Y. 1993 *Climate dynamics*, **9**, 1–15.

Smith, L.A. 1988 *Phys. Lett.*, **A 133**, 283–288.

Smith, L.A. 1992 *Physica D*, **58**, 50–76.

Smith, L.A. 1994 *Philos Trans. Roy Soc. London*, **A** (in the press).

Sonett, C.P., Giampapa, H.S. and Matthews, M.S. (eds.) 1991 *The Sun in time.* Tucson: University of Arizona Press.

Soward, A.M. (ed.) 1983 *Stellar and planetary magnetism.* New York: Gordon and Breach Publishers.

Sreenath, N. 1993 *Systems representation of global climatic change.* New York, Springer Verlag.

Stevenson, D.J. 1983 *Repts. Prog Phys.*, **46**, 455–620.

Thompson, J.M.T. and Stewart, H.B. 1986 *Nonlinear dynamics and chaos.* Chichester: John Wiley and Sons.

Thompson, P.D. 1957 *Tellus*, **9**, 275–295.

Thompson, P.D. 1988 *J. Atmos. Sci.*, **45**, 1279–1282.

Turcotte, D.L. 1993 *Fractals and chaos in geology and geophysics.* Cambridge: University Press.

Webster, P.J. and Keller, J.L. 1975 *J. Atmos Sci.*, **32**, 1283–1300.

Weiss, N.O. 1990 *Philos. Trans. Roy. Soc. London*, **A 330**. 617–625.

Weiss, N.O. 1993 pp. 219-229 in *The cosmic dynamo.* F. Krause *et al.* (eds.). The Netherlands: International Astronomical Union.

Weiss, N.O. 1994 *Philos. Trans. Roy. Soc. London*, **A 443** (in the press).

White, A.A. 1988 *Meteorol. Mag.*, **117**, 54–63.

White, A.A. 1990 *Meteorol Mag.*, 119, 1–9.

Willebrandt, J. and Anderson, D.L.T. (eds.) 1993 *Modelling ocean climate interactions.* New York: Springer Verlag.

Williams, G.P. 1979 *J. Atmos. Sci.*, **36**, 932–968.

CHAPTER 10

# CHAOTIC MODULATION OF THE SOLAR CYCLE

S.M. Tobias & N.O. Weiss
Department of Applied Mathematics and Theoretical Physics
University of Cambridge, Cambridge CB3 9EW, UK.

**Abstract**

The 11-year sunspot cycle corresponds to an aperiodic magnetic cycle, punctuated by grand minima when magnetic activity almost disappears. The record of solar activity is apparently chaotic and is consistent with the observed behaviour of other stars of similar age and mass. The crucial order parameter is the rotation rate $\Omega$, which decreases as a star grows older, owing to magnetic braking. We construct a simple third-order model of a stellar dynamo in which, as $\Omega$ increases, there is a transition from a non-magnetic state to one with periodically reversing fields, followed by quasiperiodic modulation of the basic cycle and then by chaos. This model reproduces the essential features of solar and stellar magnetic cycles and, since it relies on normal form equations, its behaviour is generic and robust.

## 1. Introduction

The Sun's magnetic activity varies cyclically, with a well-defined mean period of about 11 years. At the beginning of a new cycle, sunspots (which are dark owing to partial inhibition of convection by a strong magnetic field) appear at latitudes around $\pm 30°$; then the zones of activity expand and drift towards the equator, where they die away as the new cycle starts again at higher latitudes. Thus solar activity, as measured by the total area occupied by sunspots or by the sunspot number (as arbitrarily defined by Wolf) varies as shown in Figure 1*a*. The sunspot cycle is apparent, though its behaviour is not strictly periodic. Individual active regions are typically oriented parallel to the equator, with oppositely-directed magnetic fields in leading and following regions. The sense of these fields is opposite in the two

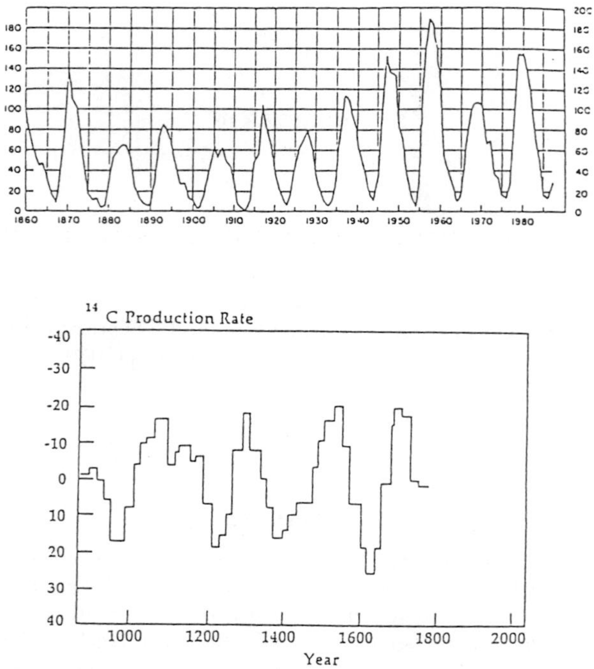


Fig. 1. Solar magnetic activity. (*a*) Variations of the sunspot number from 1860 to 1987, showing the irregular 11-year activity cycle. (After Foukal 1990.) (*b*) Modulation of the cycle over 1000 years: the $^{14}$C production rate is enhanced during episodes of reduced activity, such as the Maunder minimum in the 17th century. (After Stuiver & Braziunas 1988.)

hemispheres and reverses at sunspot minimum. So the magnetic cycle actually has a 22-year period, with waves of activity that drift towards the equator.

Sunspot records show unambiguously that there was a dearth of spots in the late 17th century (Ribes & Nesme-Ribes 1993). This episode of reduced activity – the Maunder minimum – can also be detected in proxy records. Galactic cosmic rays, whose incidence is modulated by magnetic fields in the solar wind, form radioactive isotopes in the earth's atmosphere. Anomalies in the abundances of $^{14}$C (in trees) and $^{10}$Be (in polar ice cores) show that grand minima of solar activity have occurred irregularly over the past 8000 years, with a characteristic timescale of 200 years. Figure 1*b* shows the envelope of solar activity, as derived from the $^{14}$C record, over the last millennium. The sample of magnetic variations can be enlarged by studying other lower-main sequence stars, which also show the effect of varying parameters such as the angular velocity $\Omega$ (Wilson 1994; Weiss 1994). Magnetic activity, which can be measured by monitoring $Ca^+$ emission (Baliunas *et al.* 1995), depends on $\Omega$,

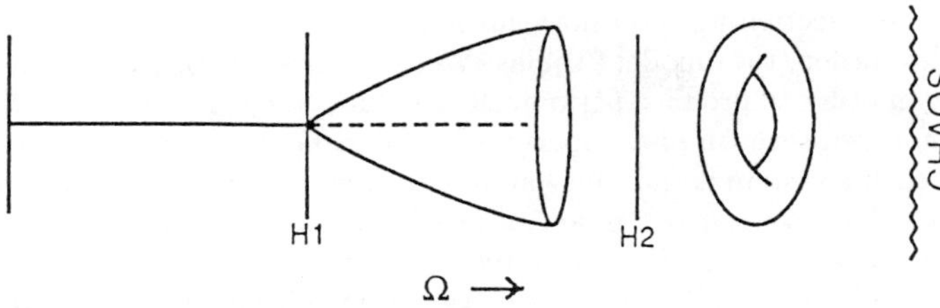


Fig. 2. Schematic Bifurcation diagram for a stellar dynamo, showing the onset of cyclic activity at a Hopf bifurcation (H1), followed by quasiperiodic modulation after a secondary Hopf bifurcation (H2) and a transition to chaos, as the angular velocity $\Omega$ is increased.

and stars spin down owing to magnetic braking. Slow rotators exhibit cycles like the Sun's and about 30% of them are quiescent, as if undergoing grand minima.

Solar activity is aperiodic but the sunspot record, extending over less than 300 years, is too short to distinguish between stochastic effects and deterministic chaos (Weiss 1990; Morfill, Scheingraber & Sonett 1991). Indeed, the solar cycle has been successfully mimicked with a stochastic ARMA model (Barnes, Sargent & Tryon 1981). However, it is reasonable to assume that we are dealing with a chaotically modulated oscillator subject to weak stochastic perturbations. Predicting future behaviour by extrapolation from the sunspot record has become a standard exercise (Tong 1990) though recent attempts have been conspicuously unsuccessful (Wilson 1994).

It is generally accepted that solar and stellar magnetic fields are produced by dynamo action. The toroidal field, which emerges in sunspots, is created from a poloidal field by differential rotation; cyclonic eddies then succeed in generating a reversed poloidal field. Simple linear models show that a magnetic instability can lead to the growth of an oscillatory field, whose amplitude is eventually limited by effects of the nonlinear Lorentz force. Both self-consistent dynamo models (where the relevant partial differential equations are solved numerically, after adopting laminar diffusivities) and parametrized mean-field ($\alpha\omega$) dynamos provide examples of aperiodic oscillations. Moreover, the behaviour of mean-field dynamos depends on the dynamo number, $D \propto \Omega^2$. Parker (1979) proposed that there are two different convective states in a rotating star, only one of which is unstable to magnetic perturbations, and that the system moves from this state towards the other, which is hydrodynamically unstable, so giving rise to modulation. The observed changes in rotation during the Maunder minimum (Ribes & Nesme-Ribes 1993) provide evidence to support this hypothesis. Thus we expect to find the bifurcation sequence sketched in Figure 2, with an initial Hopf bifurcation, leading to periodic magnetic cycles, followed by a secondary Hopf bifurcation that gives rise to periodic modulation and then by a transition to chaos.

In the next section we show how this pattern of behaviour can be studied with the aid of a low-order (toy) model (Tobias, Weiss & Kirk 1995). A third-order system is necessary in order to produce periodically modulated cyclic activity, with trajectories that lie on a two-torus in phase space; it is also sufficient for chaos. The advantage of studying such a system is that one can rely on normal form theory: although the low-order model has no *predictive* power, it provides examples of *generic* behaviour which can be expected to occur in real systems governed by nonlinear partial differential equations. In §3 we compare power spectra obtained from low order models with those obtained from proxy records and in the final section we summarize the significance of our results.

## 2. A simple third–order model

It is clear from the above discussion that modulation of the solar cycle is a real effect, which requires an explanation. We take as given that the solar cycle itself is generated by the action of a dynamo operating in or at the base of the solar convection zone. The solar dynamo problem is one of vast complexity, encompassing as it does the systematic generation of magnetic fields in a highly turbulent, self-gravitating, rotating fluid. Even with the most powerful computers available there is no hope of conducting an honest calculation that includes all the physics (and even if there were it would be difficult to gain any insight from the results). Instead many calculations have been carried out using simpler parametrized models in the hope of illuminating processes which might be important in the generation of the solar cycle. Many of these 'illustrative' models are constructed on the basis of *mean field electrodynamics.* This theory is applicable in situations where there is a separation of scales in the turbulence, and the average effect of the small-scale motions in generating the large-scale magnetic field can be calculated using certain closure approximations (see Moffatt 1978 for a review). It is not clear that these approximations are relevant to the Sun, but they do provide a starting point from which to construct simple models. These models vary in complexity and reproduce the 22-year solar cycle with varying degrees of success (see Brandenburg & Tuominen 1990 for a succinct review). None of them, however, reproduce the correct form of modulation of the basic cycle or indeed the appearance of grand minima. They are also open to the criticism that any success in reproducing behaviour of the solar cycle is due to tuning of parameters rather than to capturing the underlying physics. Thus it is important to look for generic results, which are found in all these models and cannot therefore be regarded as model-dependent, in constructing a scenario for the modulation of the solar cycle. The most robust feature of these investigations is that dynamo action leading to cyclically varying fields sets in at a supercritical Hopf bifurcation from a purely hydrodynamic (non-magnetic) state as the rotation of the star (and hence the nondimensional dynamo number) passes through a critical value. We shall use this fact as a starting point in the construction of our model.

The approach we follow is one which we believe to be applicable to many systems where little is known about the details of the physics but where some observational

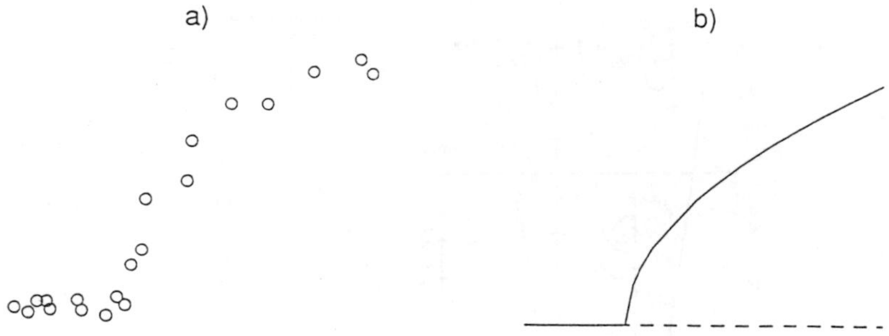


Fig. 3. Imperfect and perfect Hopf bifurcations. (*a*) Sketch showing the variation of $||\mathbf{B}||$ with $\Omega$ that might be inferred from observations. (*b*) Idealised description, showing the variation of $||\overline{\mathbf{B}}||$ with $\Omega$, with a Hopf bifurcation from the hydrodynamic subspace.

results are available. We shall exploit bifurcation theory, whose power lies in the fact that the properties of systems containing certain bifurcations can be inferred from the behaviour of normal forms containing those bifurcations. Since the magnetic fields are generated from a turbulent convecting state, the technique of deriving normal forms by a centre manifold reduction is not, however, appropriate. Instead we have to develop an idealized version of our problem, with a bifurcation structure that is amenable to analysis.

Consider a star in which no magnetic field is present. Turbulent convection, which is highly nonlinear, would still dominate and the solution of the relevant equations would evolve along a 'purely hydrodynamic' manifold in the phase space of the system. Simply put, this means that if the star starts with no magnetic field it will end up with no magnetic field – no matter how turbulent the convection. It is now reasonable to assume that a statistically steady state can be found lying in this manifold. This state corresponds to an invariant subspace of the hydrodynamic subspace (e.g. a fixed point, periodic orbit or more complicated attractor). In reality all stars possess some magnetic field (no matter how small); if the rotation rate of the star is small we do not, however, expect to find the type of large-scale field that leads to the production of systematic activity, but rather small-scale disordered fields. To examine stellar magnetic cycles we must therefore be able to distinguish between these two types of field. We do this by defining a norm $||\mathbf{B}||$, such that for weakly rotating stars $||\mathbf{B}||^2 \ll 1$. If it were possible now to observe the magnetic field in a set of stars with similar properties but varying angular velocity, we would find that dynamo action leading to large-scale magnetic fields would set in at an imperfect Hopf bifurcation at $\Omega = \Omega_c$ as shown schematically in Figure 3.

This imperfect bifurcation is not easily analysed, owing to the presence of small-scale fluctuations, so we proceed by transforming the bifurcation to a perfect bifurcation. We project the magnetic and hydrodynamic behaviour onto a few large-scale modes (averaging over the small-scale fluctuations). This enables us to obtain solutions which lie in a magnetohydrodynamic phase space, with a purely hydrodynamic

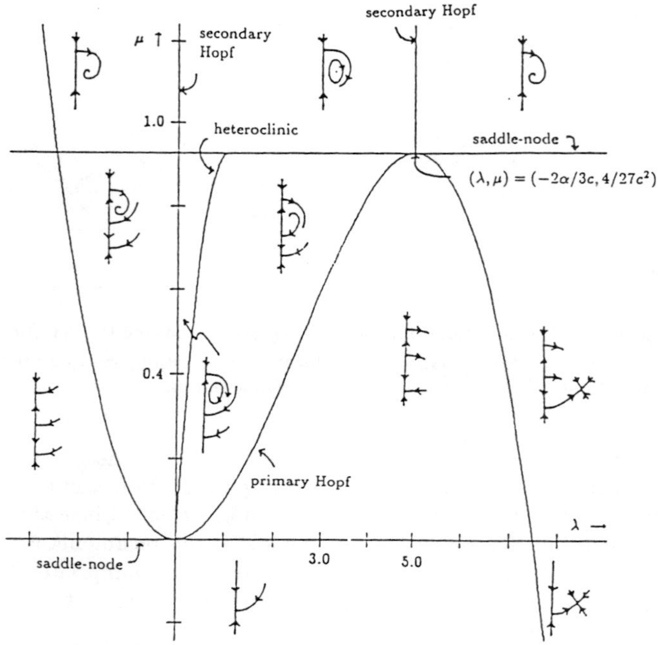


Fig. 4. Two parameter bifurcation diagram in the $\lambda\mu$-plane for the axisymmetric system, corresponding to equations (4)–(6) with $d = 0$. Bifurcation sets are indicated, together with phase portraits in the $rz$-plane. (After Kirk 1991.)

subspace. This subspace includes all solutions with $||\overline{\mathbf{B}}|| = 0$ where $\overline{\mathbf{B}}$ is the averaged (and hence large-scale) field. In this subspace we can envisage the existence of an attracting steady state which is stable to hydrodynamic perturbations. This state may however be unstable to magnetic disturbances and dynamo action can set in at a Hopf bifurcation at $\Omega = \Omega_c$ (see Figure 3).

Our model is the most drastic simplification of the above scenario. We collapse the entire hydrodynamic subspace onto the real line (the $z-$axis) and the hydrodynamically stable solution becomes a fixed point on that line. In our model the $z-$axis is therefore invariant, as initially non-magnetic solutions stay non-magnetic. We require a further two dimensions to allow for the Hopf bifurcation to cyclic magnetic activity, so we take $x$ as representing the toroidal field in the star whilst $y$ represents the poloidal field. We describe the interaction between the hydrodynamic manifold and the Hopf bifurcation using the normal form for a saddle-node/Hopf bifurcation. This is the simplest generic codimension two bifurcation describing the dynamics, but it can be shown (Tobias *et.al.* 1995) that the specific choice of normal form is unimportant.

We now take the standard unfolding of the normal form and truncate at quadratic order to obtain

$$\dot{r} \;=\; \lambda r + a_1 r z, \tag{1}$$

$$\dot{z} = \mu + b_1 r^2 + b_2 z^2, \tag{2}$$

$$\dot{\phi} = \omega, \tag{3}$$

where $r^2 = x^2 + y^2$ is a measure of the magnetic energy of the solutions and $\phi = \arctan(y/x)$ is the phase of the cycle; $\lambda$ and $\mu$ are the order parameters to be varied and $a_1$, $b_1$ and $b_2$ are constants. We remove two coefficients by rescaling, and ensure that the Hopf bifurcation is supercritical by setting $a_1 = a > 0$ and $b_1 = b_2 = -1$ (see Guckenheimer & Holmes 1986). This system now contains a secondary Hopf bifurcation to motion on a two-torus (i.e. modulation of the basic cycle frequency) but this bifurcation is degenerate (as the system is integrable) producing a family of nested tori at $\lambda = 0$. This degeneracy is unphysical, and may be removed by the addition of a higher order cubic term. We choose to add a term $cz^3$ to the $\dot{z}$ equation to achieve this. In theory the dynamics of the system may be sensitive to the particular choice of cubic term, but a thorough investigation has proved that this is not the case, providing that the choice of term does not change the invariance of the $z$-axis. The addition of this cubic term now introduces two other fixed points on the $z$-axis for certain parameter values. These appear in a line of saddle-node bifurcations at $\mu = 4/27c^2$. As required, the degeneracy of the secondary Hopf bifurcation has been removed by the cubic term. Now the two-torus is born at $\lambda = 0$, $\mu > 0$ and grows in size until it collides with and becomes heteroclinic to the two saddle-foci on the $z$-axis. The two parameter bifurcation set for this system is shown in Figure 4. Now there is a region of parameter space where motion on a two-torus leads to modulation of the basic cycle frequency. Thus quasiperiodic behaviour appears *generically* as a consequence of the supercritical Hopf bifurcation from the hydrodynamic manifold.

We finish by introducing another cubic term to break the axisymmetry of the system around the $z$-axis. This is required to introduce the possibility of chaotic dynamics in the system and also to distinguish the mechanism for generating poloidal and toroidal fields. We, therefore, choose to add a term proportional to $r^2z$ to the equation for the generation of toroidal field only. This leaves the $z$-axis invariant and yields the system

$$\dot{r} = \lambda r + arz + dr^2 z \cos\phi, \tag{4}$$

$$\dot{z} = \mu - r^2 - z^2 + c\,z^3, \tag{5}$$

$$\dot{\phi} = \omega - dzr \sin\phi, \tag{6}$$

which has been studied in detail by Kirk (1991, 1993).

The effect of this term is to replace the heteroclinic bifurcation with a heteroclinic tangle and to introduce horseshoes (and hence chaotic dynamics) into the flow. The derivation of the model is now complete and it remains to display the results of integrations of this system which we have formulated by the contraction to a normal form approach.

We can illustrate the behaviour of the system most easily by considering how solutions to the equations change as a single parameter is varied. We think of this parameter as being the rotation rate of the star which, as noted in §1, is an important

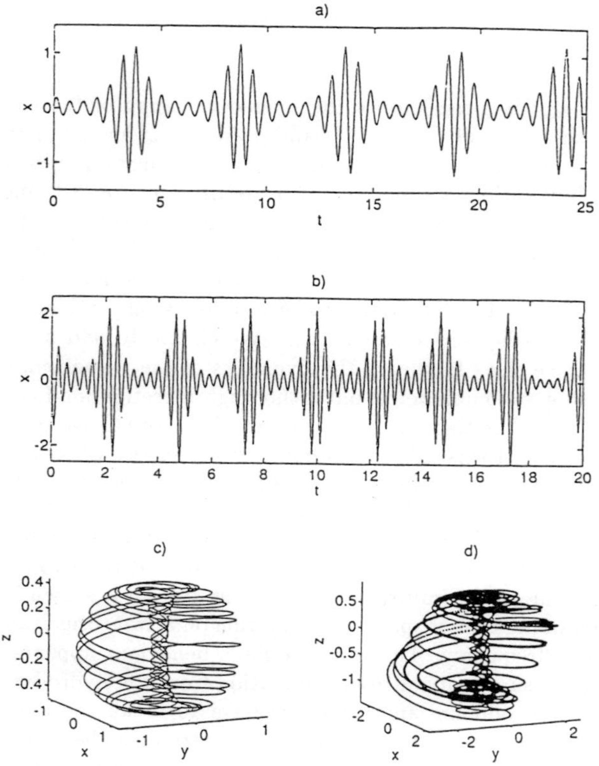


Fig. 5. Quasiperiodic and chaotic modulation of activity. ($a$) The poloidal field $x$ as a function of time $t$ for a quasiperiodic (periodically modulated) solution. ($b$) The same, but for a chaotically modulated solution. ($c$) Perspective view of the two-torus, corresponding to case $a$. ($d$) The chaotic attractor for case $b$.

parameter in determining stellar activity. We therefore consider moving along a path in the $\lambda - \mu$ plane as our order parameter $\Omega$ is varied. Our path is chosen so as to reproduce the bifurcation sequence displayed in Figure 2. It starts at $\lambda = -\infty$ near the line $\mu = 0$, where there are three fixed points on the $z$-axis. As $\Omega$ is increased the path passes through the primary Hopf bifurcation where a stable periodic orbit is born, as the stable fixed point loses stability to cyclically varying fields. A secondary Hopf bifurcation occurs as $\Omega$ is increased further and trajectories now lie on a two-torus, so the basic cycle period is modulated on a longer timescale, as in Figure 5 $a,c$. Further increase in $\Omega$ leads to the breakdown of the torus and the onset of chaotic behaviour shown in Figure 5 $b,d$. The heteroclinic bifurcation is then approached by the parameter curve and the chaotic attractor has now grown until trajectories nearly collide with the fixed points on the $z$-axis. The magnetic activity is now heavily modulated and 'switches off' regularly. Finally the parameter path is chosen to approach the $\mu$-axis so that solutions return to being quasiperiodic as $\Omega \to \infty$.

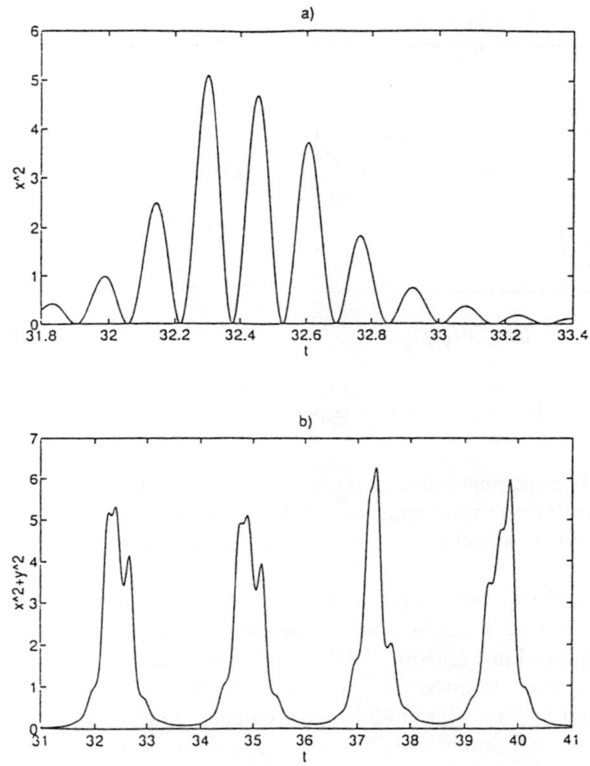


Fig. 6. Computed activity in the model system. ($a$) Variation of toroidal energy ($x^2$) with time, showing modulation of the basic cycle. ($b$) Variation of total magnetic energy ($r^2$) with time, showing the envelope of the activity cycle. These should be compared with those in Figure 1.

Clearly there are many other parameters that may be varied in this problem. These may change the specific behaviour but qualitatively it remains the same and the results are robust. In order to show that our model is capable of reproducing similar behaviour to the record of solar magnetic activity we may tune our parameters. Our best model is shown in Figure 6 where we have computed the activity cycle and the total magnetic activity of our model. This figure should be compared with Figure 1. All the behaviour described above occurs close to a codimension two point and is therefore robust. Indeed, the same behaviour may be found in a simple model which involves only one fixed point on the $z$-axis as discussed in detail by Tobias *et al.* (1995).

## 3. Frequency analysis

There has been a growing industry in analysing the solar cycle data (and more recently the $^{14}C$ and $^{10}Be$ records) by constructing power spectra from the time series in

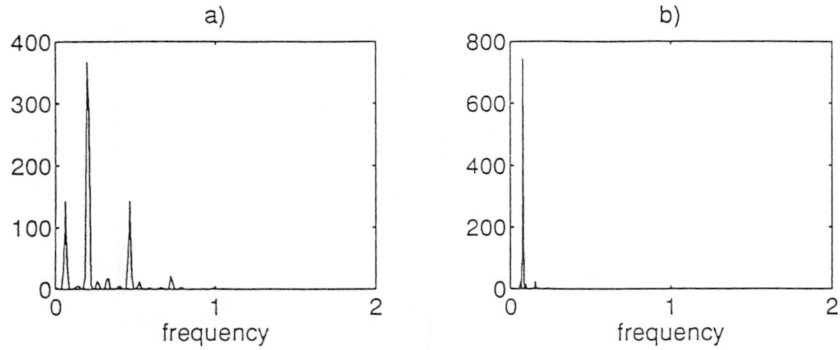


Fig. 7. Power spectra ($a$) for $x(t)$, showing clearly the cycle frequency with sum and difference frequencies; ($b$) for $z(t)$, showing a peak corresponding to the modulation frequency.

order to pull out frequencies that contribute to the signal. It is therefore interesting to perform a similar analysis using our chaotic signals; then we can discover what significance should be attached to the periodicities derived from isotopic abundance records.

We have computed power spectra from many of our numerical solutions (using the Welch method). The most interesting results come from analysing a chaotic solution and they are displayed in Figure 7. Analysis of a time series of the $x$-coordinate pulls out the dominant cycle frequency ($f$) easily as shown in Figure 7$a$. The time series is not long enough to yield the modulation frequency ($f_m$) as well, but the existence of sum and difference frequencies ($f^{\pm} = f \pm f_m$) can be ascertained. The modulation frequency itself can be found in the power spectrum of the time series for $z$ shown in Figure 7$b$. These results clearly show that it is sometimes possible to pick out individual frequencies from a highly aperiodic signal and that these frequencies may indeed have some physical significance. However, Figure 8 shows two power spectra derived from time series for a closely related problem. These take the form of a number of peaks contained within a broad envelope. This is a common feature of spectral analysis of time series which are finite. Indeed the spectra resemble more closely those derived from abundance records. It is not now apparent to which, if any, of the peaks we may attach physical significance. The nature of the attractors for these problems is similar when viewed in phase space, but is very different in Fourier space. Since this holds for simple low order models of the solar cycle, it must therefore be even more dangerous to ascribe physical significance to the presence or absence of individual peaks in power spectra derived from abundance records.

## 4. Conclusion

We have relied on bifurcation theory in order to derive a simple model system, without explicitly referring to the dynamo equations. This approach displays the underlying mathematical structure of a low-order model that succeeds in capturing the essential features of the solar cycle. Since equations (4)–(6) are normal-form equations, the

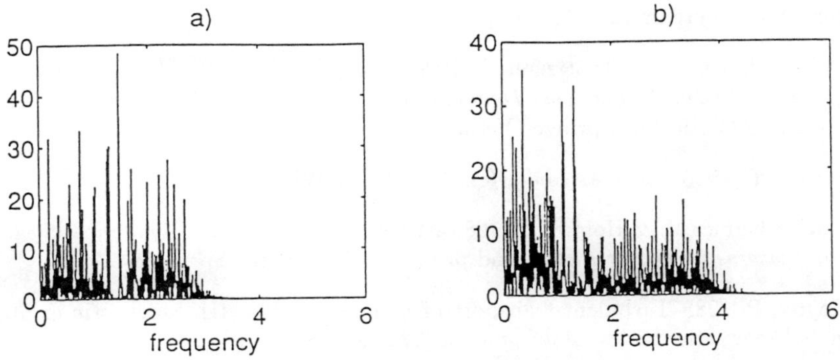


Fig. 8. Power spectra for a closely related problem, with an additional term $bz$ in equation (6). ($a$) For moderate values of $b$ the cycle frequency can still be identified, but ($b$) for higher values the frequency is no longer recognizable.

properties of their solutions are robust and likely to be found for higher order systems too. Given a supercritical Hopf bifurcation from an invariant hydrodynamic subspace, we could indeed predict the existence of chaotic modulation without knowing the details of the physics.

There is of course a different approach. The same equations can be derived by physical arguments, based on the processes that occur in a dynamo (Tobias *et al.* 1995). Although this is the simplest explanation of the observed modulation, other mechanisms have also been studied. Grand minima have, for instance, been ascribed to stochastic (Hoyng 1988) or chaotic (Spiegel 1994) switching. Alternatively, the modulation may depend on properties of the nonlinear dynamo itself, which can only be represented by a higher-order model (Weiss, Cattaneo & Jones 1984). Our procedure demonstrates, however, that it is possible to obtain generic behaviour without needing to know details of the dynamo process. It will be interesting to see whether a similar approach can be used in future to explore the connection between solar magnetic activity and climatic variations.

## Acknowledgements

The work described here was carried out in collaboration with Vivien Kirk and we thank her for guiding us through the bifurcations. SMT is grateful to SERC and PPARC for a studentship.

## References

Baliunas, S.L. & 26 others 1995 Chromospheric variations in main-sequence stars. II. *Astrophys. J.*, in press.

Barnes, J.A. Sargent, H.H. & Tryon, P.V. 1980 Sunspot cycle simulation using random noise. In *The Ancient Sun* (ed. R.O Pepin, J.A. Eddy & R.B. Merrill), pp. 159-163.

New York: Pergamon.

Brandenburg, A. & Tuominen, I. 1990 The solar dynamo. In *The Sun and Cool Stars: Activity, Magnetism, Dynamos* (ed. I. Tuominen, D. Moss & G. Rüdiger), pp. 223–233. Berlin: Springer-Verlag.

Foukal, P. 1990 *Solar Astrophysics.* New York: Wiley

Guckenheimer, J. & Holmes, P. 1986 *Nonlinear Oscillations, Dynamical Systems and Bifurcations of Vector Fields* (2nd printing). New York: Springer.

Hoyng, P. 1988 Turbulent transport of magnetic fields. III. Stochastic excitation of global magnetic modes. *Astrophys. J.* **332**, 857–871.

Kirk, V. 1991 Breaking of symmetry in the saddle-node–Hopf bifurcation. *Phys. Lett. A* **154**, 243–248.

Kirk, V. 1993 Merging of Resonance Tongues. *Physica* **66D**, 267–281.

Moffatt, H.K. 1978 *Magnetic field generation in electrically conducting fluids.* Cambridge University Press.

Morfill, G.E., Scheingraber, H & Sonett, C.P. 1991 Sunspot number variations: stochastic or chaotic? In *The Sun in Time* (ed. C.P. Sonett, M.S. Giampapa & M.S. Matthews), pp. 30–58. Tucson: University of Arizona Press.

Parker, E.N. 1979 *Cosmical Magnetic Fields: their Origin and their Activity.* Oxford: Clarendon Press.

Ribes, J.C. & Nesme-Ribes, E. 1993 The solar sunspot cycle in the Maunder minimum AD 1645 to AD 1715. *Astron. Astrophys.* **276**, 549–563.

Spiegel, E.A. 1994 Chaos and order in the solar dynamo. In *Lectures on Solar and Planetary Dynamos* (ed. M.R.E. Proctor & A.D. Gilbert), pp. 246–265. Cambridge University Press.

Stuiver, M. & Braziunas, T.F. 1988 The Solar component of the atmospheric $^{14}$C. In *Secular Solar and Geomagnetic Variations in the last 10,000 Years* (ed. F.R. Stephenson & A.W. Wolfendale), pp. 245-266. Kluwer.

Tobias, S.M., Weiss, N.O. & Kirk, V. 1995 Chaotically modulated stellar dynamos. *Mon. Not. Roy. Astron. Soc.*, in press.

Tong, H. 1990 *Nonlinear Time Series: A Dynamical Systems Approach.* Oxford: Clarendon Press

Weiss, N.O. 1990 Periodicity and aperiodicity in solar magnetic activity. *Phil. Trans. Roy. Soc. Lond. A* **330**, 617–625.

Weiss, N.O. 1994 Solar and stellar dynamos. In *Lectures on Solar and Planetary Dynamos* (ed. M.R.E. Proctor & A.D. Gilbert), pp. 59–95. Cambridge University Press.

Weiss, N.O., Cattaneo, F. & Jones, C.A. 1984 Periodic and aperiodic dynamo waves. *Geophys. Astrophys. Fluid Dyn.* **30**, 305–341.

Wilson, P.R. 1994 *Solar and Stellar Activity Cycles.* Cambridge University Press.

CHAPTER 11

# FRACTAL NATURE IN EARTHQUAKE PHENOMENA AND ITS SIMPLE MODELS

Mitsuhiro Matsuzaki
Faculty of Business Management
Kansai Jogakuin Women's College
Shijimi, Miki, HYOGO 673-05, JAPAN

**Abstract**

Earthquake prediction is one of the most important problems for countries on a plate margin with high seismic activity. The earthquake is a typical example of a common phenomena which has several kinds of fractal features. We introduce the 'stick-slip model', which can explain the fractal features of seismic phenomena to the earthquakes in Japan and discuss about the predictiblity of the destructive earthquakes.

## 1. Fracal Nature in Earthquakes

### *1.1. Introduction*

Earthquake is a fracture process of the crust. It has the features of both micro-scale and macro-scale, for example, phase transition of rock-forming minerals, micro crack of rocks, fault slippage at the active faults or plate boundaries. Therefore we can say that the earthquake is one of the most typical example of wide scale phenomena.

There are several kinds of scale invariance or fractal features in the earthquake phenomena, which are represented by a power law form, for example, the magnitude-frequency distribution, the spatial distribution of hypocenters, and the frequency of aftershock occurrences. These interesting features of seismic phenomena are investigated in this chapter.

### *1.2. Seis Distribution of Earthquakes*

The magnitude-frequency distribution of earthquakes is described by

$$\log N(>M) = a - bM \tag{1}$$

where $M$ is the magnitude of an earthquake and $b$ is a number nearly equal to 1.0 (Kanamori and Anderson, 1975). In this formula $N(>M)$ denotes the number of earthquakes having magnitude lager than $M$. This is one of the most famous formula which is called as Gutenberg-Richter's law. An example is shown in Fig.1 for the eartthquakes in Japan. The magnitude $M$ is proportional to the logarithm of the seismic energy $E$ as $M = \frac{2}{3}\log E$. Combining (1) with this, we can get the following formula for the distribution of energy

$$\log N(>M) = -\frac{2b}{3}\log E + const. \tag{2}$$

Since the magnitude is proportional to the area $S$ of earthquake source size (Utsu, 1969)

$$\log S = M - 3.7 \tag{3}$$

the Gutenberg-Richter's law can be also described as

$$\log N(>S) = -b\log S + const. \tag{4}$$

(Kanamori and Anderson, 1975).

The size of an earthquake can be estimated from the fault length. Since the fracture area-fault length relationship can be described as $\log S = 2\log L$, we can get the following formula from (4)

$$\log N(>L) = -2b\log L + const. \tag{5}$$

Thus the size-frequency distribution of earthquakes is given in the power law form.

*1.3. Spatial and Time Distribution of Earthquakes*

The spatial distribution of hypocenters of earthquakes is obviously heterogeneous. It has been confirmed to be fractal as firstly demonstrated by Kagan and Knopoff (1980) and Kagan (1981). They studied the spatial distribution of earthquake hypocenters and epicenters, using the world wide as well as local catalogues. They found that the correlation functions, that describe the number of events per unit volume or area at a distance $R$ from any earthquake, can be expressed as follows

$$\log N(R) \propto \alpha \log R \tag{6}$$

over a certain range of the distance $R$. the exponent $\alpha$ corresponds to the fractal dimension $D$ as in $D = d - \alpha$, where $d$ is the spatial dimension ($d = 2$ for an epicenter distribution and 3 for a hypocenter distribution).

The number of events per unit volume occurring within distance $R$ from any earthquake is given by

$$\log N(< R) \propto D \log R + const. \tag{7}$$

$D$ in two-dimensional space ranges from 1.0 for the world wide catalogue, 1.1 for California and 1.3 for Japan. Sadovskiy et al. (1984) also found a self-similar feature or the earthquake distribution and obtained a value of fractal dimension $D$ as 1.4 for the world wide catalogue and 1.6 for a local Soviet catalogue. Yoshida and Mikami (1986) reported $D = 2.7$ for the aftershock hypocenter distribution in Nagano Japan, which corresponded to $D = 2.7$ for the epicenter distribution.

Since the spatial distribution of hypocenters is fractal, earthquakes are not a purely random phenomenon but should have a kind of interaction. Especially, a mainshock and its aftershocks have strong interaction. An interaction of earthquake occurrences has been shown as the power law decrease of aftershocks. The decrease of aftershock is

$$P(t) = \frac{K}{t + c}, \tag{8}$$

where $P(t)$ represents a probability of aftershock occurrences at time $t$ after the main shock , $K$ is a constant proportional to the seismic energy of the mainshock and $c$ is a constant introduced to avoid the divergence at $t = 0$. This formula shows the decrease of aftershocks of which asymptotic form $1/t$ (Omori's law).

The power law decrease of aftershocks indicates a scale invariance on time of earthquake activities. In the meantime mainshocks have been considered as random phenomena, having no interactions. If we consider mainshocks and aftershocks simultaneously, we obtain the same power law relationship as (8) that is derived without mainshocks. Recently this topic is attracting much attention. Ogata (1988) showed that the auto correlation of earthquake occurrences follows a power law relationship as

$$logC(t') = -b\mathrm{log}t' + const, \tag{9}$$

where $C(t')$ is the correlation function of earthquake occcurrences for time interval $t'$.

Experimental studies of rock fracturing indicate that such power law relations emerge as fracture progresses. Hirata (1987), through rock fracture experiments, found that bursts of acoustic emissions(AE) decayed exponentially during the early stage of the experiments and then tended to decay by a power law over the last stage of fracturing. Hirata et al. (1987), also through rock fracture experiments, discovered that the spatial distribution of AE was random during the primary creep; however it gradually organized to become a fractal as the creep advanced, and the fractal dimension $D$ defined by a power coefficient of the relation:

$$\langle n(< R) \rangle \propto R^D \tag{10}$$

decreased. Here, $\langle n(< R) \rangle$ is the average number of AE within a radius $R$.

### *1.4. A Vewpoint of the Earthquake Phenomena*

Since long-range interactions are required to yield power laws, these experimental results suggest that both temporal and spatial interactions among the microcracks in the rock samples increased as the fracturing process progressed. Thereby, a self-similar fractal structure become self-organized during the fracture process. The term self-organization specially indicates that the organized state appears irrespetive of the detail on the initial conditions and on the control parameters.

Earthquakes occur repeatedly along active faults, which indicates that the crust is in a steady state of fracturing. It is surmised that the scale-invariant features of earthquakes are formed by the self-organization of the fault systems in the crust. Then it should be asked as to what kind of self-organization brings forth the fractal structure. There are many kinds of phenomena in nature which are scale-invariant or fractal (Mandelbrot, 1982). For example, critical phenomena are known to have many scale-invariant relations. Critical phenomena are found in phase transitions (Stanley, 1971) and percolation (Staufer, 1979).

Percolation, which will be defined below, has many features in common with the phase transition. During second-order phase transitions and at the critical point of percolation, various kinds of scaling relations are known. They are given in the form of power laws with the behavior of the system usually described by the critical exponents. For an example in the context of the present chapter, the number $n_s$ of clusters having size $s$ at the percolative critical point $p_c$ is given by

$$n_s(p_c) \propto S^{-\tau} \tag{11}$$

Here a cluster is defined as a group of neighboring sites with the same property (e.g., occupied sites in percolation or upward spins in the magnetic phase transition). The power law form indicates that there is no characteristic length scale and the system is spatially self-similar. The critical exponent $\tau$ describes the size-frequency relationship.

The critical state in laboratory experiments is brought forth by fine tuning the initial or boundary condition, such as temperature and probability. In contrast, there appears to be a number of systems which evolve naturally toward a critical state (Bak and Tang, 1989).

Takayasu *et al.*(1988) proposed a fractal model of river branching and pointed out that when the input of and the outflow of the river balanced, the system is in the critical state and becomes self-similar. Bak et al. (1987, 1988) proposed a cellular automaton model of self-organized criticality, which was a simple analogue of the earthquake occurrence.

## 2. Models of Earthquake

### *2.1. Go-game Model*

The earthquake phenomena show many power law behaviors. Generally speaking, the power law behaviors are commonly found in the critical state. Therefore, the earthquake is also considered as one of the phenomena in the critical state. In order to explain such a criticality many models have been proposed. We will study the fundamental physics of the models from this section.

The fracturing process of an earthquake was modeled by Otsuka (1972) as percolation process on a two-dimensional lattice during the critical state, which is called "Go-game model" in Fig.2. A square lattice is considered, and each site is supposed to have a state value of 1 (fractured) or 0 (not fractured) and a fracture probability $p$ to be occupied, independent of its neighbors. The system starts with its initial state in which all sites have uniform state value 0, then the state values of one site is chosen at random and changes to 1. This is the first step. In the second step, the state values of four nearest neighbor sites changes with a probability $p$. This process corresponds to the fracture propagation. If all the nearest neighbor sites are converted by unfractured and or fractured sites, the fracture propagation stops. In this model the size of the cluster of fractured sites is considered to be an earthquake source area.

In Go-game model, there is an parameter, i.e., the probability $p$ of fracture propagation. In the case of small $p$, the fractured area tends to be localized and for large $p$, fractured area is extended to a large region. Therefore, there would be a critical value $p_c$. This model is equivalent to the usual site percolation model and therefore it shows a phase transition. At the critical point of the phase transition, the size distribution of clusters (size of the fractured area) follows a power law. Percolation theory is concerned with clusters which are groups of neighboring occupied sites (Stauffer, 1979). For $p$ above the critical probability $p_c$, an infinite cluster exists, and the lattice is forced to undergo percolation. The system at $p_c$ is in a critical state, has no characteristic length scale, and is spatially self-similar. The percolation process may be interpreted in a different manner to describe the spreading of forest fires. A site in the lattice is chosen where a fire originates, with the probability of spreading fire to neighboring sites given by $p$. With a probability of $1-p$, the fire will not spread. All lattice sites in the next surrounded shell are tested for a probability $p$ to determine if the fire spreads or not. Then the next shell is tested with the process being repeated until the fire stops spreading. This process is called the growth method of percolation (Leath, 1976), and the result is equivalent to ordinary percolation.

As was studied above, the scaling relationship $n_s \propto s^{-\tau}$ of (11) occurs at the critical state of percolation. If a site is chosen at random, then the probability that the site belongs to a cluster of size $s$ is proportional to $S \times n_s$, since there exist $s$ ways to select a site within the size $s$ cluster. Equivalently, if a site is specified and a cluster is allowed to grow, then the growth method produces clusters of size $s$ with a probability $S \times n_s$. Therefore the size-frequency relationship at the critical state is given by

$$n_s(p_c) \propto s^{\tau+1} \qquad (12)$$

through the growth method, instead of (11) in ordinary percolation.

It should be pointed out that the size-frequency in seismology implies the size distribution of earthquakes in space, and therefore it corresponds to the latter relation (Hirabayasi et al., 1989). In a manner similar to the model that simulates the spreading of forest fires, Go-game model simulates the spread of fracturing by the growth method of percolations.

Now, we will apply the cellular automaton model (Ito and Matuzaki, 1990) to the earthquake process. We then modify the model so that every earthquake may trigger aftershocks. Then we introduce some mechanical models and discuss about the predictability of destructive earthquake along the active faults.

### *2.2. Cellular Automaton Model*

Takayasu and Matuzaki (1988) extended the one-dimensional stick-slip model of Burridge and Knopoff (1967) to a multidimensional mechanical model and found that it has an order-disorder phase transition, above which the entire system slips together as one cluster. The behavior of the system depends on the strength of the coupling. In the critical state, the size distribution of clusters follows the power law similar to that of earthquakes. Their two-dimensional mechanical model is similar to the cellular automaton model of Bak et al. (1987, 1988) which evolves into a self-organized critical state.

In the two-dimensional mechanical model, the friction elements and interconnecting springs are arranged in a square lattice and are located on a constantly moving surface. Each element is interconnected with the four surrounding elements by springs and is also connected to a rigid support by a flat spring as Fig.3. The force acting on a given element is the sum of the forces of the springs. Each element adheres to the moving surface while the force is less than a threshold value, that is, the frictional force. The element slips when the force exceeds the frictional force and the force that was acting on it is distributed to the surrounding elements through connecting springs. The system evolves, under the following procedure consisting of three steps.

First, the normalized force $F(i,j)$ acting on each unit block $(i,j)$ continuously increases as $\frac{d}{dt}F(i,j) = v(i,j)$, until the force on at least one $F(i,j)$ exceeds 1.0, that is, the normalized friction force. Here, $v(i,j)$ is related to the strength of the flat spring and the velocity of the constantly moving surface.Second when $F(i,j) = 1$ the block $(i,j)$ slips and $F(i,j)$ reduces to 0. Since block $(i,j)$ is connected by springs with the four surrounding blocks $(i\pm1, j\pm1)$, the slippage of the $(i,j)$ block creates an increase of force in the $(i\pm1, j\pm1)$ blocks by certain rate $\Delta F$. If $F(i\pm1, j\pm1) + \Delta F$ exceeds 1.0, then the slippage extends to the neighboring blocks of $(i\pm1, j\pm1)$. For all $F(i\pm1, j\pm1) + \Delta F \geq 1$, let $F(i\pm1, j\pm1) = 0$ and add $\Delta F$ to every blocks surrounding the $(i\pm1, j\pm1)$ blocks except the one which had already slipped. This step is then repeated. Third, if all the $F(i,j)$ are less than 1.0, then the process returns to the first step. The total number of friction elements which have slipped in the second step gives the size of an earthquake. We will study this model in later section.

In the present cellular automaton model, a two-dimensional regular lattice, $n \times n$, is considered. Each site corresponds to a friction element in the mechanical models. Each bond represents the interaction with the neighboring sites, corresponding to the connecting springs in the mechanical model. The state value $z(i,j;t)$ at a time

$t$, corresponding to the force $F(i,j)$ in the mechanical model, takes discrete values; $0, 1, 2, \ldots 7$, in the cellular automaton model. The threshold value $z_c$ corresponds to the frictional force in the mechanical model and is assigned an arbitrary value of 3. The following procedure again consisting of three steps is followed as in the mechanical model.

First, the state value $z(i,j;t)$ increases by a speed of $1/n^2$, until at least one $z(i,j;t)$ exceeds the threshold value $z_c$. Because is discrete, an alternative procedure is actually taken such that one random site $(i,j)$ among the $n^2$ sites is chosen and $z(i,j;t)$ is increased by 1 so that

$$z(i,j;t+1) = z(i,j;t) + 1 \tag{14}$$

The first step in the procedure represents the increase in the tectonic force due to plate motion. Second, when $z(i,j) \geq 4$, the site $(i,j)$ "slips" (release the state value) and $z(i,j)$ decrease by 4. Thus $z(i,j;t^+) = z(i,j;t) - 4$ only for $z(i,j;t) \geq 4$. When the site $(i,j)$ "slips" $z$ of the four surrounding states $(i{\pm}1, j{\pm}1)$ increase by $\Delta z$. In the mechanical model, $\Delta F$ is a control parameter. However, it is expected from the elasticity theory of dislocation that $\Delta z$ should be close to 1 or that $z(i,j;t) - z(i,j;t^+)$ is close to $\Delta z$ times the number of nearest neighbors. A dislocation (fault) is formed when an earthquake occurs and causes strain around the fault. According to the elasticity theory of dislocation, the strain field generated by a dislocation on an infinitesimal surface element is equivalent to that generated by a double-couple force applied on the surface element in the absence of a dislocation (Stketee, 1958).

This may be interpreted for the present cellular automaton model in the following way. The state such that the force applied on a given site $(i,j)$ as $z(i,j;t) = 4$ is equivalent to the state that the force is free at $(i,j)$, resulting in $z(i,j;t^+) = 0$; however, the surrounding sites $(i{\pm}1, j{\pm}1)$ are each increased by 1. In elasticity theory, a double-couple force heterogeneous strains in its surroundings. In the present model, however, it is assumed for simplicity that the force freed from a given site $(i,j)$ is distributed homogeneously among the four nearest sites $(i{\pm}1, j{\pm}1)$. Thus it follows that $z(i{\pm}1, j{\pm}1; t^+) = z(i{\pm}1, j{\pm}1; t) + \Delta z$ with $\Delta z = 1$.

Although the elasticity theory of dislocation has been successful in expanding the crustal deformation caused by great earthquakes, difficulty arises in stopping a fracture in the model once it has started. In order to accomplish this, plasticity is required. Similarly, if the sum of $z$ is kept constant in the second step of the procedure, the same problem occurs. Hence a probability $1-p$ representing plasticity is introduced. Step two of the procedure is thus taken for $z(i,j;t) \geq 4$ as

$$z(i,j;t^+) = z(i,j;t) - 4 \tag{15a}$$

$$z(i{\pm}1, j{\pm}1; t^+) = z(i{\pm}1, j{\pm}1; t) + 1 \tag{15b}$$

with a probability $p$. The probability $p$ should be close to a value of 1.0.

Takayasu and Matsuzaki (1988) excluded blocks that had already slipped once from applying the distribution of the force described in (13b), however, multiple slippage of a unit is allowed to occur in the present cellular automaton model. As long as the size of an earthquake is defined as the number of sites which have slipped, it does not make a difference whether multiple slippage is allowed or not. Then this step is repeated.

In seismic phenomena, the time scale of the first step of the procedure may be 100∼1000 years, while that of the second step of the procedure is, in general, of the order on seconds Therefore $t^+$ in (13a) is set equal to $t$, and the time increase is neglected during the second step of the procedure.

Third, if all $z(i,j)$ are less than 4, return to the first step.

The procedure is essentially equivalent, with the exception of minor points, to that of the mechanical model of Takayasu and Matsuzaki (1988) at phase transition. The present model differs from that of Bak et al. (1987, 1988) only in the consideration of the probability $p$. In the latter, the boundary is fixed as inactive. When a fracture extends to the boundary, the total sum of $z(i,j)$ starts to decrease and thus the fracture will stop. Since the fixed boundary condition result in a significant system-size dependency, a periodic boundary condition is used and the probability $p$ is introduced. The behavior of the present model when $p$ is close to 1 is essentially the same as that of Bak et al. (1987,1988) with a large system size.

The sum of $z(i,j)$, $\theta(t) = \sum_{i,j} z(i,j;t)$, is chosen as the state of variable of the system. When starting from an initial condition where $z(i,j;t=0) = 0$ for all $(i,j)$, it can be seen in Fig.4 that $\theta$ gradually increases and then reaches a steady state, In the steady state, an earthquake originates when the energy $z$ of one site exceeds the friction force $z_c$, with fracturing propagating to the neighboring sites. As a site slips, the sum of the force decreases with a rate proportional to $1-p$, the propagation is finally halted, and the event (earthquake) ends.

Sites which "slip" during this event from a cluster, and the number of "slipped" sites $S$ corresponds to the fault area or the size of an earthquake. The decrease of $\theta$ or the number of fracturing event would represent the energy released by an earthquake. After one seismic event in this model, $\theta$ starts to increase through the first step of the procedure, and another earthquake is initiated at another site.

In order to show the change of $\theta(t)$ in the steady state in more detail, a part of graph Fig.4-*d* is expanded and shown as Fig.4-*e*. Each vertical drop of $\theta$ in Fig.4-*e* corresponds to the energy release by an earthquake in this model. The "steady" state with the same $p$ does not depend on the initial condition (compare *b* with *d* in Fig.4). The total sum of the state values (forces) $\theta$ at the "steady" state depends only slightly on the probability $p$ (compare *a*,*b* and *c* in Fig.4). If $p$ is significantly less than 1 or plasticity is significant (case *c* in Fig.4), then large earthquakes rarely occur. When $p = 1.0$ or plasticity is nil (case *a* in Fig.4), a large earthquake, once it starts, will continue indefinitely. As the case with $p = 0.999$ simulates the size distribution of earthquakes very well, the results with $p = 0.999$ are shown in the following of the present chapter.

Fig.5 demonstrates the change of cluster patterns when the system starts from a totally inactive state, gradually reaching the "steady" state. At the beginning of the stage, active sites are randomly distributed, forming small clusters. During the "steady" state, active sites tend to form a few large clusters. The size distribution of the clusters during the "steady" state has a power law similar to that of earthquakes. This will be discussed in the next section. Fig.6 shows the size-frequency relation of clusters for the present model with a value of $p = 0.999$. The cluster size in the model corresponds to the size of fractured zone $S$ for earthquakes. The relation is given also in the form of a power law. When $p$ is significantly less than 1, then the size frequency relation for large size deviates from the power law. Tang and Bak (1988a) have found that the relationship obeys a power law, such as

$$N(S) \propto S^{-\tau+1} \tag{16}$$

where $N(S)$ is the number of size-$S$ clusters, obtaining $\tau = 2.0$ for the cellular automaton model. The present result indicates $N(S) \propto S^{-1.06}$; thus $\tau = 2.06$. It should be pointed out that the cluster size distribution in this cellular automaton model is calculated by the so-called growth method of the percolation problem (Leath, 1976). It has been explained that this method produces clusters of size $S$ with a probability proportional to $S \times n(S)$. Here $n(S)$ is the number of size-$S$ clusters in an infinite lattice and is given by $n(S) \propto S^{-\tau}$ (Stauffer, 1979). This is why the exponent in (16) is written as $-\tau+1$. Since (1) was obtained for earthquake occurring over wide areas, $(\tau - 1)$ in (16) is compared with $b$ in (4) when the present model is applied to the earthquake occurrence. The comparison gives the estimate that $b = \tau - 1 = 1.06$. This agrees with observed values of $b$, since $b$ is usually about 1.

This magnitude-frequency relationship of earthquakes is thus in accord with the critical behavior of the present model. The model, however, has some difficulties. First, aftershocks which are common in natural earthquakes do not occur in the model. Second, the distribution of epicenters remains homogeneous even in the steady state and does not evolve to being fractal. We need to modify the model to overcome these difficulties, especially the absence of aftershocks.

### *2.3. Decay of Aftershocks*

In order to consider aftershock occurrences, we may have two alternative hypothesis why aftershocks occur: 1 Aftershocks occur because the mainshock does not release the strain completely down to the steady state value, or 2 aftershocks occur because the main shock disturbs the strain distribution from the steady state one.

The automaton model may be modified according to the hypothesis 1 in such a manner that the state value $z$ of some sites which "slipped" during the main shock are increased after the main shock is terminated. Alternatively, the automaton model may be modified according to the hypothesis 2 in such a manner that the state value $z$ at each site is perturbed, being either added to or subtracted from by 1, after the main shock. In both cases, some sites may exceed the threshold value and aftershocks will originate there.

We can estimate without simulation the difference of the two types of aftershock occurrences by analyzing the relation behavior of the present cellular automaton model. The aftershock is a type of relaxation processes 1 and 2 occur in conditions that 1 the total some of the state values (forces) $\theta$ relaxes to the "steady" state value, and 2 the distribution of $z(i,j)$ approaches the "steady" state while $\theta$ is kept constant. Process 1 can be called the force (energy) relaxation and process 2 can be termed the entropy relaxation. Tang and Bak (1988a), using a mean-field approximation, analyzed the relaxation behavior of type 1. They predicted that the number of active sites decreases exponentially with time. Ito and Matsuzaki (1990) have analyzed, using the same mean-field approximation, the relaxation behavior of type 2. In this latter type of relaxation, the number of active sites decrease in a power law form. Presently, the number of clusters of active sites may correspond to the number of aftershocks.

The frequency $f$ of aftershocks decreases with a hyperbolic form known as Omori's law (Omori, 1984). The following form is now commonly used (Utsu, 1970):

$$f \propto \frac{k}{(t+c)^p} \tag{17}$$

where $c$ is a small constant and $p$ has a volume of approximately 1.0, ranging from 1.0 to 1.4. Because $c$ is small, $f$ for a large time $t$ can be expressed by an inverse power law. The agreement of decay formulas of (17) suggests that aftershocks may occur due to process 2 rather than process 1. Enya (1901) already proposed hypothesis 2 after Omori (1894) found the hyperbolic decay of aftershocks. Accordingly, we are going to modify the model on the line of hypothesis 2.

### *2.4. A Modified Trigger Model*

In a modified trigger model, in which each event may be a trigger of the following events, we take the following procedure.

First, the state value $z(i,j;t)$ increases by (14). Second, when $z(i,j)$ of a site exceeds the threshold value, an earthquake occurs. The rule of fracture propagation is the same as (15a) and (15b) as long as the fracture is propagating. When the propagation stops or all $z(i,j)$ become less than 4, the main shock ends. Third, instead of returning to the first step as in the original cellular automaton model, the system is perturbed in the modified trigger model. As aftershocks occur only within the fractured zone of the main shock, the area of perturbation in the model is also limited within the cluster of the main shock. The value $z$ of all sites which slipped during the main shock is perturbed; that is, $z(i,j)$ is either added to or subtracted from by 1, with an equal probability.

We should notice that the sum of $z(i,j;t)$ is not statistically perturbated. If some of these sites exceed their threshold value, then aftershocks will originate at those sites. For each aftershock, we repeat the second step until the propagation of fracture stops. when all aftershocks terminates, we repeat the third step of procedure explained above for the fractured area of every aftershock, that is to say, we regard each aftershock as a main shock which may trigger "after-aftershocks". This iterative process will be repeated until none of perturbed $z(i,j;t)$ after the third step exceeds the threshold value. Then we return to the first step.

Because the evolution rule is the same as the original cellular automaton model, the modified trigger model shares many common features with the original model. Time evolution of $\theta(t)$ given in Fig.4 and that of the active site pattern shown in Fig. 5 are essentially the same. The size-frequency relation of clusters in the modified trigger model also obeys a power law, but the exponent is larger, being about 1.7. An increase of the slope in the modified trigger model may be expected, because all aftershocks are smaller than the main shock, and thus small earthquakes occur more frequently in the modified trigger model than in the original one. The interpretation of this exponent 1.7 is complicated. The modified trigger model is no longer a simple version of the growth method of percolation, in which each growth is independent. Aftershocks in the modified trigger model are spatially correlated. This is why the exponent is not denoted by $\tau$ as in the original cellular automaton model. If we interpret, going to an extreme, that the cluster size distribution obtained for the modified trigger model corresponds directly to the size distribution of earthquakes, then the comparison gives an estimate that $b = 0.7$ from (4). Because this interpretation should give a minimum estimate for $b$, the estimate is satisfactory.

The other difficulty in the original cellular automaton model that the fractal nature of epicenters distribution does not appear is also cleared in the modified trigger model. The spatial distribution of epicenters evolves from a random to fractal structure as the system approaches the steady state, as shown in Fig.7. This behavior will discussed in the next section.

### *2.5. Spatial Clustering of Hypocenters*

In the present automaton model, an earthquake is considered as a cluster of active ("slipped") sites, and the epicenter is defined as the first site that became active. The fractal dimension $D$ (the two-point correlation dimension $D_2$) of the epicenter distribution in the present model is obtained for the 400 succeeding earthquakes by the correlation integral method (Grassberger and Procaccia, 1983). The definition of the correlation integral is

$$Ci(r) \equiv \lim_{N\to\infty} \frac{1}{N^2} \sum_{i}^{N} \sum_{i}^{N} H(r - |X_i - X_j|) \tag{18}$$

where $H(x)$ is Heaviside function. Here $N$ is the number of epicenters, and $|X_i - X_j|$ is the distance between $i$-th and $j$-th epicenters. When the distribution of points is fractal,then $Ci(r)$ behaves as a power law of distance between the hypocenters $r$, when $r$ is small, as

$$Ci(r) \propto r^{\nu} \tag{19}$$

The exponent $\nu$ is a kind of fractal dimension and is called the (two-point) correlation dimension $D_2$. Since the fractal structure in the modified trigger model changes gradually, the limit $N \to \infty$ can not be obtained. In the present study, $N$ is set to a value of 400, which is found to be large enough to estimate $D_2$, and the change of $D_2$ is examined as the system evolves. Fig.8 demonstrates that the correlation dimension decrease and approaches a value of 1.1 as the system attains the "steady" state. This behavior agrees with the results of Hirata et al. (1987), found in rock-fracturing experiments.

## 3. Stick-slip Model

### *3.1. Two-dimensional Stick-slip Model*

A simplified two-dimensional stick-slip model will be introduced in this section. The system shows a kind of dynamical order-disorder phase transition with a control parameter proportional to the strength of interaction. At the critical point, this model is consistent with the empirical laws of earthquakes such as Gutenberg-Richter's law, the spatial distribution of hypocenters, and the correlation function of the occurrence of earthquakes as results of a dynamical critical phenomenon. This model can naturally explain the locality that local magnitude distributions often deviate from Gutenberg-Richter's law. The two-dimensional stick-slip model of Matsuzaki and Takayasu (1991) consists of oscillators arranged on a square lattice. The oscillators are on the frictional surface moving with a constant velocity. Each oscillator is interconnected with the four surrounding oscillators by horizontal springs of rigidity $k$ and also connected to the rigid support by a vertical spring of rigidity $g$ (Fig.3).

While the sum of the forces from the horizontal springs and vertical spring $f$ at every oscillator is smaller than the a frictional force $f_c$, the oscillators move with a constant velocity $V$ and their spring forces increase by $Vg$. When the sum of the spring forces $f(i,j)$ at the $(i,j)$ site exceeds the threshold frictional force $f_c(i,j)$, the oscillator slips back to the position at which the total spring force is equal to 0, and the forces that were acting on it are distributed to the surrounding oscillators except the ones that have already slipped at the same time step.

In order to avoid unnecessary complication, here we assume that the spring constants $k$ and $g$ are identical for all oscillators. Normalizing the variables by the threshold frictional forces, we have the following rule

$$\frac{d}{dt}F(i,j) = v(i,j); \quad \text{If} \quad F(i,j) \geq 1$$

$$\text{then} \quad F(i,j) = 0, \quad \text{and} \quad F(i\pm1, j\pm1) = F(i\pm1, j\pm1) + d \tag{20}$$

$$\text{where} \quad F(i,j) = \frac{f(i,j)}{f_c(i,j)}, \; v(i,j) = \frac{V\cdot g}{f_c(i,j)}, \; d = k/(g+4k).$$

Because the thresholds are randomly distributed, the normalized time derivatives of forces, $v(i,j)$, distribute randomly. A slip of the oscillator at $(i,j)$ creates an increase of the force on the surrounding oscillators by $d$. If $F(i\pm1, j\pm1)$ becomes larger than 1.0 by this effect, the slip extends to the neighboring oscillators except the one which has already slipped. When all force values are less than 1.0, the slip propagation stops. A cluster is defined as a group of oscillators which slipped at one time step through this process.

In this model, the cluster size gives a measure of the fracture area of an earthquake and the force drop is considered to be proportional to the moment of an earthquake. This model has a parameter $d$ which is related to the ratio of vertical spring to the horizontal spring constant. For small $d$ , the events tend to be localized and the system belongs to a random phase. Conversely, for large $d$, the force of the slipped oscillator is nearly conserved, so the events tend to extend over a large region and the system belongs to the coherent phase (Fig.9).

These two phases are separated at a critical value of $d$. At the critical point, the model is consistent with the empirical laws of earthquakes such as Gutenberg-Richter's law, the spatial distribution of hypocenters, and the correlation function of the earthquake occurrences (Fig.10, 12, and 13) .

Although the correlation function of the time sequence of "earthquakes" by this model is consistent with that of real seismic phenomenon, it is known that aftershocks are much fewer than real earthquakes.

In order to improve the two-dimensional stick-slip model so that aftershocks may occur, we introduce a hypothesis of aftershock occurrences as follows: A man shock causes the strain decrease and the crust of the focus region may melt partially. Aftershock may be viewed as the relaxation process of the state of the crust from the partial melting state in which little strain energy can be hold, to the solid state in which much strain energy can be hold. This assumption is based on the entropy relaxation presented in the previous section. The low viscosity area corresponds to the perturbed area of the modified trigger model.In order to realize this scenario, now we modify the evolution rule as,

$$\frac{d}{dt}F(i,j) = v(i,j); \quad \text{If} \quad F(i,j) \geq 1$$

$$\text{then} \quad F(i,j) = 0, \quad \text{and} \quad F(i\pm1, j\pm1) = F(i\pm1, j\pm1) + d,$$

$$F_c(i,j) = 0.01 \qquad \text{in a certain period } t_r \tag{21}$$

For example $t_r = 100/ < v(i,j) >$ (angle brackets denote an average over $(i,j)$ in the following simulation, which is long enough to change the behavior of the system from the original two-dimensional stick-slip model. In the case the period $t_r$ is short, the behavior of the system is almost equivalent to the original two-dimensional stick-slip model. This case corresponds to the situation that the crust of the fault region rarely melts. In the case of the period $t_r$ is long, the behavior of the system is very different from that of the original two-dimensional stick-slip model. This case corresponds to the situation that aftershocks occur for a long period because the state of the fault region does not return to the steady state. The maximum period in real seismic phenomena is estimated as 10000 days (Ogata, 1987). Fig.13 shows a pattern of the aftershock occurrences after a big event of the reduced threshold model. When a main shock occurs, the slope of this curve increases because many aftershocks occur in a short period. This is caused by the decrease of thresholds of the oscillators. When the thresholds return to the original values, the interval of two nearest events becomes longer and then the next mainshock occurs. The biggest aftershock occurs at the point of (A) of this curve, after a comparatively long interval.

### *3.2. One-dimensional Stick-slip Model*

The one-dimensional version of the stick-slip model described in previous section is applied for active fault planes facing two crustal blocks. Frictional boxes on a constantly moving crustal block is connected to the adjacent crustal block by a leaf spring which constant is , and also connected to the two nearest neighbor boxes by coil springs which constant is.

The evolutional rule of this system which normalized by the thresholds is described as follows, just like two-dimensional model.

$$\frac{d}{dt}F(i) = v(i); \quad \text{If} \quad F(i) \geq 1$$

$$\text{then} \quad F(i) = 0, \quad \text{and} \quad F(i \pm 1) = F(i \pm 1) + d \tag{22}$$

When we assume that spring constants $g$ and $k$ are identical in each spring, the temporal behavior of this system is determined by a parameter $d = k/(g + 2k)$, which is related to the ratio of leaf and coil springs, just like the two-dimensional model. For small $d$, a large part of the force drop cannot distribute to the surrounding oscillators and events tend to be localized. this feature corresponds to the creaping of the fault. conversely, for large $d$, the force drop of the slipped oscillator is distributed to the connecting oscillators, and eventually extends over a large region.

If we change the value of model parameter $d$ in the range from 0.2 to 0.3, we can get the following result. A model with smaller $d$ value result in a number of small events, while a model with larger $d$ value results in only one great event, comparable to the system size, confirming the above atatement. For $d = 0.23$, the results exhibit the results exhibit a few large events similar to those obtained by the historical earthquake analysis. It is very interesting to note that the two-dimensional stick-slip model with value of $d = 0.23$ generate a number of events whose size distribution consistent to the Gutenberg-Richter's law (Matsuzaki and Takayasu, 1991). Although the one-dimensional stick-slip model studied here couldn't exhibit fractal features of the earthquake phenomena, it will provide rich information for seismic couplings between adjacent fault segments (Hung and Turcotte, 1990).

### *3.3. The Caterpillar Model*

Earthquake prediction is one of the most important problems for countries on a plate margin with high seismic activity. Now, we introduce "stick-slip model", which can explain the fractal features of seismic phenomena to the earthquakes in Japan and discuss about the predictability of the destructive earthquakes.

Kanaori et al. (1991a,b, 1993) investigated space time patterns of destructive earthquake occurrences in Central Japan and proposed a new model of seismic cycles. In this model, block boundaries or tectonic lines are defined as linking majour active faults distributed in the area (Fig.14). In this Figure, we can find a sliding block and some rotating blocks. If we assume this block rotation model, we can see the seismic coupling between these block boundaries.

The space-time patterns of destructive earthquakes with magnitude 6.4 or grater have been analyzed by relocating the postulated epicenters to the corresponding block boundaries or tectonic lines. This study revealed that the destructive earthquake in that area have clear periodicities of active and quiet intervals with recurrence time of about 1000 years and the duration of active periods are several hundreds years. We can also find the seismic coupling between West Nankai Trough and other block boundaries, which is the empirical feature of seismic events along West Nankai Trough occurring in 30 years after the earthquake HKFL (Hanaore-Kongo Fault Line) and in order of 100 years after the events along the easter block boundaries (Fig.15).

We arrange the one-dimensional stick-slip model to the Block rotation model of Southwest Japan which studied by Kanamori et al. (1991a,b, 1993). We made three circles of stick-slip chain and one pair of straight stick-slip chain to explain the seismic coupling in this area (Fig.16). To avoid complication, we enclosed HKFL to West Nankai Trough and also enclosed East Nankai Trough to the boundaries of three rotating blocks.

The simulated earthquake activities are shown in Fig.17 which agrees the space time patterns of real historical destructive earthquakes in this area. It is noted that the spatial pattern of rupure zones of large events are stable for a long period. This suggests that the block boundary lines are able to be classified into several segments which generates earthquakes of quite similar sizes in each active periods. This observation should be consistent with the characteristic earthquake model. We also be able to find out the seismic coupling between the block boundaries. The recurrence time of the modeled seismic events at the West Nankai Trough is about 150 or 180 years, consistent with the observations. According to this caterpillar model, the coupling between these block boundaries are strong and stable for a long time. Tis result show us the predictability of the earthquakes is not so small if we only see the destructive earthquakes in long time.

The majour conclusion of this simulation is that the seismic coupling of large events can be represented by this caterpillar model. The recurrence time of active periods is stable, while spatial patterns of seismic events along the block boundaries should vary in long time scale. The caterpillar model studied here should provide a physical basis to the empirical approaches for destructive earthquake prediction.

## 4. References

Bak, P., C. Tang, and K, Wiesenfeld, 1987 Self-organized criticality, *Phys, Rev. A*, **38**, 364 -374.

Bak P., C. Tang, and K. Wiesenfeld, 1987 Self-organized criticality: An explanation of noise, *Phy s. Rev. Lett.*, **59**, 381-384.

Bak, P., and C. Tang, 1989 Self organized criticality, *Phys. Today*, **42**, s27-s28.

Burridge, R., and L. Knopoff, 1967 Model and theoretical seismicity, *Bull. Seismo. Soc. Am.*, **57**, 341-371.

Enya, O., 1901 On aftershocks (in Japanese), *Rep. Earthquake Invest Comm. location*, **35**, 35-36.

Grassberger, P., and I. Procaccia, 1983 Measuring the strangeness of strange attractors, *Physica D* , **9**, 189-208.

Hirabayashi, T., K.Ito, and J. Kigami, 1989 "Goishi" model, revisited (in Japanese), *J. Seismol. Japan*, **42**, 235-237.

Hirata, T., T. Sato, and K. Ito, 1987 Fractal structure of spatial distribution of microfracturing in rock, *Geophys. J. R. Astron. Soc.*, **90**, 369-374.

Hirata, T., 1987 Omori's power law aftershock sequences of microfracturing in rock-fracture experiment, *J. Geophys. Res.*, **92**, 6215-6221.

Huang, J., and D. L. Turcotte, 1990 Evidence for chaotic fault interactions in the seismicity of the an Andreas fault and Nankai trough, *Nature*, **348**, 234-236.

Ito, K and M. Matsuzaki, 1990 Earthquakes as self-organized critical phenomena.*J. Geophys. Res.*, **95**, 6853-6860.

Kagan, Y. Y., & L.Knopoff, 1980 The spatial distribution of earthquakes: The two-point correlation function.*Geophys.J.R.Astron.Soc.*, **62**, 303-320.

Kagan, Y. Y., 1981 Spatial distribution of earthquakes: The three point moment function.*Geophys. J. R. Astron. Soc.*, **67**, 697-717.

Kanamori,H., and D.L.Anderson, 1975 Theoretical basis of some empirical relations in seismology.*Bull. Seismol. Soc. Am.*, **65**,1095.

Kanamori, Y., Kawakami, S. & Yairi, K., 1991a The block structure and Quaternary strike-slip block rotation of central Japan. *Tectonics*, **11**, 47-56.

Kanamori, Y., Kawakami, S. & Yairi, K., 1991b Space-time distribution patterns of destructive earthquakes in central Japan: activity intervals and locations of earthquakes.*Eng. Geol.*, **31**, 209-230.

Kanaori, Y., Kawakami, s. & Yairi, K., 1993 Space-time correlations between inland earthquakes in central Japan and great offshore earthquakes along the Nankai trough: implications for destructive earthquake prediction.*Eng. Geol.*, **33**, 289-303.

Leath, P. L., 1976 Cluster size and boundary distribution near percolation threshold, *Phys. Rev. B*, **14**, 5046-5055.

Mandelbrot, B. B., 1982 *The Fractal Geometry of Nature*, **460**pp., W. H. Freeman, New York.

Matsuzaki, M., & Takayasu, H., 1991 Fractal features of the earthquake phenomenon and a simple mechanical model, *J. Geophys. Res.*, **96**, 19925-19931.

Ogata, Y., 1988 Statistical models for earthquake occurrences and residual analysis for point process, *J. Am. Stat. Assoc.*, **83**, 9-27.

Omori, F., 1894 On the aftershock of earthquakes, *J. Coll. Sci. Imp. Univ. Tokyo*, **7** , 111-200.

Otsuka, M., 1972 A chain-reaction-type source model as a tool to interpret the magnitude-frequency relation of earthquakes, *J. Phys. Earth*, **20**, 35-45.

Sadovskiy, M, A., T. V. Golubeva, V. F. Pisarenko, and M. G. Shnirman, 1984 Characteristic dimensions of rock and hierarchical properties of seismicity, *Izv. Acad. Sci. USSR Phys. Solid Earth, Engl. Transl.*, **20**, 87-96.

Stauffer, D., 1979 Scaling throe of percolation clusters, *Phys. Rep.*, **54**, 1-74.

Stanley, H. E., 1971 Introduction to Phase Transitions and Critical Phenomena, *Clarendon, Oxford.*

Stketee, J. A., 1958 Some geophysical applications of the elasticity theory of dislocations, *Can. J. Phys.*, **36**, 1168-1198.

Takayasu, H., I. Nishikawa, and H. Tasaki, 1988 Power law distribution of aggregation systems with injection, *Phys. Rev. A*, **37**, 3110-3117.

Takayasu, H., and M. Matsuzaki, 1988 Dynamical phase transition in threshold elements, *Phys. Lett. A*, **131**, 244-247.

Tang, C., and P. Bak, 1988a Critical exponents and scaling relations for self-organized critical phenomena, *Phys. Rev. Lett.*, **60**, 2347-2350.

Utsu, T.,1969 Aftershock and earthquake statistics (I). *J.Fac.Sci.Hokkaido Univ.*, **3,1** 29-195.

Utsu, T., 1969 Aftershocks and earthquake statics (I), *J. Fac. Sci. Hokkaido Univ., Ser. VII*, **3**, 129-195.

Utsu, T., 1970 Aftershocks and earthquake statistics (II)*J. Fac. Sci. Hokkaido. Univ., Ser. VII*, **3**, 197-266.

Yoshida, A., and N. Mikami, 1986 Temporal and spatial distribution of aftershocks: A preliminary report(in Japanese), *Mathematical Seismology* (ed. M. Saito), pp. 98-108, Institute of Statistical Mathematics, Tokyo.

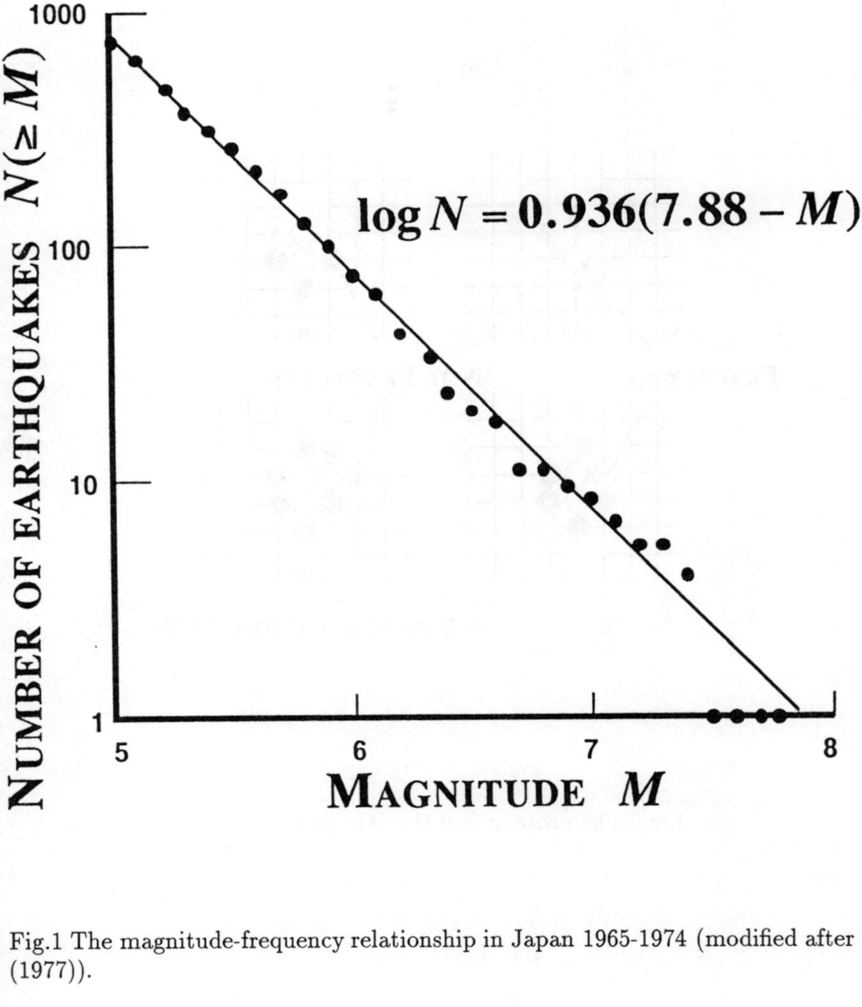


Fig.1 The magnitude-frequency relationship in Japan 1965-1974 (modified after Utsu (1977)).

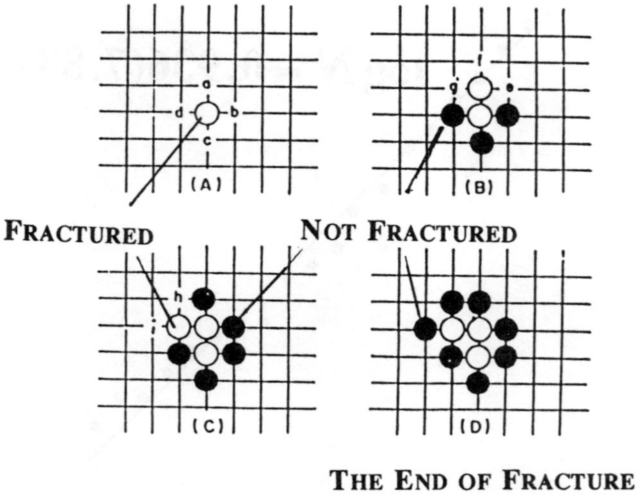


Fig.2 The procedure of the "Go game model".

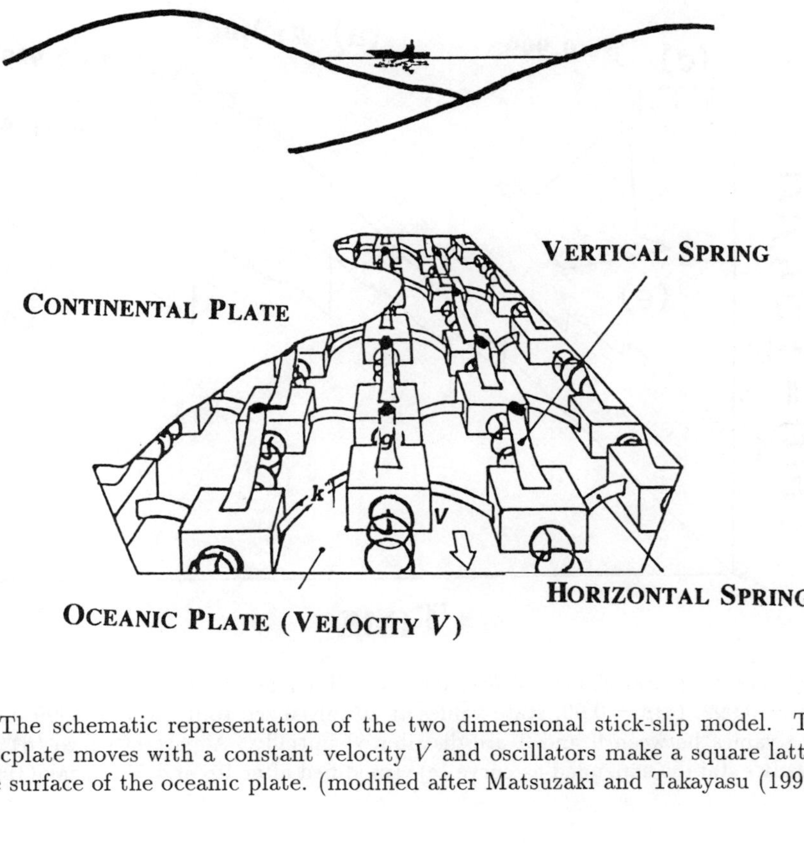


Fig.3 The schematic representation of the two dimensional stick-slip model. The oceanicplate moves with a constant velocity $V$ and oscillators make a square lattice on the surface of the oceanic plate. (modified after Matsuzaki and Takayasu (1991))

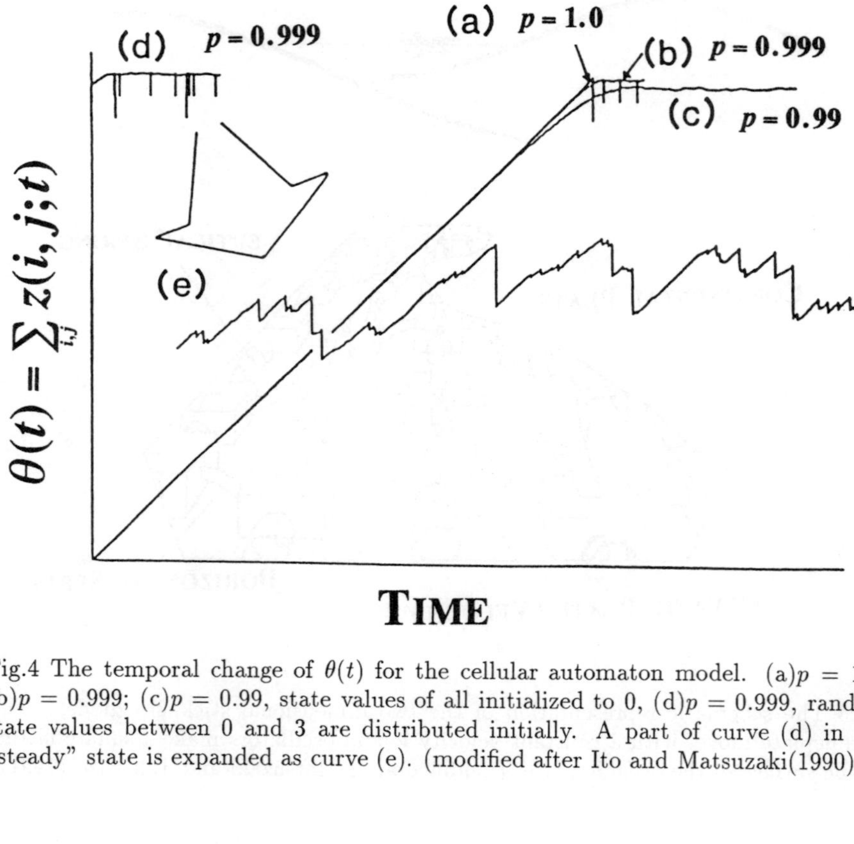


Fig.4 The temporal change of $\theta(t)$ for the cellular automaton model. (a)$p = 1.0$; (b)$p = 0.999$; (c)$p = 0.99$, state values of all initialized to 0, (d)$p = 0.999$, random state values between 0 and 3 are distributed initially. A part of curve (d) in the "steady" state is expanded as curve (e). (modified after Ito and Matsuzaki(1990))

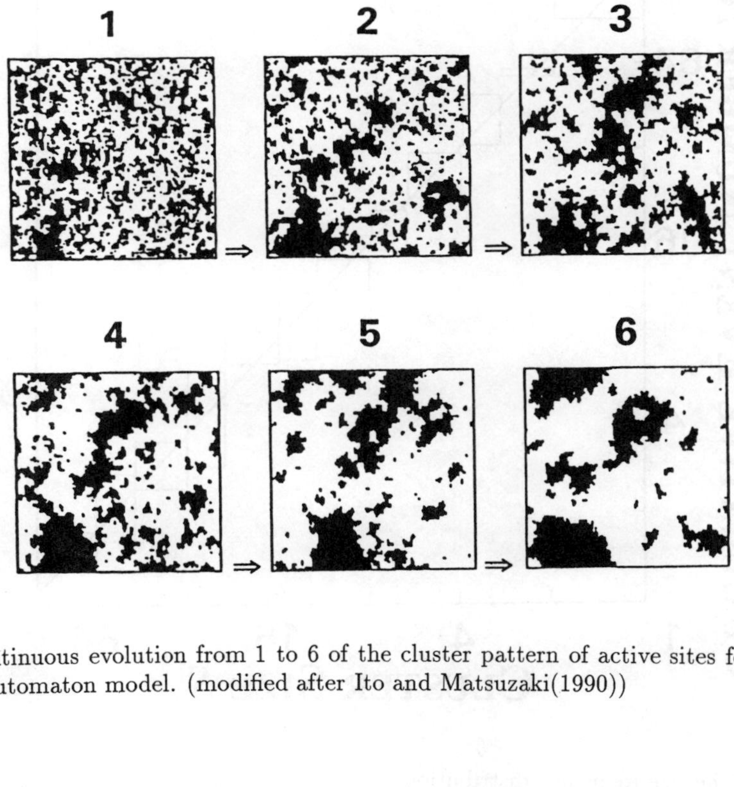


Fig.5 Continuous evolution from 1 to 6 of the cluster pattern of active sites for the cellular automaton model. (modified after Ito and Matsuzaki(1990))

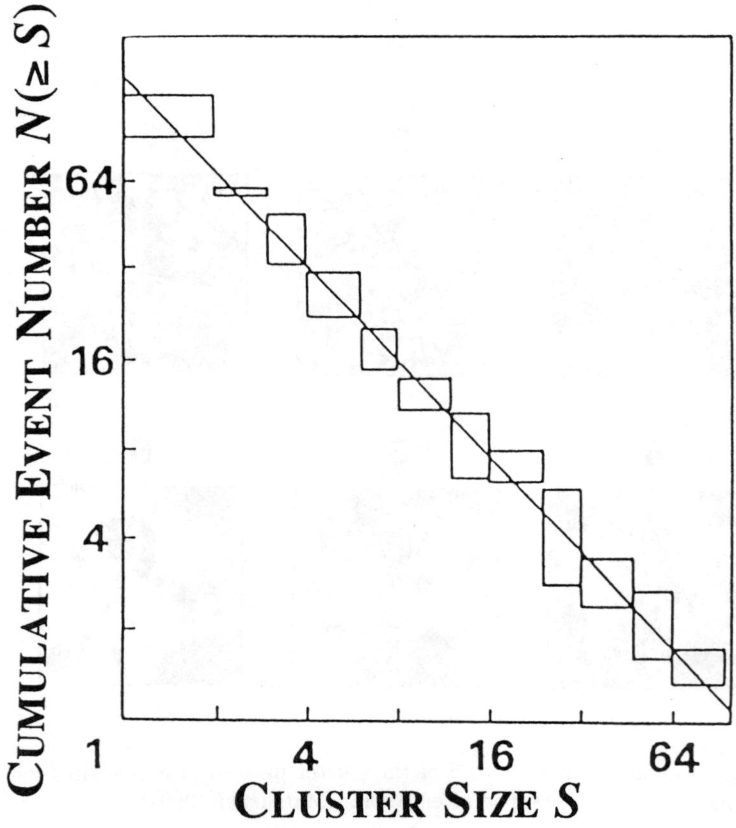


Fig.6 The size-frequency distribution of the cellular automaton model. The straight line has the slope-1.06. (modified after Ito and Matsuzaki(1990))

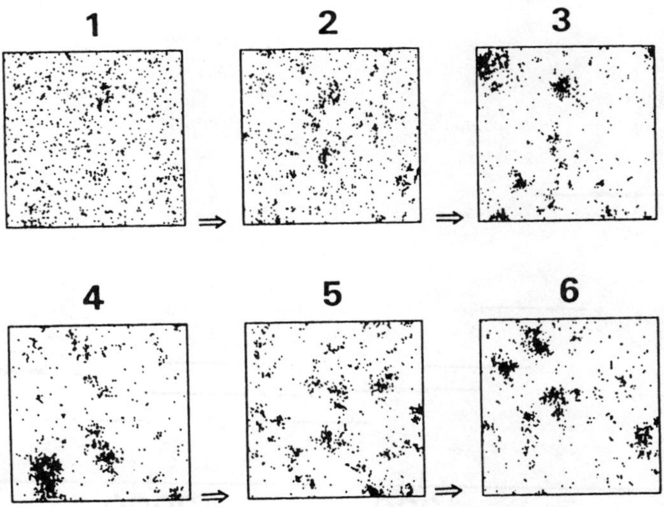


Fig.7 Continuous evolution from 1 to 6 of the epicenter distribution for the modified trigger model. (modified after Ito and Matsuzaki(1990))

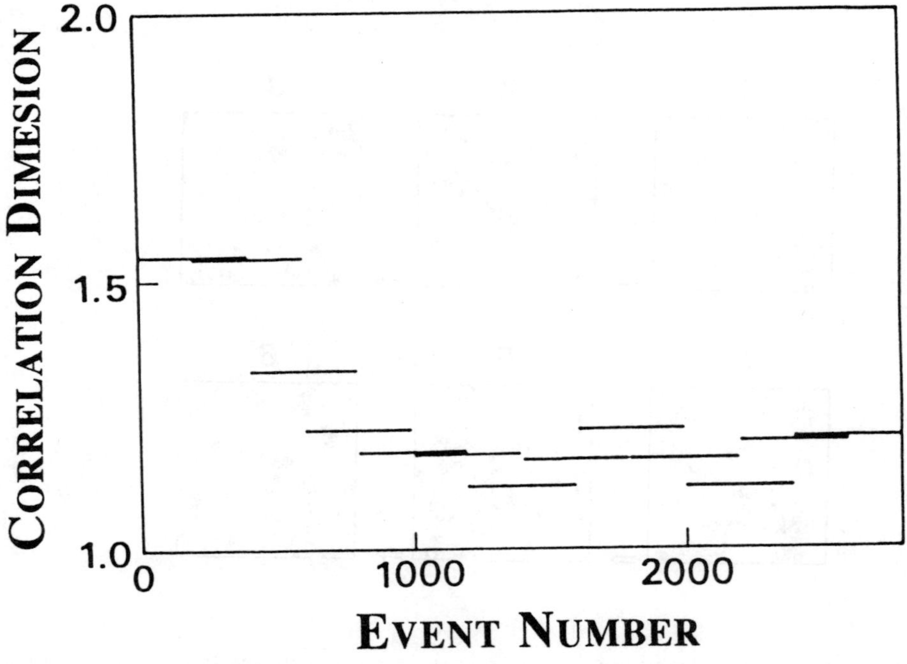


Fig.8 The change of the correlation dimension for epicenter distributions in the modified trigger model. (modified after Ito and Matsuzaki(1990))

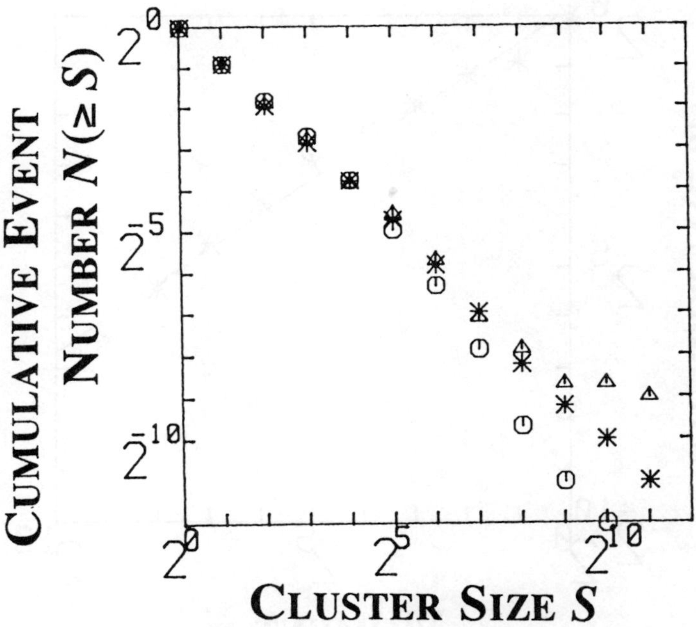


Fig.9 Cumulative distributions of the cluster size of two-dimensional stick-slip model for three typical $d$ values in log-log scale; triangle $d = 0.24 > d_c$; asterisk $d = 0.23 = d_c$; circle $d = 0.22 < d_c$. (modified after Matsuzaki and Takayasu(1991))

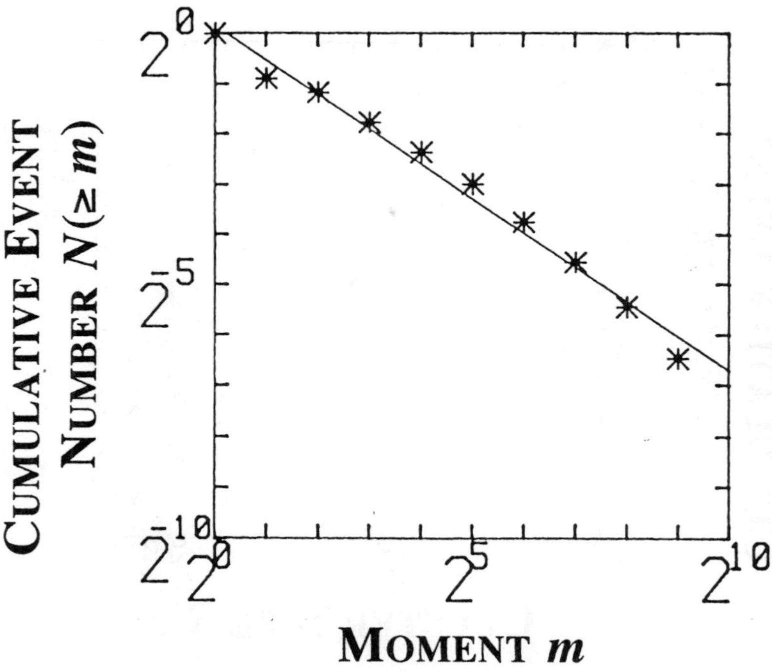


Fig.10 Cumulative distribution of the moment of two-dimensional stick-slip model in log-log scale at the critical state. The exponent is about 0.67. (modified after Matsuzaki and Takayasu(1991))

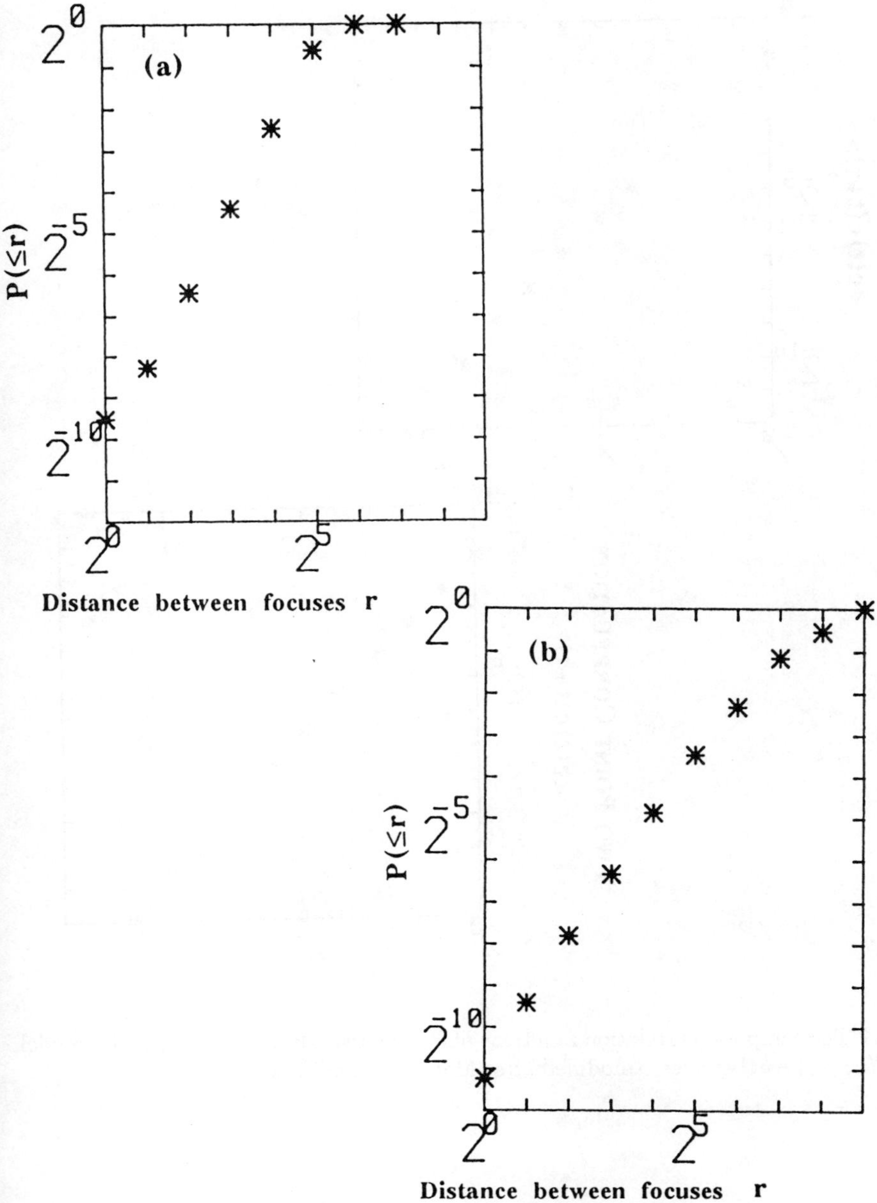


Fig.11 The spatial distributions of the focuses of (a) the two-dimensional stick-slip model and (b) real earthquakes. (modified after Matsuzaki and Takayasu(1991))

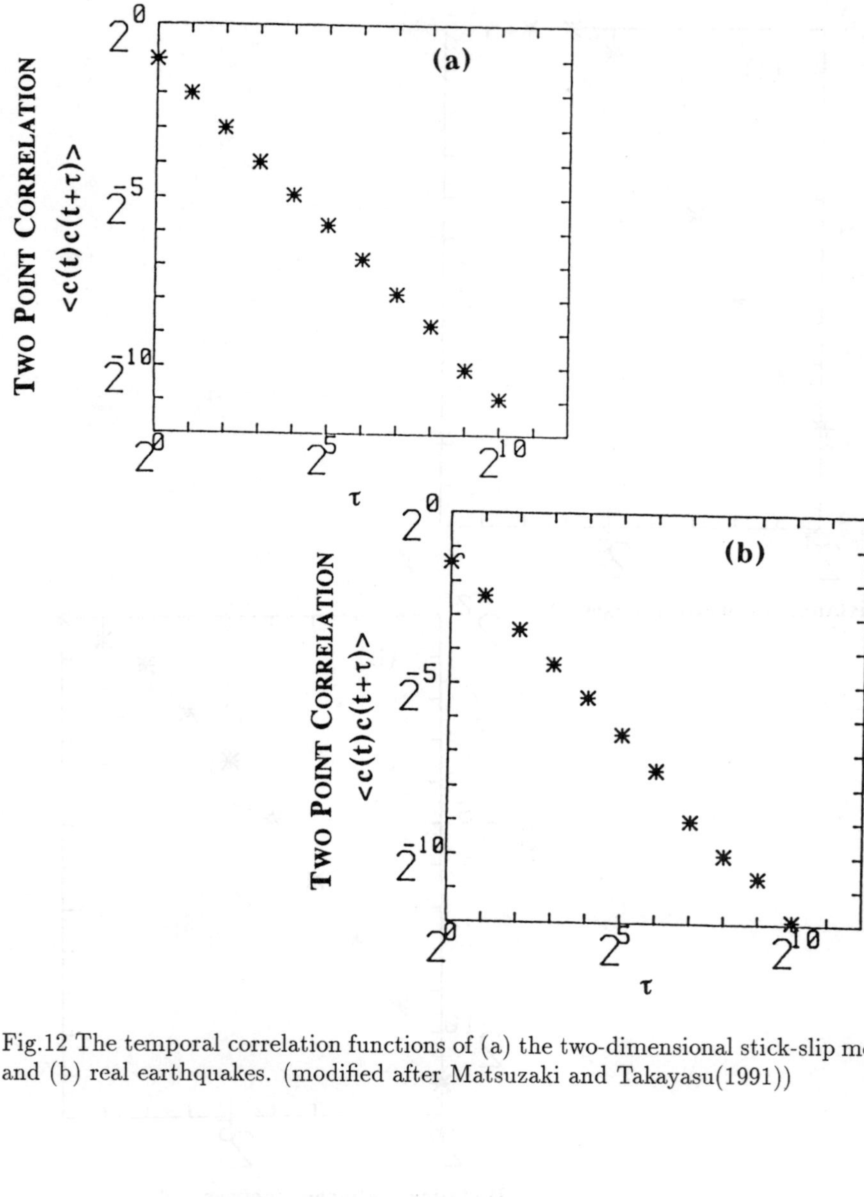


Fig.12 The temporal correlation functions of (a) the two-dimensional stick-slip model and (b) real earthquakes. (modified after Matsuzaki and Takayasu(1991))

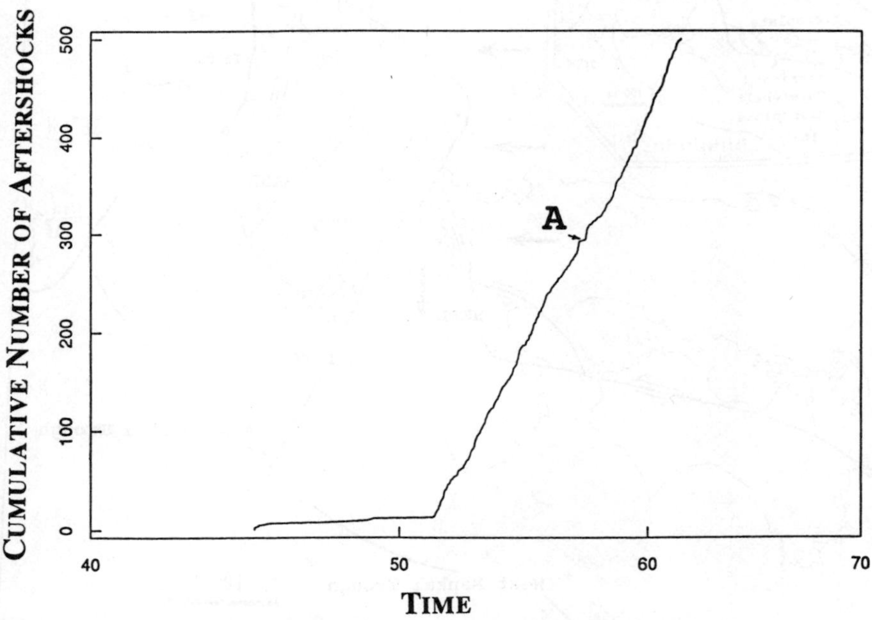


Fig.13 Cumulative number of aftershock after a large event, whose cluster size is comparable to the system size. Note that the time axis is in log scale.

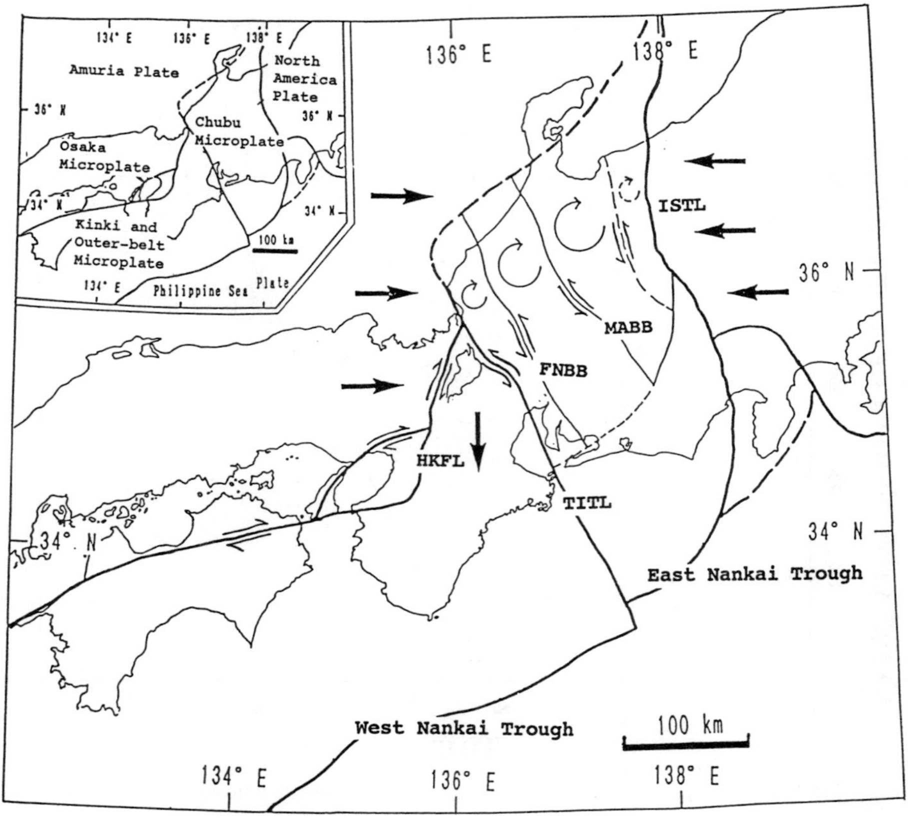


Fig.14 The block rotation model of Southwest Japan. This area is under the pressure due to the plate motion and the rotation of the three or four blocks makes the coupling of the seismic phenomena. (modified after Kanaori et al.(1993))

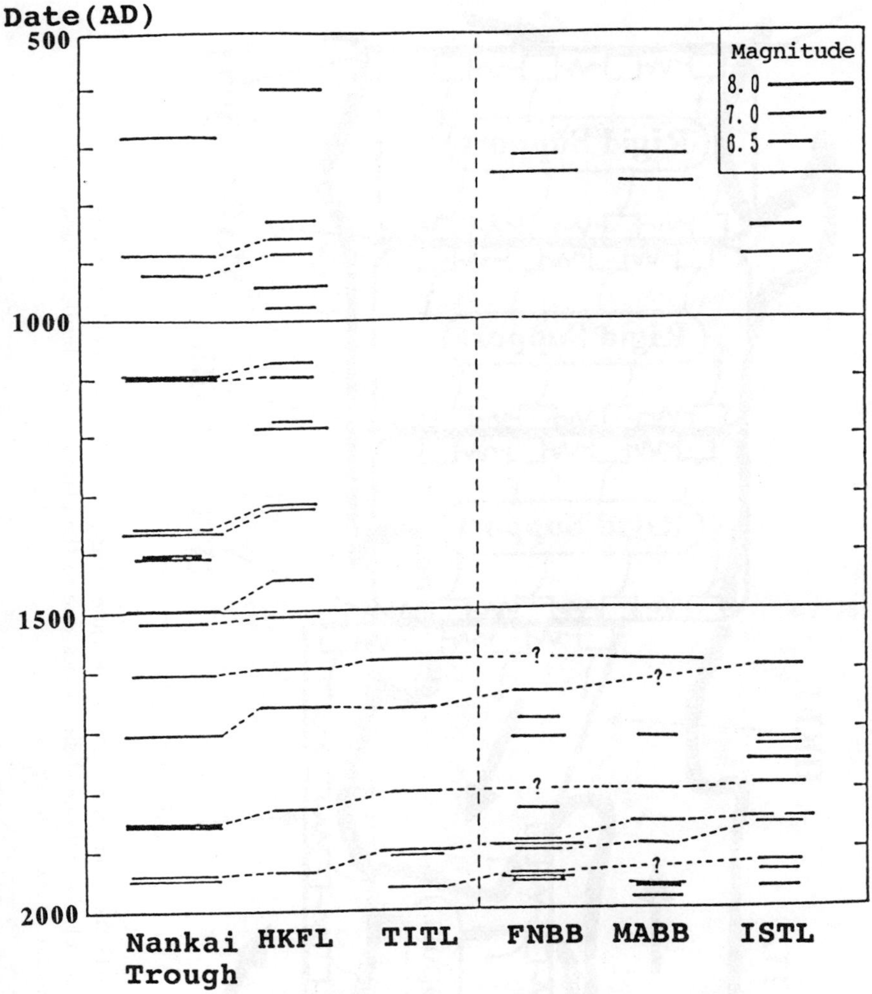


Fig.15 The space-time patterns of the Southwest Japan estimated from the historical earthquakes. Each event is relocated along the block boundaries and the length of each line corresponds to the magnitude of each event. We can find the seismic coupling in this area. (modified after Kanaori et al.(1993))

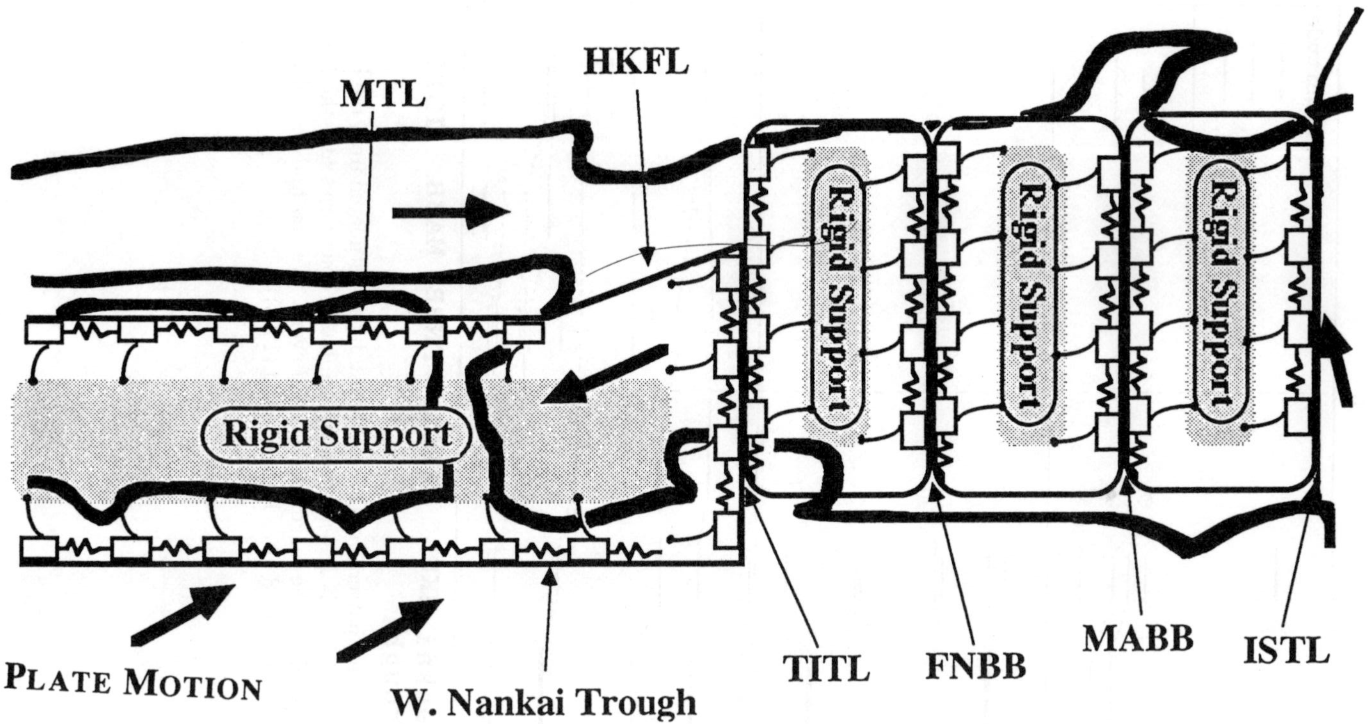


Fig.16 The schematic representation of the caterpillar model. In this model, we make three rotating blocks and one sliding block due to the plate motion. Inside of each block is thought as rigid support.

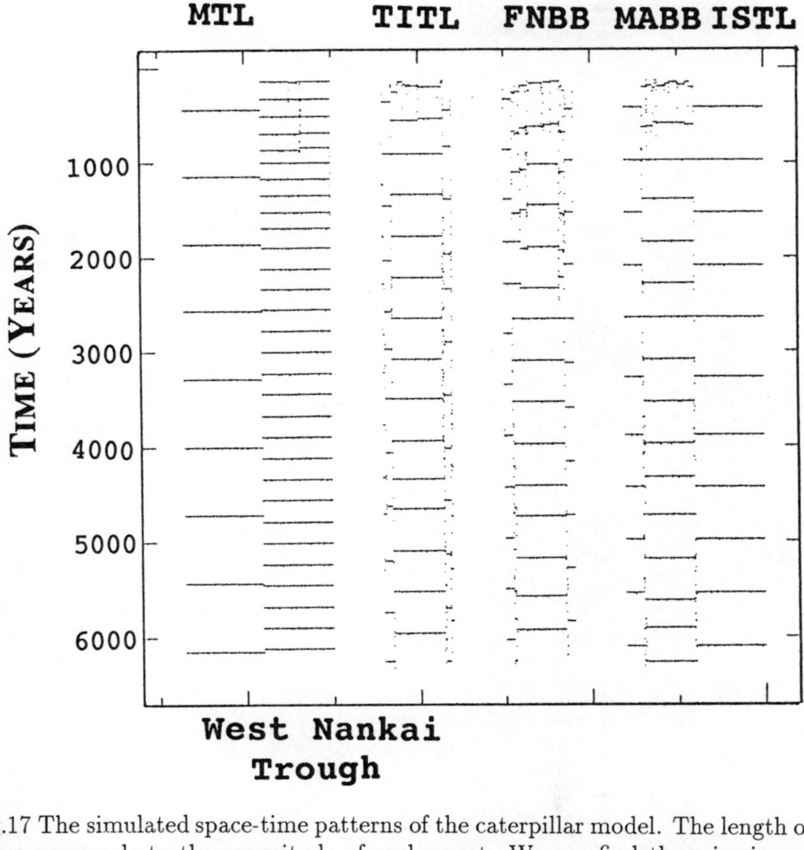


Fig.17 The simulated space-time patterns of the caterpillar model. The length of each line corresponds to the magnitude of each event. We can find the seismic coupling and stable periodicity.

CHAPTER 12

# SINGULAR VECTORS AND THE PREDICTABILITY OF WEATHER AND CLIMATE

T. N. Palmer[1], R. Buizza[1], F. Molteni[1], Y.-Q. Chen[2] and S. Corti[3]
[1]European Centre for Medium-Range WeatherForecasts
Shinfield Park, Reading, Berkshire RG2 9AX, U.K.
[2]Joint Institute for the Study of the Atmosphere and Ocean
Department of Atmospheric Sciences
University of Washington, AK40, Seattle
Washington 98195, U.S.A.
[3]Department of Physics, University of Bologna, via Irnerio
Bologna 461-40126, Italy

**Abstract**

The local instability properties of a chaotic system are determined by the singular vectors and singular values of the dynamical evolution operator, linearized about a finite trajectory portion of the integral curves of the nonlinear equations. Knowledge of these quantities allows an assessment of the reliability of a finite-time forecast from a chaotic system.

After a brief study of the Lorenz model, singular vector analysis is applied to study three predictability issues in atmosphere–ocean dynamics. The first concerns the predictability of weather forecasts of a few days, and singular vector calculations are made from a large-dimensional numerical weather prediction model using an iterative Lanczos algorithm. The second concerns the predictability of El Niño on seasonal to interannual timescales. Here singular vector calculations are made using a coupled ocean–atmosphere model of the tropical Pacific region. Finally we show results from a multi-decadal integration of a medium-resolution quasi-geostrophic model, and discuss the possible relevance of singular vector analysis for the problem of climate change.

## 1. Introduction

The global instability characteristics of a chaotic system are, of course, determined by its positive Lyapunov exponents. They measure the asymptotic exponential divergence of initially close trajectories on the system's strange attractor; its climate.

Such attractors are not amorphous. The Lorenz (1963 *a*) attractor, for example, comprises two distinct régimes or 'butterfly wings'. This structure in turn induces inhomogeneity in the local phase-space divergence of trajectories. Knowledge of the local-phase space instability of a chaotic system allows an assessment of the reliability of a finite-time forecast from a chaotic system. Inhomogeneity of local phase-space predictability also affects the system's variable response to a fixed external forcing.

In this paper, we shall explore the use of singular vector analysis as a means of quantifying the local instability properties of a chaotic system. The singular vectors in question are of the dynamical evolution operator, linearized about a finite trajectory portion of the integral curves of the nonlinear equations. Unlike the eigenvectors of that operator, the singular vectors can usually be chosen to form an orthonormal basis for linear perturbations. The corresponding singular values give local divergence rates, and the singular vector with largest singular value represents the linear perturbation with largest growth over the trajectory portion. Unlike the Lyapunov exponents, the singular values do not necessarily describe exponential growth.

In § 2 we briefly describe some of the formalism of singular vector analysis and outline the close relations to more familiar energy-growth concepts: normal mode instability, upscale quasi-turbulent transfer, and the already-mentioned Lyapunov exponents.

In § 3, after a brief discussion of these concepts in the Lorenz model, we apply the analysis to three prediction problems in atmospheric–ocean dynamics. The first relates to forecasting the predictability of weather on timescales of a few days using complex numerical weather prediction models. The second relates to an analysis of the predictability of the El Niño phenomenon, during which equatorial Pacific sea surface temperatures warm by several degrees and disrupt the normal weather patterns over much of the globe on timescales of a few seasons. The El Niño is an irregular event, occurring every three or four years on average. Its prediction requires coupled ocean–atmosphere models. Finally, we discuss the possible relevance of singular vector analysis for the problem of climate change.

## 2. Basic formalism

We start by considering a general ($M$-dimensional) nonlinear evolution equation,

$$\mathrm{d}\boldsymbol{x}/\mathrm{dt} = \boldsymbol{A}[\boldsymbol{x}]. \tag{1}$$

For example, for the application of this analysis in § *3.1*, (1) represents a complex numerical weather prediction model used at the European Centre for Medium-Range Weather Forecasts (ECMWF) (Simmons *et al.* 1989).

Consider a small perturbation $\boldsymbol{x}'$ of the state vector $\boldsymbol{x}$. For sufficiently short time intervals, its evolution can be described by the linearized approximation

$$\mathrm{d}\boldsymbol{x}'/\mathrm{dt} = \boldsymbol{A}_l \boldsymbol{x}' \tag{2}$$

of (1). $\boldsymbol{A}_l \equiv (\mathrm{d}\boldsymbol{A}/\mathrm{d}\boldsymbol{x})|_{\boldsymbol{x}(t)}$ is the linear evolution operator evaluated on the nonlinear trajectory $\boldsymbol{x}(t)$.

Equation (2) can be written in the integral form

$$\boldsymbol{x}'(t) = \boldsymbol{L}(t, t_0)\boldsymbol{x}'(t_0). \tag{3}$$

The operator $\boldsymbol{L}(t, t_0)$ is referred to as the forward tangent propagator; it maps small perturbations along the (nonlinear) trajectory from an initial time $t_0$ to some future perturbation quantities. For the application to weather prediction, if $\boldsymbol{x}(t_0)$ is the typical error in the initial conditions for a weather forecast, then (2) and (3) hold for approximately three days of integration time.

We now define an inner product $(\boldsymbol{x}; \boldsymbol{y})$ based on total perturbation energy. By using (3) the perturbation norm at time $t$ is given by

$$||\boldsymbol{x}(t)||^2 \equiv (\boldsymbol{x}(t); \boldsymbol{x}(t)) = (\boldsymbol{x}(t_0); \boldsymbol{L}^*\boldsymbol{L}\boldsymbol{x}(t_0)), \tag{4}$$

where $\boldsymbol{L}^*$ is the adjoint of $\boldsymbol{L}$ with respect to the energy inner product.

Unlike $\boldsymbol{L}$ itself, the operator $\boldsymbol{L}^*\boldsymbol{L}$ (sometimes referred to as the Oseledec operator (Abarbanel *et al.* 1991)) is symmetric. Hence its eigenvectors $\boldsymbol{\nu}_i(t_0)$ can be chosen to form a complete orthonormal basis in the $M$-dimensional tangent space of linear perturbations, with real eigenvalues $\sigma_i^2 \geq 0$ (see, for example, Noble & Daniel 1977), i.e.

$$(\boldsymbol{L}^*\boldsymbol{L})\boldsymbol{\nu}_i(t_0) = \sigma_i^2 \boldsymbol{\nu}_i(t_0). \tag{5}$$

At future time $t$, these eigenvectors evolve to $\boldsymbol{\nu}_i(t) = \boldsymbol{L}\boldsymbol{\nu}_i(t_0)$, which in turn satisfy the eigenvector equation

$$(\boldsymbol{L}\boldsymbol{L}^*)\boldsymbol{\nu}_i(t) = \sigma_i^2 \boldsymbol{\nu}_i(t). \tag{6}$$

From (4) and (5),

$$||\boldsymbol{\nu}_i(t)||^2 = (\boldsymbol{\nu}_i(t_0); \boldsymbol{L}^*\boldsymbol{L}\boldsymbol{\nu}_i(t_0)) = \sigma_i^2. \tag{7}$$

Because any $\boldsymbol{x}(t)/||\boldsymbol{x}(t_0)||$ can be written as a linear combination of the set $\boldsymbol{\nu}_i(t)$, it follows that

$$\max_{\boldsymbol{x}(t_0)\neq 0} (||\boldsymbol{x}(t)||/||\boldsymbol{x}(t_0)||) = \sigma_1. \tag{8}$$

Following the terminology of linear algebra, the $\sigma_i$, ranked in terms of magnitude, are here called the singular values of the operator $\boldsymbol{L}$ and the vectors $\boldsymbol{\nu}_i(t)$ are called the singular vectors of $\boldsymbol{L}$. Maximum energy growth over the time interval $t - t_0$ is therefore associated with the dominant singular vector: $\boldsymbol{\nu}_1(t_0)$ at initial time, and $\boldsymbol{\nu}_1(t)$ at optimization time.

Singular vector analysis is a generalization of classical normal mode instability analysis. This can be made explicit by linearizing about a stationary solution, so that normalized eigenvectors $\boldsymbol{\xi}_i$ of $\boldsymbol{A}_l$ with eigenvalues $\mu_i$ give rise to modal solutions $\boldsymbol{\xi}_i \exp[(t-t_0)\mu_i]$ of (2). The integral operator $\boldsymbol{L}(t, t_0)$ can be written as $\exp[(t-t_0)\boldsymbol{A}_l]$, with eigenvectors $\boldsymbol{\xi}_i$ and eigenvalues $\exp[(t-t_0)\mu_i]$.

For application to atmosphere–ocean dynamics, the linear evolution operators associated with realistic stationary basic state flows are generally not normal ($\boldsymbol{L}^*\boldsymbol{L} \neq \boldsymbol{L}\boldsymbol{L}^*$) because of vertical and horizontal shear. In this circumstance, normalized eigenvectors $\boldsymbol{\eta}_i$ and eigenvalues $\theta_i$ of the adjoint operator $\boldsymbol{L}^*$ satisfy the biorthogonality condition

$$(\mu_i - \theta_i^{\mathrm{cc}})(\boldsymbol{\eta}_i, \boldsymbol{\xi}_i) = 0, \tag{9}$$

where cc denotes complex conjugate. This condition ensures that the eigenvalues of an eigenvector/adjoint eigenvector pair that are not orthogonal, must form a complex conjugate pair. The magnitude of the inner product $(\boldsymbol{\eta}_i, \boldsymbol{\xi}_i)$ for such eigenvector pairs depends on the angle, $\alpha_i$, they subtend in phase space.

If an initial disturbance comprises a linear combination of the eigenvectors $\boldsymbol{\xi}_i$ so that

$$\boldsymbol{x}(t) = \sum_i c_i \boldsymbol{\xi}_i e^{\mu_i (t-t_0)}, \tag{10}$$

then from the biorthogonality condition (9)

$$c_i = (\boldsymbol{\eta}_i, \boldsymbol{x}(t_0))/(\boldsymbol{\eta}_i, \boldsymbol{\xi}_i). \tag{11}$$

From (10), the fastest growing normal mode will ultimately dominate the linear combination. Hence for sufficiently long optimization times, the dominant singular vector at optimization time will correspond to the most unstable normal mode. (Because the singular values are real, whilst the normal mode eigenvalues are complex, there is an arbitrary phase factor that has to be defined to make this correspondence precise.)

To maximize the contribution of the first normal mode at optimization time, $c_1$ in (10) should be as large as possible. If $\boldsymbol{x}(t_0)$ equals $\boldsymbol{\xi}_1$ then $c_1 = 1$, which could be highly sub-optimal. If fact, if $\boldsymbol{x}(t_0)$ projects onto $\boldsymbol{\eta}_1$, then $c_1$ is maximized and is given by the projectibility factor $1/(\cos \alpha_1)$ (Zhang 1988).

Hence, for indefinitely long optimization time, the dominant singular vector, at initial time, is determined by the first adjoint eigenvector, whereas the dominant singular vector at optimization time is determined by the first normal mode itself. The singular value will depend on both the e-folding time of the dominant normal mode and its projectibility. For finite optimization time the dominant singular vectors will no longer project onto individual normal mode solutions (and their adjoints), and the amplitude of finite-time instabilities need not be bounded by properties of the dominant normal modes alone.

If, instead of linearizing about a single stationary point on the climate attractor, we consider the other extreme of linearizing about a (time-evolving) trajectory portion which is sufficiently long to approximately cover the entire attractor, then according to the multiplicative ergodic theorem of Oseledec (1968) the singular values again have exponential dependence $\exp[(t-t_0)\lambda_i]$ on optimization time, where the Lyapunov exponents, $\lambda_i$, do not themselves depend on position on the attractor.

The corresponding set of singular vectors $\boldsymbol{\nu}_i(t)$ can be referred to as the Lyapunov vectors. Note that although the Lyapunov exponents are themselves global quantities, the Lyapunov vectors are still defined locally, and thus vary with position on the attractor, and hence with time.

Singular vectors are, in general, not modal. For the application in § *3.1*, their shape evolves not only in geographical space but also in the spectral distribution of energy. This spectral evolution describes, in a linear context, the upscale energy transfer associated with turbulent processes. To study this upscale energy transfer more explicitly we introduce a spectral projection operator $\boldsymbol{P}_{[n_1,n_2]}$, where $[n_1, n_2]$ denotes the total wavenumber interval $n_1 \leq n \leq n_2$. $\boldsymbol{P}_{[n_1,n_2]}$ is defined as

$$\boldsymbol{P}_{[n_1,n_2]}\boldsymbol{x}_n = \begin{cases} \boldsymbol{x}_n & if\ n \in [n_1, n_2], \\ 0 & otherwise. \end{cases} \tag{12}$$

Here $\boldsymbol{x}_n$ is the wavenumber $n$ component of the spherical harmonic expansion of the (atmospheric) state vector. If we wish to find perturbations, initially constrained to be in $[n_3, n_4]$, with maximum energy in $[n_1, n_2]$, these are given by the singular vectors of $\boldsymbol{P}_{[n_1,n_2]}\boldsymbol{L}\boldsymbol{P}_{[n_3,n_4]}$.

When systems with a large number of degrees of freedom (e.g. $O(10^4)$ or more) are considered, the eigenvalue problem (5) and (6) cannot be solved using conventional matrix algorithms. However, iterative techniques provide an alternative possibility if the adjoint propagator has been coded. The power method, whereby a random initial vector is operated on repeatedly by $\boldsymbol{L}^*\boldsymbol{L}$, is an example. A more sophisticated technique such as the Lanczos algorithm (Strang 1986) is required if more than the largest singular vector is required.

## 3. Applications of singular vector analysis

### *3.1. The Lorenz model*

The first application of the above formalism is to the prototype dissipative chaotic system: the Lorenz (1963 *a*) model

$$\dot{X} = -\sigma X + \sigma Y, \tag{13 a}$$

$$\dot{Y} = -XZ + rX - Y, \tag{13 b}$$

$$\dot{Z} = XY - bZ, \tag{13 c}$$

Figure 1 shows the familiar Lorenz attractor, superimposed on which are three ensembles of points integrated from initial conditions on different parts of the attractor. The dispersion of the ensemble identifies the origin as a region of particular unpredictability.

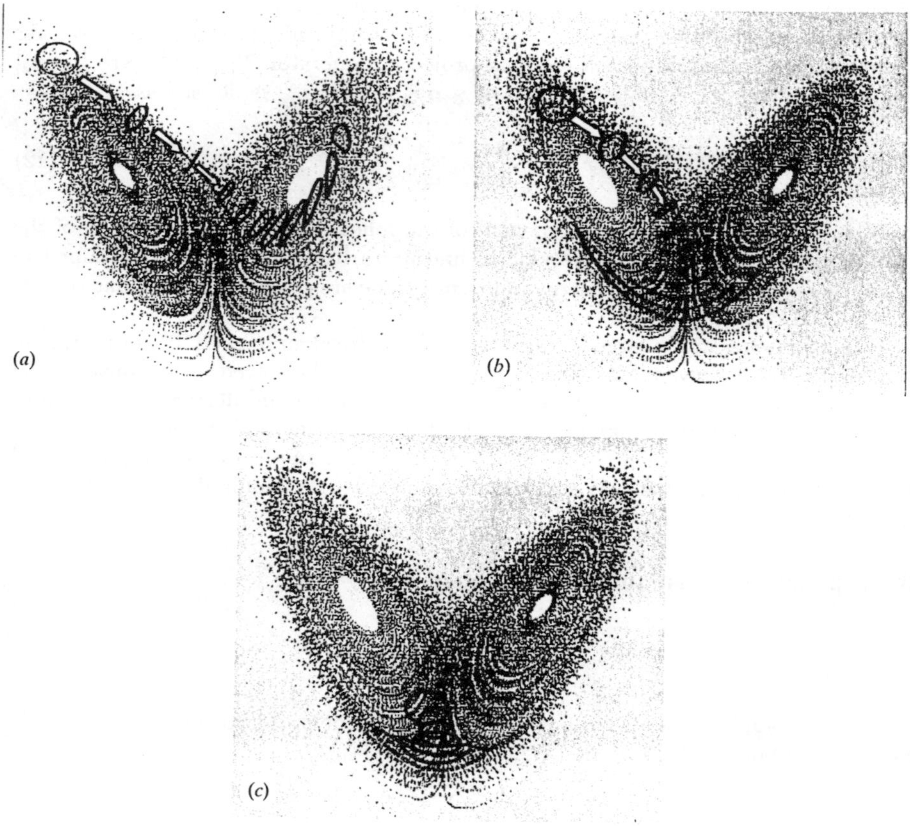


Fig. 1. Phase-space evolution of an ensemble of initial points on the Lorenz attractor, for three sets of initial conditions.

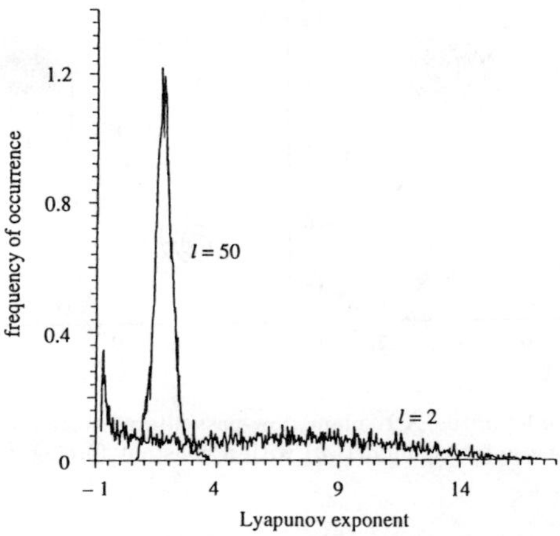


Fig. 2. The distribution of the largest singular value, $\sigma$, (shown as its exponent $\ln \sigma$ for the Lorenz model (13) for two trajectory lengths $L = 2$, $L = 50$. Each distribution is normalized to unity. From Abarbanel *et al.* (1991).

Singular values for the Lorenz model have been computed by a number of authors (see, for example, Mukougawa *et al.* 1991; Abarbanel *et al.* 1991; Trevisan 1993). Figure 2 shows the distribution of exponents of dominant singular values for two choices of the trajectory length. For relatively long trajectory portions, the distribution of singular values is relatively narrow and is clearly asymptoting to the appropriate largest Lyapunov exponent (here about 1.5). For short trajectory portions the distribution of maximum exponents is broad, varying from negative values (i.e. decaying singular vectors) to values over one order of magnitude greater than the fastest growing Lyapunov exponent.

Figure 3*a* shows the state vector probability density function (PDF) of the Lorenz model computed from a long integration. It is effectively symmetric with maxima corresponding to the centroids of the butterfly-wing régimes. In figure 3*b* a constant forcing $F = 1/\sqrt{2}$ is applied to the right-hand side of (13 *a*, *b*). The system continues to be chaotic, though the PDF is now biased to one of the régimes; however, the phase space position of the régime centroids remains essentially unchanged. This behaviour can be understood by noting that the influence of $F$ on the state vector is greatest near the origin where the dominant singular value is large (see, for example, Palmer 1993).

These simple properties of the Lorenz model have some correspondence with more complex models of the atmosphere and oceans; see below.

### *3.2. A numerical weather prediction model*

We discuss here results from a complex primitive equation numerical weather

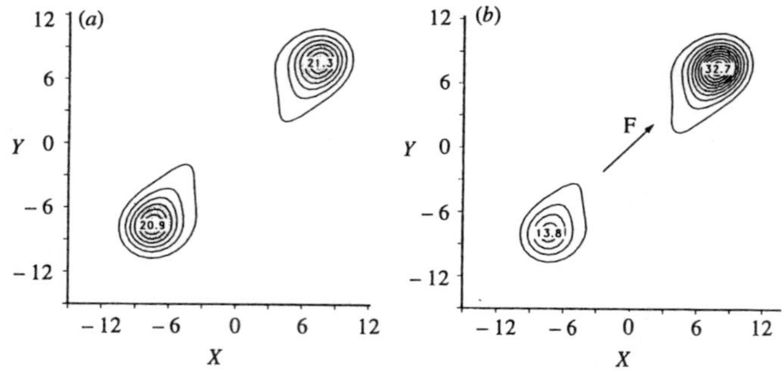


Fig. 3. PDF of the Lorenz model in the $XY$ plane, low-pass filtered to remove oscillations around a régime centroid ($a$) from the unforced model ($b$) with a constant forcing $F_X = F_Y = 1/\sqrt{2}$, $F_Z = 0$.

prediction model used at ECMWF (Simmons *et al.* 1989). The adjoint of this model has been coded (Courtier *et al.* 1991), and estimation of the first 20 or so singular vectors, optimized over periods of a few days, are possible using an iterative Lanczos technique. See Buizza & Palmer (1994) for details. The singular vector calculations described here are made using 19 vertical levels with triangular truncation at either wavenumber 21 or wavenumber 42.

We discuss some of the characteristics of atmospheric singular vectors from a particular case study. During the period under question (9–12 January 1993), there was a vigorous east–west oriented flow across the north Atlantic associated with strong baroclinity. (During this period the oil tanker Braer was grounded on the Shetland Island during heavy storms (Mansfield 1993).)

Figure 4 shows the dominant singular vector calculated using the primitive equation model, for a three day trajectory portion made from initial conditions on 9 January 1993, at three levels in the atmosphere (200 hPa, 700 hPa and 850 hPa) at initial and optimization time. The figure illustrates features which bear qualitative resemblance to idealized normal-mode baroclinic instability (Charney 1947; Eady 1949): the disturbance clearly amplifies at is propagated through the region of maximum baroclinity (where north–south temperature gradients are strongest), and the disturbance shows evidence of westward tilting phase with height, consistent with a northward flux of heat.

On the other hand, the figure also clearly illustrates the non-modal nature of the disturbance and is qualitatively similar in some respects to idealized non-modal optimal baroclinic excitation described by Farrell (1989). At initial time the disturbance is localized near the north Atlantic jet entrance region; at optimization time the disturbance has propagated downstream to Europe. At initial time, maximum disturbance amplitude is located in the lower troposphere, whilst at optimization time maximum amplitude is located in the upper troposphere at the level of maximum winds (understandable qualitatively using wave-activity conservation concepts;

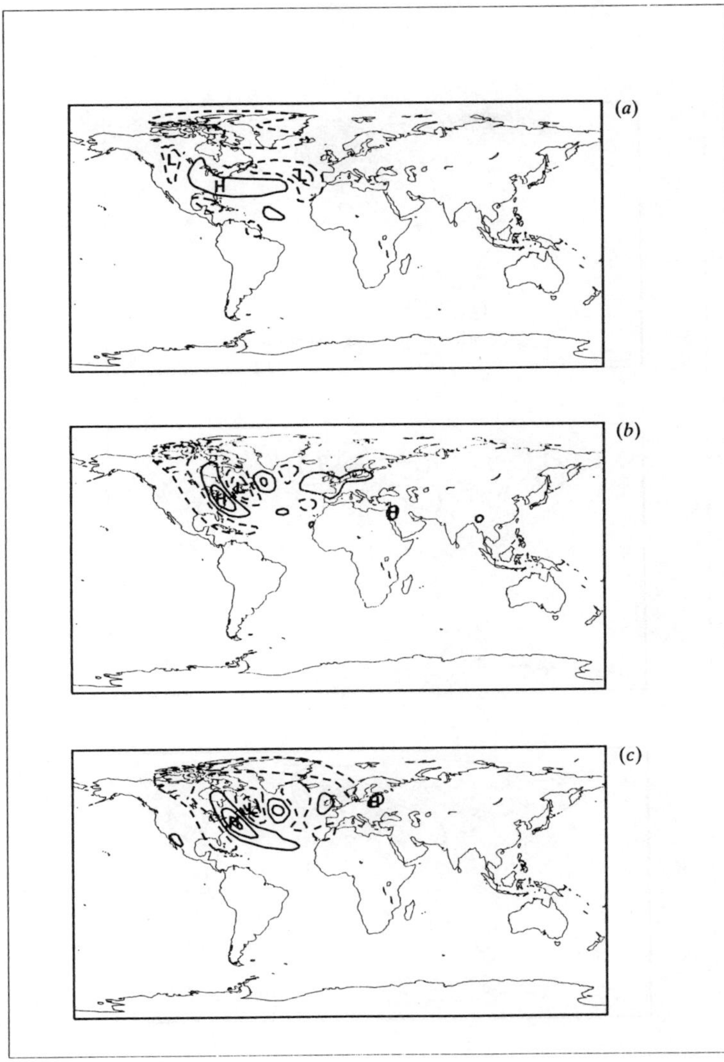


Fig. 4. $(a)$–$(c)$ For description see next page.

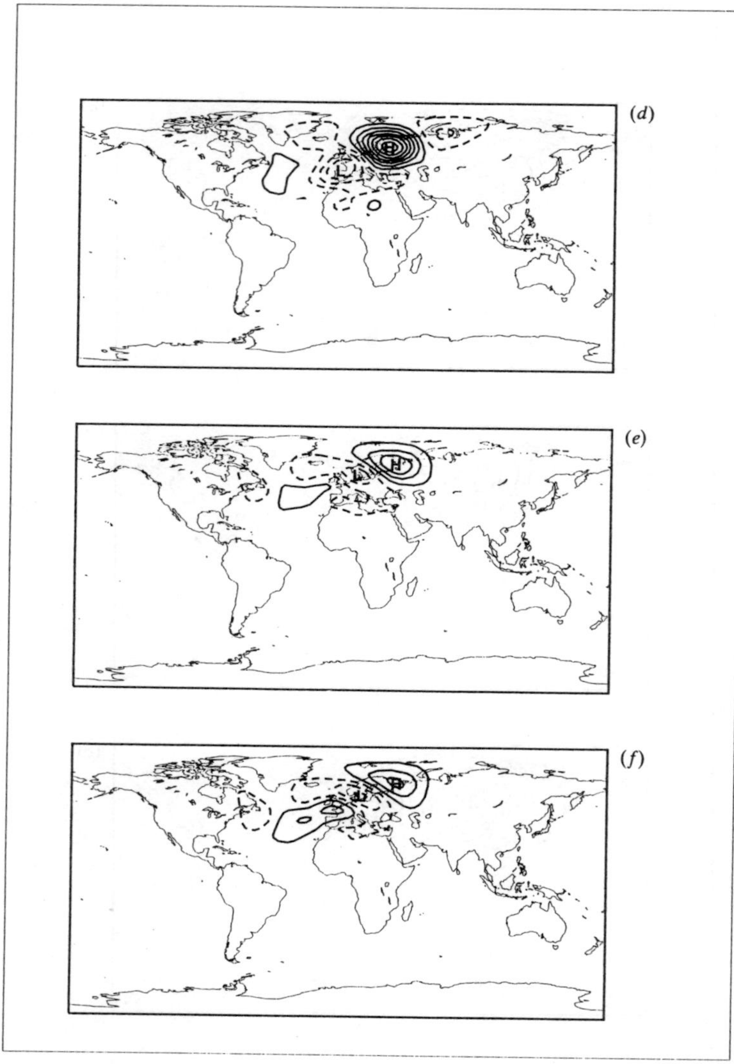


Fig. 4. Streamfunction of the dominant atmospheric singular vector calculated using a primitive equation numerical weather prediction model for a three-day trajectory portion made from initial conditions of 9 January 1993 at three levels (200 hPa $(a)$, $(d)$, 700 hPa $(b)$, $(e)$, 850 hPa $(c)$, $(f)$). Panels $(a)$–$(c)$ show initial time. Panels $(d)$–$(f)$ show optimization time. Contour interval at optimization time is 20 times larger than at initial time.

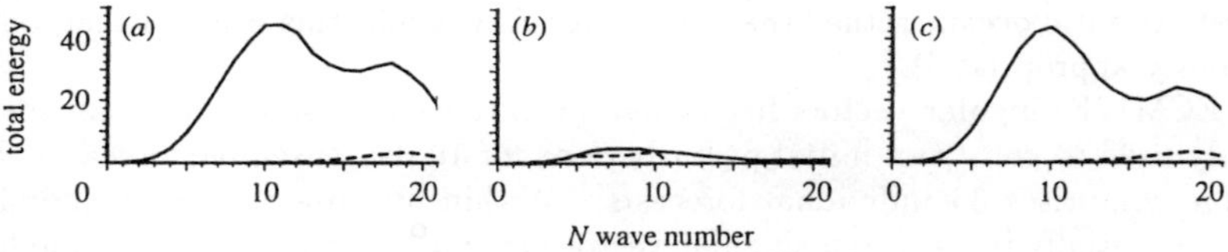


Fig. 5. Energy distribution of three-day singular vector from 9 January 1993 as a function of (total) wavenumber. Dashed lines show initial time with values multiplied by 20 in order to be visible. Solid lines show optimization time. ($a$) For singular vector shown in figure 4. ($b$) For singular vector with energy optimized for wavenumbers 0–10, and constrained to wavenumbers 0–10 at initial time. ($c$) For singular vector with energy optimized for wavenumbers 0–10, and constrained to wavenumbers 11–20 at final time.

see Buizza & Palmer 1994). Finally, the horizontal scale of the initial disturbance is noticeably smaller at initial time than at optimization time.

This final aspect of non-modality is explored further in figure 5 which shows the energy distribution of the singular vector at initial and final time, as a function of total wavenumber. Figure 5$a$ shows the spectral distribution of the disturbance shown in figure 4 peaking near the truncation limit at initial time (dashed line) and at about wavenumber 10 at optimization time. This upscale energy transfer can occur because the basic state (unlike those in many idealized calculations) is itself an unrestricted solution to the equations of motion, and in particular contains scales comparable with those in the disturbance field. This allows triad interactions between two disturbance scales and a third trajectory scale.

Figure 5$b$, $c$ shows the spectral distribution of two further singular vector calculations made using this trajectory. In these further calculations, the spectral projection operator has been applied both at initial and optimization time. For both calculations the operator at optimization time maximizes energy between wavenumbers 0 and 10. The initial perturbation is constrained at wavenumbers 0–10 in figure 5$b$ and at wavenumbers 11–20 in figure 5$c$.

The results are quite dramatic. Constraining the perturbation to have the same energy distribution in wavenumber space at initial and final time (which a normal mode solution, if it exists, must have), severely restricts perturbation growth. On the other hand, constraining the perturbation at initial and final time to have energy in non-overlapping wavenumber intervals hardly restricts energy growth at all.

The result of this calculation illustrates the butterfly effect in its original sense (Lorenz 1963 $b$), i.e. that small-scale initial disturbances will ultimately have an overwhelming influence on large-scale disturbances. This is in addition to the commonly perceived meaning of the butterfly effect that small-amplitude initial disturbances will ultimately have an overwhelming effect on large-amplitude disturbances. (In fact, Lorenz refers to the influence on the weather of a flap of a sea-gull's rather than a butterfly's wings. In view of the location of the initial singular vectors over the west Atlantic ocean, rather than, say, over Amazonia, this original metaphor is fortuitously appropriate!)

At ECMWF singular vectors from these primitive equation calculations are now routinely used to construct initial perturbations for 10-day ensemble forecasts. Each ensemble comprises 33 individual forecasts. Within an ensemble, each individual forecast essentially integrates a singular vector beyond the range for which the linear approximation is valid (about 3 days). Figure 6 shows the ensemble dispersion of two ensemble forecasts made one week apart. The first example shows large dispersion, the second small. From a practical point of view a weather forecaster would be warned that any single deterministic prediction from the first example is unlikely to be reliable, whereas any prediction from the second example is likely to be relatively skilful. The skill of the operational ECMWF forecast is also shown for these two cases confirming the ensemble predictions of reliability. Using these ensemble predictions, ECMWF now routinely issues dynamically based probabilistic weather forecasts out to 10 days.

From a meteorological point of view, both cases in figure 6 involved a transition in flow type. In some sense these mimic ensemble predictions (figure 1 $c, a$) in the Lorenz model, which also showed unpredictable and predictable régime transitions.

### *3.3. A coupled tropical ocean–atmosphere model*

There is currently much excitement in the climate community about the feasibility of making predictions about the natural year-to-year fluctuations of the climate system. Much of this excitement has stemmed from the demonstration that the El Niño/Southern Oscillation phenomenon is predictable up to a year or so in advance using relatively simple numerical models of the atmosphere and ocean coupled together (Zebiak & Cane 1987). El Niño is the name given to an irregular oscillation in the sea surface temperature of the tropical Pacific ocean. The Southern Oscillation denotes a corresponding atmospheric perturbation in the surface pressure field above the equatorial Pacific. According to Münnich *et al.* (1991), long-term variability of the El Niño/Southern Oscillation phenomenon is intrinsically chaotic (independent of the chaotic nature of weather itself).

It is known that during years in which the El Niño phenomenon is active, much of the world's weather patterns are significantly disturbed, leading to drought in some parts of the world, floods in other parts and to overall global warming (IPCC 1992). For example, the warming in surface temperature averaged over the Northern Hemisphere is related to the existence of atmospheric teleconnections arising from exitation of atmospheric planetary-scale wave patterns by anomalous latent-heat release over the El Niño area closely related to the patterns shown in figure 8. These teleconnections lead to a bias in extratropical régime (or weather-pattern) statistics, similar to that shown in figure 3$b$ for the Lorenz model.

The skill of El Niño forecasts made with coupled ocean–atmosphere models is seasonally dependent (Cane *et al.* 1986). Typically, seasonal forecasts beginning in spring tend to be less skilful than forecasts beginning, for example, in autumn. Blu-

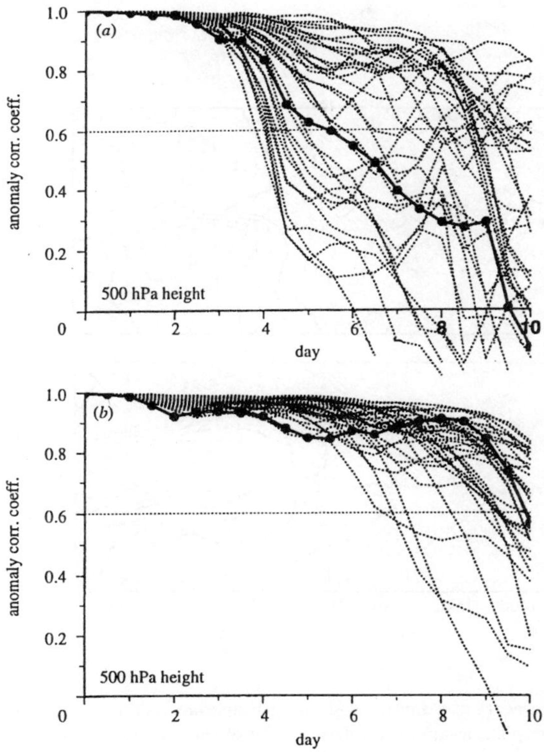


Fig. 6. Ten-day ensemble forecast dispersion as measured by anomaly correlation between 500 hPa geopotential height of control and perturbed forecasts, over Europe (light lines). Skill of 10-day control forecast (heavy line), (*a*) forecasts from 30 October 1993, (*b*) forecasts from 13 November 1993.

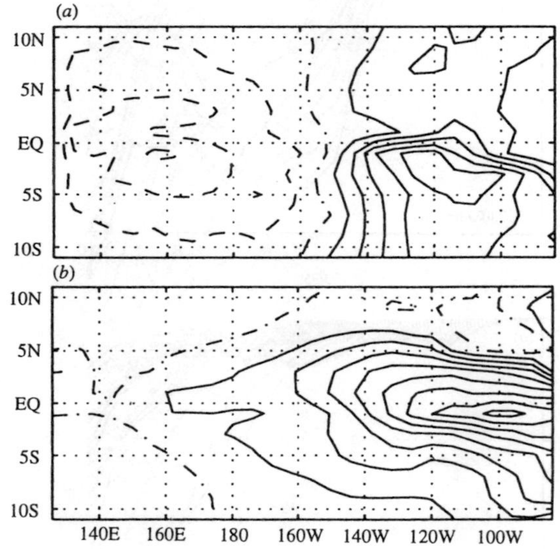


Fig. 7. Sea surface temperature of a dominant six-month singular vector at initial, $(a)$, and optimization, $(b)$, time from a coupled ocean–atmosphere model of the tropical Pacific. Starting conditions are in April, and the basic state is constructed from climatology.

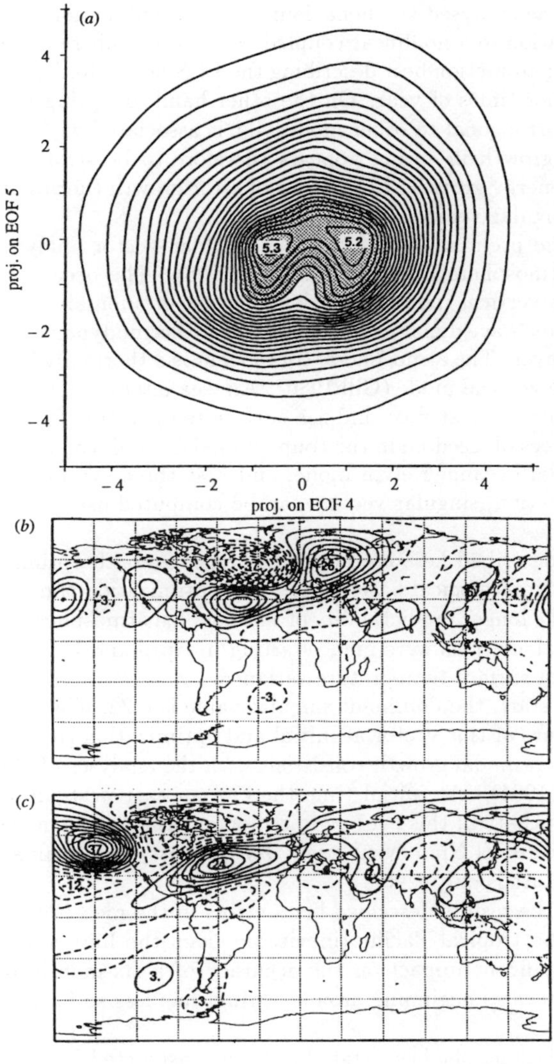


Fig. 8. ($a$) Two-dimensional cut of the PDF from 100 consecutive winters of a multi-decadal integration in a T21L3 quasi-geostrophic model. ($b$), ($c$) The empirical orthogonal functions from the model, used to define the axes in ($a$). The PDF is bimodal along the axis given by ($b$).

menthal (1991) has analysed the behaviour of the eigenfunctions of a linear Markov model approximation to a nonlinear coupled ocean–atmosphere model. He finds that in summer the eigenfunction best describing the El Niño oscillation has a larger eigenvalue than at other times of year. On the other hand, in spring this El Niño eigenfunction is least orthogonal to other modes, i.e. is associated with large projectibility (see § 2). Error growth in spring appears therefore to be strongly associated with non-self-adjoint energy growth. Both types of error growth (normal and non-normal) are implicit in singular vector analysis.

We show some preliminary results of a singular vector analysis applied to the coupled ocean–atmosphere model of Battisti (1988). The ocean component of this model is a single vertical mode tropical Pacific basin anomaly model, governed by linear shallow water wave dynamics. The nonlinear thermodynamics are only active in a surface mixed layer. The atmospheric component is a thermally forced steady linear model with single vertical mode (Gill 1980). Air–sea interactions are nonlinear, given by surface wind stress, heat flux, and sea surface temperature (SST). The number of independent degrees of freedom in the coupled model is reduced to 420 by considering only the equatorial oceanic Kelvin mode, and first three symmetric Rossby modes. With such a reduction, singular vectors can be computed using conventional matrix algorithms.

Calculations described here are performed for six-month optimization times, linearized about a climatological basic state. The optimization time is greater than in § *3.2* because of the longer predictability of this coupled atmosphere–ocean oscillation. Singular vector calculations were made starting in April and October, and are based on maximizing sea surface temperature variance.

For the April start, the dominant singular value is 6.22. The structure of the sea surface temperature of this vector at initial and optimization time is shown in figure 7. It illustrates non-modal growth consistent with the analyses of Blumenthal (1991) and Xue *et al.* (1994). At initial time there is an east-west dipole in sea surface temperature perturbation; this evolves to a monopole structure in the eastern Pacific similar to the mature El Niño event itself. By contrast, the dominant singular value for the October start is only 1.96.

According to the results described here, errors in the measurement of sea surface temperature in the tropical Pacific (inevitable given the limited data in that area) would have a significant impact on the accuracy of a six-month forecast of the El Niño event for forecasts beginning around springtime; less so for forecasts beginning in autumn.

In these calculations, the basic state has been constructed from climatology. More generally the predictability of tropical Pacific sea surface temperature is a function of the actual ocean–atmosphere state, particularly with respect to the El Niño oscillation. As with the calculations in § *3.2*, these oceanic singular vectors can be readily generalized to include observed variations in basic-state flow. Such calculations are in progress.

*3.4. Climate change*

One atmospheric forecast which has attracted more attention worldwide by governments and the public alike, is of a trend to warmer temperatures as anthropogenic greenhouse gases, such as $CO_2$, increase. Such warming necessarily occurs in a one-dimensional radiative equilibrium atmosphere without dynamics. However, one can ask the question of how chaotic nonlinear dynamics may compromise such radiative equilibrium calculations.

As discussed in § *3.1*, for the Lorenz model, the impact of a small externally imposed forcing on a dynamical system is potentially greater on parts of the attractor where singular values are large. On the other hand the response to this forcing, given by changes in the system PDFs, may be largely determined by the values of the PDFs at régime centroids, well removed (in phase-space) from regions of maximum instability. Whether the response to the imposed forcing is an increase in the frequency of one régime rather than another, may depend quite strongly on the local attractor geometry in the regions of strong instability.

In principle it would be possible to run the primitive equation model above over long climatic periods of time, and obtain detailed knowledge of the structure of the climate attractor, and associated estimates of the singular-vector structure of this attractor. However, this is overwhelmingly too computationally expensive at present. On the other hand, excessive truncation of the atmospheric equations of motion (say to less that 100 degrees of freedom) produces results which may depend excessively on quasi-resonant properties of low-order models.

However, it is now feasible to run quasi-geostrophic models with reasonable resolution over century periods of time, and compute local singular vector structures. Figure 8*a* shows the PDF from a 100 consecutive winter sample of a 1200 perpetual winters integration of a 3-level T21 quasi-geostrophic model (Marshall & Molteni 1993). Singular vectors from this model can be estimated using conventional matrix eigenvector techniques (Molteni & Palmer 1993). The PDF is estimated in a phase-space plane spanned by two of the dominant empirical orthogonal functions of the model. It can be seen that the PDF during this chosen period is bimodal along an axis that corresponds to the intensity of the north Atlantic jet (figure 8*b*). These two model régime centroids would generate different surface temperature fields, both over the Atlantic and Europe. (In fact the calculations described in § *3.2* were made when the atmosphere was in one of these régimes with strong north Atlantic westerlies.) The impact of a small external forcing would lead to either an increase or decrease in one régime over another. Which one increases may depend crucially on the geometry of the attractor at some possible unstable saddle point between the régimes. As such, the fact that one régime is associated with warmer temperatures than the other is not directly relevant in determining whether its PDF increases or decreases. Calculation of singular vectors associated with finite trajectory portions that make up the PDF in figure 7*a* are currently under way.

These general remarks also apply to considerations of changes to global temperature. For example, as mentioned above, global-mean temperature tends to be warmer than average during years in which the El Niño event is active. If the impact of enhanced greenhouse gases was to alter the frequency of El Niño events, then over decadal and longer timescales, this would contribute to a component of global climate change. We have briefly discussed the structure and growth of singular vectors in the tropical Pacific that lead to El Niño. Projection of radiative forcing from enhanced $CO_2$ onto these geographically localized singular vectors (particularly during the northern spring and summer) may be critical in determining this component of global temperature change.

These considerations are, of course, implicit in the calculations of climate change made with our best comprehensive global circulation models (IPCC 1992). These models consistently predict global warming, qualitatively similar to the radiative equilibrium calculations. The models are our best hope at quantifying climate change, and their predictions must therefore treated seriously. On the other hand, such models are not free of systematic error; not least they are generally poor in simulating the observed statistics of the El Niño/Southern Oscillation phenomenon. Analysis of the processes by which atmosphere–ocean dynamics influence climate is an important component of research on climate change.

## 4. Conclusions

We have described the formalism of singular vector analysis which allows one to study local phase-space instability of inhomogeneous chaotic attractors. Using adjoint models and iterative eigenvector solvers, we are able to apply the formalism to large-dimensional systems. In this paper we have applied the method to the study of the predictability of weather and climate. However, in principle, the techniques may also be of use in studies of the predictability of other large-dimensional chaotic systems as occur, for example, in the natural sciences, engineering, economics and elsewhere.

## References

Abarbanel, H. D. I., Brown, R. & Kennel, M. B. 1991 Variation of Lyapunov exponents on a strange attractor. *J. Nonlinear Sci.* **1**, 175–199.

Battisti, D. S. 1988 Dynamics and thermodynamics of a warming event in a coupled tropical atmosphere–ocean model. *J. atmos. Sci.* **45**, 2889–2919.

Blumenthal, B. 1991 Predictability of a coupled ocean–atmosphere model. *J. Climate* **4**, 766–784.

Buizza, R. & Palmer, T. N. 1994 The singular vector structure of the atmospheric general circulation. *J. atmos. Sci.* (In the press).

Cane, M. A., Zebiak, S. & Dolan, S. 1986 Experimental forecasts of El Niño. *Nature, Lond.* **321**, 827–832.

Charney, J. G. 1947 The dynamics of long waves in a baroclinic westerly current. *J. Meteor.* **4**, 135–163.

Courtier, P., Freydier, C., Geleyn, J.-F., Rabier, F. & Rochas, M. 1991 The Arpege project at Météo, France. In *Proc. ECMWF Seminar on Numerical methods in atmospheric models*, vol. 2, pp. 193–231. Reading, U.K.: ECMWF.

Eady, E. T. 1949 Long waves and cyclone waves. *Tellus* **1**, 33–52.

Farrell, B. F. 1989 Optimal excitation of baroclinic waves. *J. atmos. Sci.* **46**, 1193–1206.

Gill, A. E. 1980 Some simple solutions for heat-induced tropical circulation. *Q. Jl R. met. Soc.* **106**, 447–462.

IPCC 1992 *Climate change, the supplementary report to the IPCC scientific assessment.* Cambridge University Press.

Lorenz, E. N. 1963 *a* Deterministic nonperiodic flow. *J. atmos. Sci.* **20**, 130–141.

Lorenz, E. N. 1963 *b* The predictability of hydrodynamic flow. *Trans. New York Acad. Sci. 2* **25**, 409–432.

Mansfield, D. A. 1993 The storm of 10 January 1993. *Met. Mag.* **122**, 140–146.

Marshall, J. & Molteni, F. 1993 Toward a dynamical understanding of planetary-scale flow régimes. *J. atmos. Sci.* **50**, 1792–1818.

Molteni, F. & Palmer, T. N. 1993 Predictability and finite-time instability of the northern winter circulation. *Q. Jl R. met. Soc.* **119**, 269–298.

Mukougawa, H., Kimoto, M. & Yoden, S. 1991 A relationship between local error growth and quasi-stationary states in the Lorenz system. *J. atmos. Sci.* **48**, 1231–1237.

Münnich, M., Cane, M. A. & Zebiak, S. E. 1991 A study of self-excited oscillations in a tropical ocean–atmosphere system. Part II: Nonlinear cases. *J. atmos. Sci.* **48**, 1238–1248.

Noble, B. & Daniel, J. W. 1977 *Applied linear algebra.* Prentice-Hall.

Oseledec, V. I. 1968 A multiplicative ergodic theorem. Lyapunov characteristic numbers for dynamical systems. *Trudy Mosk. Mat. Obsc.* **19**, 197.

Palmer, T. N. 1993 Extended-range atmospheric prediction and the Lorenz model. *Bull. Am. met. Soc.* **74**, 49–65.

Simmons, A. J., Burridge, D. M., Jarraud, M., Girard, C. & Wergen, W. 1989 The ECMWF medium-range prediction models development of the numerical formulations and the impact of increased resolution. *Meteorol. atmos. Phys.* **40**, 28–60.

Strang, G. 1986 *Introduction to applied mathematics.* Wellesley-Cambridge Press.

Trevisan, A. 1993 Impact of transient error growth on global average predictability measures. *J. atmos. Sci.* **50**, 1016–1028.

Xue, Y., Cane, M. A., Zebiak, S. E. & Blumenthal, M. B. 1994 On the prediction of ENSO. A study with a low order Markov. *Tellus.* (In the press.)

Zebiak, S. E. & Cane, M. A. 1987 A model El Niño–Southern Oscillation. *Mon. Wea. Rev.* **115**, 2262–2278.

Zhang, Z. 1988 The linear study of zonally asymmetric barotropic flows. Ph.D. thesis, University of Reading, U.K.

CHAPTER 13

# PREDICTION AS A CRITERION FOR CLASSIFYING NATURAL TIME SERIES

George Sugihara
Scripps Institution of Oceanography
University of California at San Diego
La Jolla, U.S.A. 92093-0202

**Abstract**

During the past decade, there has been a growing trend to view time series as products of dynamical systems. This viewpoint has proven to be particularly useful in stimulating debate and insight into the nature of the underlying generating mechanisms. It has provoked investigations into such questions as: 1) are the underlying dynamics stable or are they chaotic? 2) to what extent does the complexity of a particular time signature come from the nonlinear interaction of variables versus high dimensional (stochastic) effects? 3) how important is observational noise as a source of complexity in a time series?

Here I review some of the issues concerning the use of prediction in the detection of nonlinearities and possible chaos. In particular, I discuss the phenomenon of 'high dimensional' or 'stochastic chaos,' and illustrate it with a simple example. It is suggested that this phenomenon is likely to be the most ubiquitous form of chaos in nature. With stochastic chaos it is shown how recent attempts to measure meaningful Lyapunov exponents for natural data (eg. Ellner 1991, Turchin & Taylor 1992, Hastings et. al. 1993) are fundamentally flawed. Moreover, when observational noise is convolved with process noise (noise which is folded into the dynamics themselves), computing Lyapunov exponents for the real system will be difficult. Such problems pave the way for more operational definitions of dynamic complexity (cf. Sugihara 1994, Yao & Tong, this volume).

Aside from its use in the characterization of chaos, I attempt to show how nonlinear forecasting can be used more broadly in pragmatic classification problems. Here I review a recent example of nonlinear forecasting as it is applied to classify human heart rhythms. In so doing, I introduce a method for characterizing nonlinearity using 'S-maps' and a novel method for analyzing multiple short time series with composite attractors.

## 1. Introduction

Suppose we are given a time series such as the one shown in figure 1*a* where we have few preconceptions. These are data for the Scripps Pier diatoms that were extracted from a record that W.E. Allen collected over a 20 year period (Tont 1981). This is an especially difficult time series to interpret since it represents counts of organisms taken from very small samples (a five gallon bucket) of a patchy sea where the potentially confounding mechanisms of population growth, physical transport and mixing among others, are taking place. What kinds of information might one hope to uncover from such a series?

From a dynamical systems viewpoint, there are three main potential sources of complexity here. The first is observational noise or measurement error. The samples are small and as such may not be representative for these populations. To the extent that this is true, the apparent complexity that we see in the time series may be largely a consequence of sampling variability. The second and related source of complexity is the dimensionality of the system, or the possible action of many variables. Thus, the apparent multitude of modes, the irregular ups and downs, may be a consequence of a multitude of effects blended together in additive fashion by superposition. Explaining complexity in a time series in terms of the action of a large number of variables characterizes the traditional linear approach to understanding complexity in time series. The third possible source of complexity that we see here may be the multiplicative or nonlinear interaction of these variables. That is, rather than attributing what is seen here to the action of a large number of variables, the focus is on the nonlinear interaction of possibly a small number of variables. Herein, such nonlinear behavior, both chaotic and stable, shall be called 'dynamic complexity.'

Real time series such as this one probably contain a mixture of these effects, however, in the past decade, there has been a special focus in the natural sciences to uncover evidence for dynamic complexity (eg. Ascioti et. al. 1993, Lefebvre et. al. 1993, Tsonis 1993, Greshenfeld & Weigend 1993). The compelling reasons for this interest are that both nonlinearity and chaos hold out the promise of producing complex time signatures with relatively simple (low dimensional) underlying dynamics, and the acknowledgment that such nonlinearity may allow improved short-term predictability while at the same time it can place limits on long-term forecastability. Although a major motive in this search is the possibility of finding a low dimensional model, as we shall see, many of the cases where nonlinear dynamics appear to operate also have high dimensional or stochastic characteristics.

I shall now review some of the issues in this search, particularly as they apply to uncovering nonlinearities in biology where the issues differ in a fundamental way from those in the conventional signal-processing literature.

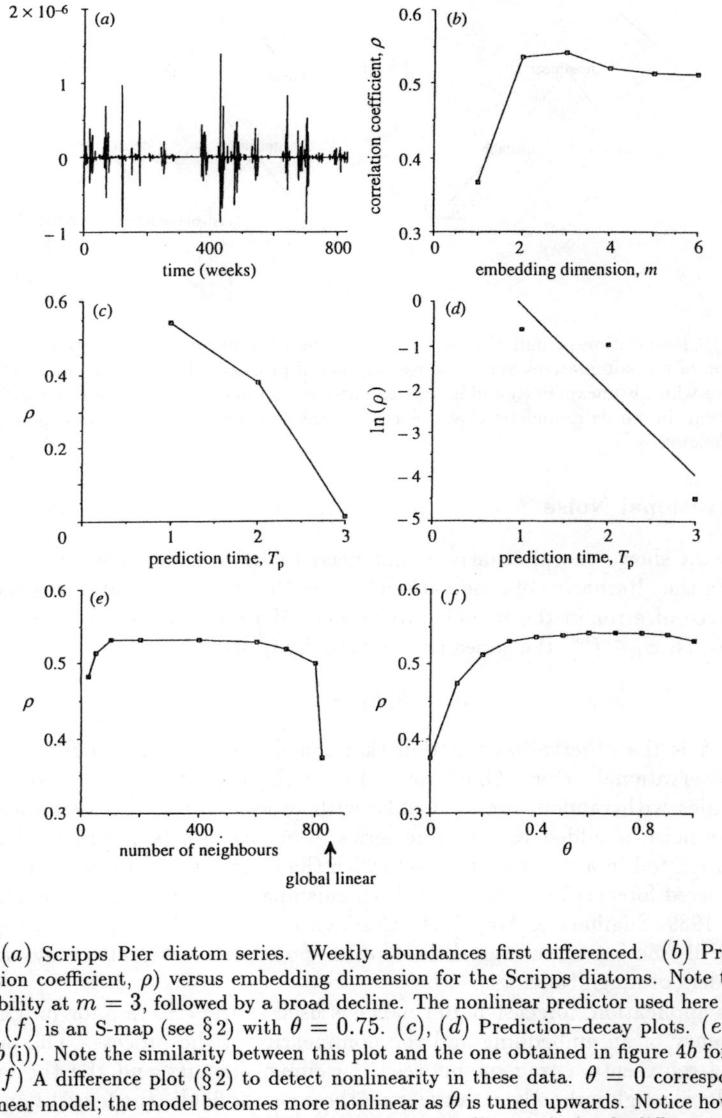


Fig. 1. $(a)$ Scripps Pier diatom series. Weekly abundances first differenced. $(b)$ Predictability (correlation coefficient, $\rho$) versus embedding dimension for the Scripps diatoms. Note the peak in predictability at $m = 3$, followed by a broad decline. The nonlinear predictor used here and in $(c)$, $(d)$ and $(f)$ is an S-map (see § 2) with $\theta = 0.75$. $(c)$, $(d)$ Prediction–decay plots. $(e)$ DVS plot (see § 1 $b$ (i)). Note the similarity between this plot and the one obtained in figure 4$b$ for stochastic chaos. $(f)$ A difference plot (§ 2) to detect nonlinearity in these data. $\theta = 0$ corresponds to the global linear model; the model becomes more nonlinear as $\theta$ is tuned upwards. Notice how markedly predictability is improved as the model beco mes more nonlinear. A simple difference test (§ 2$a$) for the $H_0 : \rho_{\text{linear}} = \rho_{\text{nonlinear}}$ suggests rejection of $H_0$ at $p < 0.0005$. (After Sugihara & May 1990.)

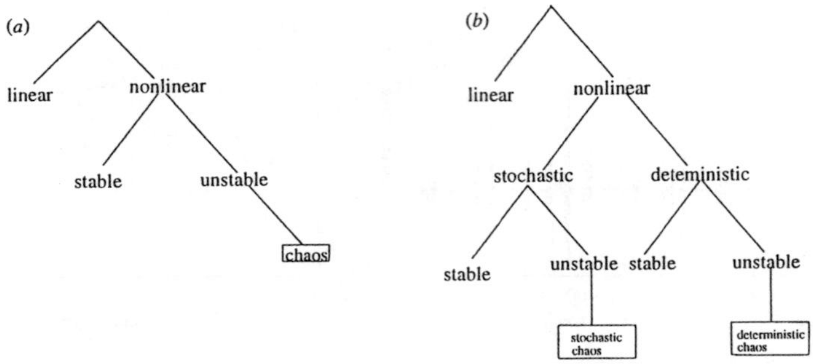


Fig. 2. (*a*) Classification of null alternatives under the assumption of observational noise. (*b*) Classification of null alternatives under the assumption of process noise. Process noise is that part of the system which is unexplained and is approximated as IID noise. Strictly speaking with process noise there can be no deterministic chaos, except in the limiting case (no process noise) or in a noise-free skeleton.

## 2. Observational Noise

Figure 2*a* shows the alternatives that need to be considered when one is only considering the alternative of observational noise; that is, if observational noise is the major source of error in the modeled dynamics. Here, given the noise-free process, $\boldsymbol{S}_t = \boldsymbol{F}(\boldsymbol{S}_{t-1})$, $\boldsymbol{S}_t \in R^m$, the series that is actually observed is

$$X_t = h(\boldsymbol{S}_t) + \epsilon_{\text{obs}}, \tag{1}$$

$X_t \in R^1$, $h$ is the observation function that maps points in $R^m$ to $R^1$, and $\epsilon_{\text{obs}}$ is the IID observational error. One essentially overlays the time series produced by the dynamics with random noise. Unlike with process noise, the assumption here is that the noise is added to the time series after the fact (superimposed), rather than incorporated in a way that interacts with the dynamics. Most of the early work which involved forecasting focused on distinguishing chaos from measurement noise (Casdagli 1989, Sugihara & May 1990, Casdagli et. al. 1992). This is an especially important distinction in ecological data where population estimates may vary over several orders of magnitude.

In this application, forecast performance is used as the criterion for judging both the sufficiency of an embedding and the nonlinearity of the process. The optimum embedding represents a tradeoff between low dimensionality and the the one that produces local neighborhoods with the smoothest vector fields (the fewest non-unique points or crossing vectors). This will correspond to the embedding that gives the best predictability sensu Sugihara & May (1990). Thus, the optimum is identified by doing a trail-and-error search to find the embedding that optimizes predictability (figure 1*b*).

Whether the underlying dynamic is linear or nonlinear is resolved here by seeing which model (linear or nonlinear) produces the best out-of-sample forecast. The basic philosophy here is that the model that produces the best out-of-sample forecast is the best dynamical description of the data. Thus a time series would be classified as nonlinear, if a nonlinear model is capable of producing significantly better forecasts (out-of-sample) than a linear model of equivalent complexity.

As far as stability versus instability is concerned, simple scaling rules have been developed for (1) to distinguish the decay in predictability that would be expected from chaotic signals (Wales 1991) versus those from linear and fractal processes contaminated with white and colored noise (Tsonis & Elsnor 1992). Decay in predictability with forecast horizon, combined with strong nonlinearity are used as evidence for the operation of chaotic dynamics. The obvious appeal of this approach is in its directness and in the relatively modest amounts of data it requires (Sugihara & May 1990). The method is easy to apply and, as explained in the next section, it can also be applied to study cases with process noise.

### *2.1. Process Noise and Stochastic Chaos*

As shown in figure 2*b*, the problem appears to become more complicated in terms of the number of alternatives that need to be considered when the decision is made to explicitly include process noise in the picture. A crucial fact, which we are not often conscious of in the exploration of real data, is that whether or not to include process noise in the model is ultimately a philosophical decision. Process noise is not objectively fixed, but is that part of the system dynamics that, as a function of our limited reckoning, remains unexplained. According to this view, the underlying system consists of an explained part (that which is modeled, or the noise-free skeleton) and an unexplained part (approximated by IID noise of some kind, $\epsilon_{\text{process}}$). Notice that process noise will generally be a function of the noise-free map that is chosen. If the underlying dynamics are high dimensional, but we choose a noise-free model which is low dimensional, then the remaining error is process noise. Thus, according to this view, the real $m$-dimensional system dynamics can only be approximated in $k$ dimensions, with the dynamics in the remaining $m - k$ dimensions approximated as noise. Hence, if the true system dynamics are

$$\boldsymbol{S}_t = \boldsymbol{F}(\boldsymbol{S}_{t-1})$$

$\boldsymbol{S}_t \in R^m$. Then in $R^k$, $k << m$, we have

$$\boldsymbol{Z}_t = \boldsymbol{J}(\boldsymbol{S}_t)$$

where $\boldsymbol{J}$ behaves like an observation function, mapping $R^m$ to $R^k$. But notice that

$$\boldsymbol{Z}_t \neq \boldsymbol{G}(\boldsymbol{Z}_{t-1})$$

where $\boldsymbol{G}$ is a diffeomorphism since the trajectories are no longer unique (may cross). Thus, we approximate the full $m$-dimensional dynamics in $k$ dimensions as

$$\boldsymbol{Z}_t \approx \boldsymbol{G}(\boldsymbol{Z}_{t-1}, \epsilon_{\text{process}}) \tag{2}$$

where $\boldsymbol{G}(\boldsymbol{Z}_{t-1})$ is the modeled part or noise-free skeleton and the dynamics in the remaining $m-k$ dimensions are approximated as IID noise of some kind. A specific form for incorporating process noise which is commonly studied is one where $\boldsymbol{Z}_t = \boldsymbol{G}(\boldsymbol{Z}_{t-1})$ is the noise-free map, and error-free (in terms of measurement error) observations on the system are approximated (although equality is used) as

$$\boldsymbol{Z}_t = \boldsymbol{G}(\boldsymbol{Z}_{t-1} + \epsilon_{\text{process}}) \tag{3}$$

In this example, process noise is added and iterated through the dynamics of a noise-free map. The invariant measure for the system (which tells us the density with which various parts of the attractor are visited) is induced by the process noise.

An interesting phenomenon that emerges with process noise is stochastic chaos (Tong 1990, Chan & Tong 1994, Rand & Wilson 1991, Sidorowich 1992). This is likely the most ubiquitous form of chaos in nature. This is illustrated by the noise-driven logistic map with a two-point cycle shown in figure 3. Here, $X_t = aX_{t-1}(1 - X_{t-1}) + \epsilon_{\text{process}}$, $a = 3.2$, $\epsilon_{\text{process}}$ is gaussian $(0, \mathit{sigma})$ with the domain boundaries (0,1) treated as reflecting barriers, to keep the function restricted to the interval (0,1). Again, the invariant measure, or the specific sequence of points visited on the noise-free map is not determined by the map itself, but by the interaction of the noise on the map. Thus, if the Lyapunov exponent, $L$, for the system is calculated from the geometric mean of the Jacobian elements, $J$, the invariant measure must be induced by the noise:

$$L = \lim_{n\to\infty} \frac{1}{n} \sum \ln(|J|)\,.$$

If the noise level is increased from zero to very low levels, there is a tendency to drift slightly away from the 2-point equilibria, initially spending more time visiting flatter portions of the map (net contraction) to yield more stable dynamics (figures 3*b*, *d*). As the noise level is increased a bit further, steeper, unstable portions of the map are visited more frequently (regions of net divergence), and the Lyapunov exponent can become positive. Thus, the system can become stochastically chaotic, although the skeleton is stable (figures 3*c*, *d*), and such behavior can be achieved at very low noise levels ($\mathit{sigma} = 0.05$). In this example, the Lyapunov exponent appears to be a smooth function of noise level (figure 3*d*).

Of special interest is that the form of chaos demonstrated by this simple example is not deterministic or low dimensional. Although the noise-free map was low dimensional, the full system is high dimensional because it must include the noise (the unexplained part). Chaos only appears when the map is driven by the noise. Thus, contrary to common vernacular, the chaos which appears in this example, neither implies low dimensionality nor determinism.

### 2.1.1. Identifying Stochastic Chaos

Strictly speaking, deterministic chaos (the right-hand limb of Figure 2*b*) does not exist for a system with process noise; only stochastic chaos is possible. It may

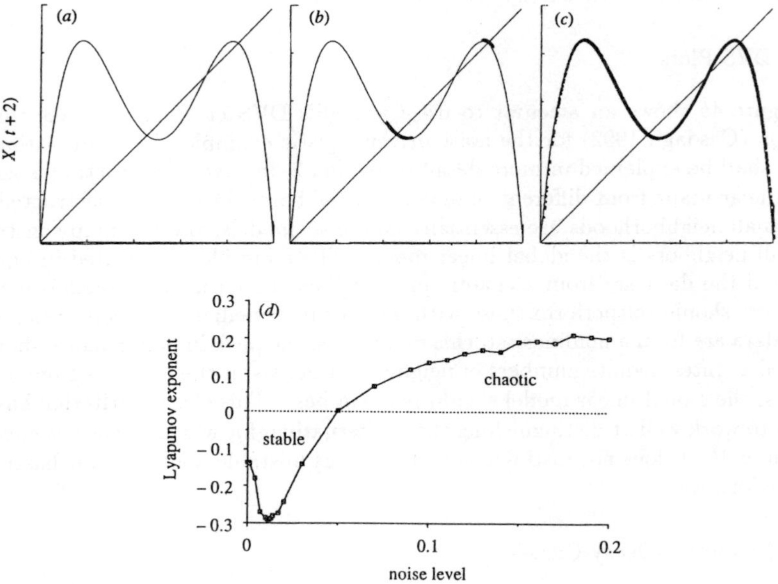


Fig. 3. Stochastic chaos induced by the action of gaussian noise on a sta ble period-2 quadratic map ($X_{t+1} = 3.2X_t(1 - X_t) + \epsilon$). $(a)$ The noise-free ($\epsilon = 0$) time-2 map. $(b)$ Shows what portions of the noise free map are hit as a very small level of noise ($\sigma = 0.005$) is introduced. Notice that because the system spends more time in the flatter more stable portions of the map, it is actually more stable than the noise-free map. $(c)$ Has a slightly higher noise level ($\sigma = 0.1$) and shows how the system spends more time in the steeper unstable portions of the map (repellor regions where the tangent $|slope| > 1$). $(d)$ Summarizes these effects showing how the Lyapunov exponent for the system (invariant measure is induced by the noise) is a s mooth function of noise level. Thus, process noise can induce chaos even if the noise-free skeleton is chosen to be stable.

exist as an abstraction for a noise-free skeleton, and as will be discussed, this may be a reasonable thing to consider in certain circumstances such as noise-filtering in engineering applications (Abarbanel et. al. 1993). However, for a natural system with process noise, the noise-free skeleton may have no objective meaning at all. Indeed, the noise-free skeleton may simply be an artifact of whatever fitting procedure is used to extract it. Thus, insofar as our objective is to learn about the real system, with process noise the difficult alternatives that one wants to distinguish are stable versus unstable nonlinear stochasticity (stochastic chaos).

### 2.1.2. DVS Plots

Figure 4*b* shows an attempt to use Casdagli's DVS criterion to make this distinction (Casdagli 1992) for the noise-driven logistic example. This forecasting test, which shall be explained in more detail in section 2, involves constructing a series of local linear maps from different numbers of neighbors. The maps constructed from very small neighborhoods are essentially nonlinear models, and the map constructed from all neighbors is the global linear model. The main idea illustrated in figure 4*a* is that if the data are from a chaotic process then the local linear models with few neighbors should outperform those with large or intermediate numbers of neighbors; if the data are from a nonlinear stochastic process, the peak in performance should be realized at intermediate numbers of neighbors; whereas, if the data are from a linear process, the global linear model should perform best. This elegant criterion has been shown to work well at distinguishing those alternatives for which it was designed, but as seen in 4*b*, it does not distinguish stable versus unstable nonlinear stochasticity in any obvious way.

### 2.1.3. Prediction-Decay Curves

Prediction-decay curves are obtained by plotting the correlation coefficient (Sugihara & May 1990, Wales 1991, Tsonis & Elsnor 1992) or rms error (Farmer & Sidorowich 1987), (both either logged or unlogged) against prediction time ($T_p$). As can be seen in figure 4*c*, stable systems will decay, however, at a rate which is considerably slower than the exponentially decaying unstable cases. Stable systems decay because the gaussian noise which is small on average, will occasionally produce a value from the tails of the distribution that will push the system into a repellor region (unstable or local divergence). Uniform noise from a distribution whose width is smaller than the smallest domain of attraction around the time-2 fixed points, will not decay. Thus, in general, one should expect steeper prediction-decay curves when the system is chaotic and shallower ones when it is stable. As can be seen in figure 4*c*, this qualitative distinction may be useful at identifying the extremes, but there is a smooth (albeit sharp with respect to increasing noise level) continuum in between, where the distinction will be blurred in any practical sense. Nonetheless, applied with care, prediction-decay curves can be a useful tool for arguing the distinction between stable and chaotic stochasticity.

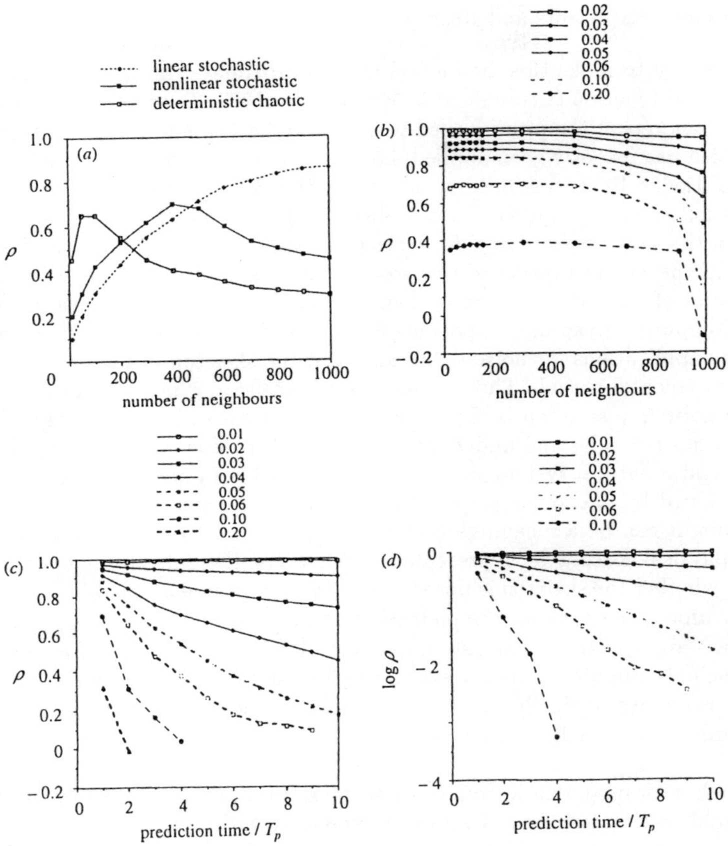


Fig. 4. The use of forecast methods to detect stochastic chaos. $(a)$ Hypothetical example of Casdagli's (1992) DVS plot to distinguish between chaotic, nonlinear stochastic, and linear stochastic alternatives. $(b)$ DVS plots for the data in figure 3, to distinguish stable (solid lin es) from unstable (dashed lines) nonlinear stochasticity. $(c)$, $(d)$ Prediction–decay curves to distinguish stable (solid lines) from unstable (dashed lines) nonlinear stochasticity. As expected (Sugihara & May 1990; Wales 1991), the stable cases exhibit slow decay while the unstable (chaotic) cases exhibit rapid exponential decay. With uniform noise instead of gaussian, the stable cases will show no decay.

2.1.4. Lyapunov Exponents and Reality

Another way to make this distinction is to approximate the noise-free map with a smooth fitted function (Crutchfield & MacNamara 1987, Tong 1990, Nychka et. al. 1992) and then repeat the process used to compute the Lyapunov exponents in figure 3*d*. This idea has recently been applied, however incorrectly, in ecology (Ellner 1991, Turchin & Taylor 1992, Hastings et. al. 1993, Hanski et al 1993) where Lyapunov exponents have been computed for the abstracted noise-free maps, rather than for the system (in some cases fitting $R^5$ surfaces to time-series containing as few as 30 points!). In the fitting procedure the process noise is assumed to have the specific additive form of (3). However, it is then ignored in the invariant measure for the system. Computing Lyapunov exponents for a noise-free skeleton may make sense in engineering applications for noise-reduction, or if the skeleton is a classical or a well defined fundamental model. Thus, if one has a transistor which is fed noisy inputs, where the noise-free skeleton is the action of the transistor, it is meaningful here to try to filter out the noise and understand the transistor. However, when one is trying to understand a natural system such as the Scripps diatom data in figure 1, and the map is obtained by phenomenological fitting so that one really does not know what it represents, it can be a meaningless exercise (eg. are the coordinates all biological or could physical forcing be represented? What are the subsystem boundaries?). Moreover, whether the skeleton is linear or nonlinear or stable or unstable can depend to an uncomfortable extent on the method of fitting that one chooses to apply to the small noisy data sets that biologists must typically work with. Thus, although one can in principle compute Lyapunov exponents for a noise-free skeleton and determine whether it is deterministically chaotic by the rigorous definition, it is uncertain what this determination actually means, when the real dynamics are assumed to contain noise.

Although with model data from our noisy logistic example it may be possible to get reasonable estimates of the Lyapunov exponent, with real data, where observational noise may be involved, this may not be possible (Yao & Tong 1994). With observational noise convolved on top of process noise we are looking at data of the form

$$Y_t = h(\boldsymbol{Z}_t) + \epsilon_{\text{obs}}$$

where again, $\boldsymbol{Z}_t = \boldsymbol{G}(\boldsymbol{Z}_{t-1} + \epsilon_{\text{process}})$, $\boldsymbol{Z}_t \in R^k$, $k << m$, and $h$ maps $R^k$ to $R^1$. The difficult problem is not in estimating some noise-free skeleton (Nychka et. al. 1992, Turchin & Taylor 1992, Hastings et. al. 1993). Rather, the difficulty, which may not have a practical solution (eg. in terms of data requirements), is in disentangling the process noise from the observational noise to reconstruct the invariant measure. One can no longer simply compute products of the Jacobians from the time series itself because the invariant measure is contaminated with observational noise. Unless there is prior knowledge of the observational noise or the process noise, it is difficult to see how an invariant measure for the system can be deduced from the data. More work needs to be done here.

*2.2. Should we care about chaos?*

The above conundrum has prompted Yao & Tong (this volume) to propose an alternative definition of sensitive dependence that appears to be more practical than Lyapunov exponents. One of the central reasons for wanting to know whether a system is chaotic is the possibility of low-dimensional determinism. This motivation is somewhat diminished by the fact and probable ubiquity of stochastic chaos, where ultimately the driving force for prediction decay might be the interaction of some lower dimensional map with the remaining high dimensional process noise.

Regardless of the fact that we still cannot compute meaningful Lyapunov exponents, the various other methods that have been developed for nonlinear analysis of time series remain useful tools for understanding nature and for constructing arguments for chaotic dynamics. These include the new criterion of Yao & Tong (this volume), nonlinear forecasting (Kravtsov & Etkin 1984, Farmer & Sidorowich 1987, Sugihara & May 1990, Casdagli 1989), and the use of null models (Brock 1986, Osborne & Provenzale 1987, Theiler et. al. 1991). Forecast methods may be especially useful since they regard any process noise as may be present as convolved in the system. The 3-way forecast criterion of Sugihara & May requires: 1) finding an optimal low-dimensional model (optimal in terms of a peak in predictability versus embedding); 2) demonstration of strong nonlinearity (nonlinear model performs significantly better than linear model); and 3) steeply declining prediction-decay curves. The emphasis is on nonlinearity and forecast decay. Operational methods such as this are easy to implement and may provide good evidence (albeit not 'proof') for the operation of chaotic dynamics narrowly defined. Alternatively, one could side-step this conundrum and take the 3-way criterion as an 'operational definition' of chaos.

## 3. Sequential Locally Weighted Global Linear Maps: SLWGLM (or 'S-map')

The idea of locally weighted maps has been discussed in various contexts in the statistics literature (Cleveland 1979, Hardle 1990, Tong 1990), by connectionists (Bottou 1993), and in the literature on dynamical systems (Crutchfield 1979, Kravtsov & Etkin 1984, Crutchfield & MacNamara 1987, Farmer & Sidorowich 1987, Casdagli 1989, 1992, Casdagli et. al. 1991, 1992, Diebold & Nason 1990, Stokbro & Umberger 1992). Such ideas have also been discussed in relation to the geometry of the attractor (Sugihara & May 1990, Mees 1991). I will now describe a predictor which is similar to several of the above techniques, but which can often give more skillful forecasts when the time signature contains a mixture of linear and nonlinear components. Moreover, like Casdagli's DVS model, this model can span the spectrum of models from global linear to local nonlinear, and as described in the next section, it can provide a simple test for nonlinearity in a time series (see also the elegant BDS criterion of Brock et. al. 1987). Essentially the S-map is an extremely general formulation of nonlinear forecasting which involves a variety of possible forms of local linear maps which portray a nonlinear manifold as piece-wise linear.

Suppose we have an embedding of an observed time series $\boldsymbol{X}_t \in R^{m+1}$, where $\boldsymbol{X}_t(0) \equiv 1$ (for the constant term in the solution of (5) below). Let the time series value $T_p$ time steps forward be $\boldsymbol{X}_{t+T_p}(1) = Y_t$. Then the forecast at $T_p$ is

$$\hat{Y}_t = \sum_{j=0}^{m} \boldsymbol{C}_t(j)\boldsymbol{X}_t(j) \tag{4}$$

For each predictee, $X_t$, one finds the SVD solution for $\boldsymbol{C}$ using historical points from the fitting set or library set (referred to below by the subscript '$i$') as follows:

$$\boldsymbol{B} = \boldsymbol{AC} \tag{5}$$

where

$$B_i = w(||X_i - X_t||)Y_i,$$
$$A_{ij} = w(||X_i - X_t||)\boldsymbol{X}_i(j)$$

and

$$w(d) = e^{-\theta d/\bar{d}} \tag{6}$$

$\theta >= 0$, $d$ is the distance between the predictee and the neighbor vector, and the scale factor, *dbar*, is the average distance between neighbors. Note that $\boldsymbol{A}$ has dimensions $n_l \times (m+1)$ where $n_l$ = size of the library. To maintain independence in the out-of-sample solution for each fitted map, I eliminate all vectors from the library whose coordinates include any of the coordinates of the predictee $X_t$ in the time series. Thus for each prediction, I construct a different exponentially weighted global linear map, where the degree of the local weighting is controlled by $\theta$. If $\theta = 0$, we have the simple global linear solution, as $\theta$ is tuned to higher positive values, the solutions become more local and hence nonlinear.

Notice that if instead of (6), the weighting function is $w(d) = 1$ for $d < k$ and $w(d) = 0$ otherwise, one recovers the local linear maps described by Casdagli (1992) and Diebold & Nason (1990). When there is a mixture of linear and nonlinear signal in the data (eg. nonlinear response to regular forcing) this weighting function can penalize distant points too severely (zero weight); thus the exponential but global fit was chosen for this particular S-map. The S-map is a very general formulation of first-order nonlinear forecasting which covers the spectrum of local linear (piecewise linear = nonlinear) to global linear maps.

### *3.1. The Difference Test: Scripps Diatoms*

S-maps can provide a useful way to characterize the degree of nonlinearity in a stationary time series in terms of the difference in forecast skill achieved with a nonlinear versus a linear model (see also the bi-spectrum of Subba Rao & Gabor 1984 and Subba Rao 1991 and the BDS test of Brock Dechert & Scheinkman 1987, Brock & Potter 1993 and Brock et. al. 1991). In figure $1f$ it is applied to the Scripps diatom data, where the difference in linear versus nonlinear predictability is shown as a function of $\theta$. Comparisons of forecast skill achieved with nonlinear versus linear models shall be called 'difference tests.' As described in Sugihara & May (1990), one can ask whether the difference between linear and nonlinear models is significant simply by comparing the two correlation coefficients. This can be done by computing a z-statistic (Fisher's z-transform for the product moment correlation coefficient) to test the null hypothesis that $\rho_{\text{nonlinear}} = \rho_{\text{linear}}$. The availability of this standard test was one of the reasons that Sugihara & May (1990) used the correlation coefficient to measure predictability. To be rigorous, before applying it, one should take care that successive residuals are roughly independent, and if not, adjustments need to be made to obtain an accurate p-value.

As can be seen, this difference is highly significant in the diatom data ($p < .0005$). This strong nonlinearity, combined with the steep prediction-decay curve (figures 1*c*, *d*) is evidence for the operation of chaotic dynamics. Notice further, how similar this DVS plot looks (figure 1*d*) to those obtained in figure 4*b* for stochastic chaos; both show a sharp decline in predictability as the global linear map is approached.

### *3.2. The Difference Test: Surrogate Measles*

Figure 5 contains another example of a difference test as applied to the New York measles data (Schaffer 1985ab, Olsen et. al. 1988) and data from a null hypothesis (Stone 1992) which questioned the characterization of these data as an example of chaotic dynamics (Schaffer et. al. 1988, 1990, Olsen et. al. 1988, Sugihara & May 1990, Sugihara, Grenfell & May 1990). This rather complicated null hypothesis incorporated much of the natural detail of the epidemic. Indeed there was some question as to its neutrality as a null hypothesis (Stone 1992). It was constructed by stitching together gaussian wave forms taken from 4 empirically fit distributions sampled in a probabilistic order that would mimic the pattern observed. As seen in figure 5*b*, it very effectively mimics the general shape and features of the original measles data (figure 5*a*). These surrogates also had a degree of predictability which was similar to the real data (using a nonlinear forecast model) and which had similarly steep prediction-decay curves, yet were not from a chaotic process. This prompted Stone to accept the null hypothesis. However, as seen in figures 5*c*, *d*, they do not have the same apparent degree of nonlinearity as the natural data. One can safely reject this null hypothesis at the $p < 0.0001$ level. The difference between linear and nonlinear predictability is an important characteristic to consider when constructing null models to study chaos (see figure 2), and is useful in distinguishing chaos from colored noise (Sugihara & May 1990).

## 4. Broader Applications of Nonlinear Forecasting in Medicine

The aim of this next section is to show how, apart from addressing the more narrow question of detecting chaos, nonlinear methods can have other important applications in the natural sciences, and in particular in biology and medicine. Specifically, it is shown how the key characteristics, nonlinearity and prediction-decay, can be useful

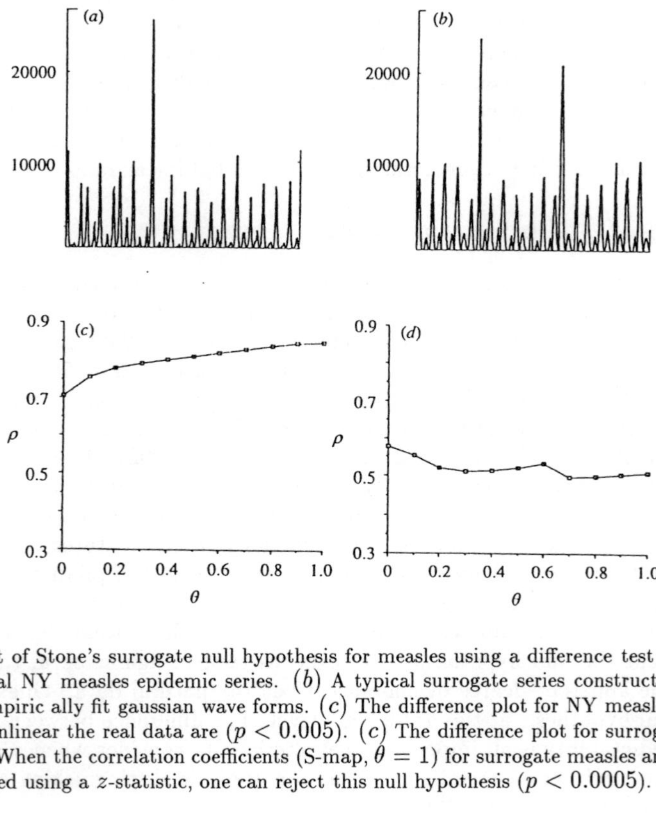


Fig. 5. Test of Stone's surrogate null hypothesis for measles using a difference test and an S-map. $(a)$ The real NY measles epidemic series. $(b)$ A typical surrogate series constructed by stitching together empiric ally fit gaussian wave forms. $(c)$ The difference plot for NY measles showing how strongly nonlinear the real data are ($p < 0.005$). $(c)$ The difference plot for surrogate data is not nonlinear. When the correlation coefficients (S-map, $\theta = 1$) for surrogate measles and the real data are compared using a $z$-statistic, one can reject this null hypothesis ($p < 0.0005$).

attributes for the pragmatic classification of medical time series.

### *4.1. Human Heart Rhythms and Forecasting for Practical Diagnosis*

An area of application where nonlinear methods and forecasting will likely have a great deal to offer is in the study of human heart rhythms and in particular the characterization and diagnosis of pathologies. This is an area where there has been some very strong theoretical work (eg. Glass & Mackey 1988), however, less has been done in the way of applying modern time series methods to analyze empirical EKG data (Lefebvre et. al. 1993, Lipshitz & Goldberger 1992, Gough 1992, Goldberger 1991, Skinner et. al. 1990, Goldberger et. al. 1990, Goldberger et. al. 1984).

Two important results from the empirical analysis of human heart rhythms have to do with differences in RR-intervals as a function of age, and those differences in heart rhythms that may be associated with heart disease. With regard to age, Kaplan et. al. (1991) found that young healthy hearts contain a high degree of variability (measured in terms of standard deviation) which becomes diminished as a normal part of the aging process. Lipshitz & Goldberger (1992) found a decrease in the correlation dimension with age. As regards health, Kleiger et. al. (1987), showed how after an acute myocardial infarction, low heart rate variability (low variance in RR-intervals) is associated with increased mortality. Lefebvre et. al. (1993) found significant differences in predictability with nonlinear forecast methods, between older patients who had suffered a heart attack and older patients who had not. They also found suggestive differences (though not significant) in the rate of prediction-decay of diseased versus healthy patients.

Here, I shall apply the S-map given in (4) to EKG data from 29 patients kindly provided by J. Lefebvre, M. Kamath, D. Goodings and M. Fallen (see Lefebvre et. al. 1993 for details). The aim of this exercise is to determine whether the degree of nonlinearity (difference test) and the rate of prediction-decay can discriminate age and health. I shall do this in two ways: by modeling each patient separately and by creating composite mappings for each group.

In all cases, I shall use first-differenced data embedded with a standard lag of 1 and an embedding dimension $m = 7$. This embedding convention was used by Lefebvre et. al. 1993 with the simplex predictor (Sugihara & May 1990), and $m = 7$ is in the plateau of optimal predictability for all cases. Moreover, as Lefebvre et. al. (1993) noted, a vector having 7 components is just long enough to span a complete respiratory cycle with respiration likely being an important controlling factor.

#### 4.1.1. Modeling Individual Attractors

Figure 6 shows illustrative examples of prediction-decay curves and difference plots constructed by comparing changes in absolute forecast error for a young versus an old healthy patient ($6c, d$) and for a healthy versus a heart-diseased patient ($6a, b$). For reasons which will be explained in the next section, absolute forecast error is a better discriminator of heart disease than the correlation coefficient. Table 1 presents the

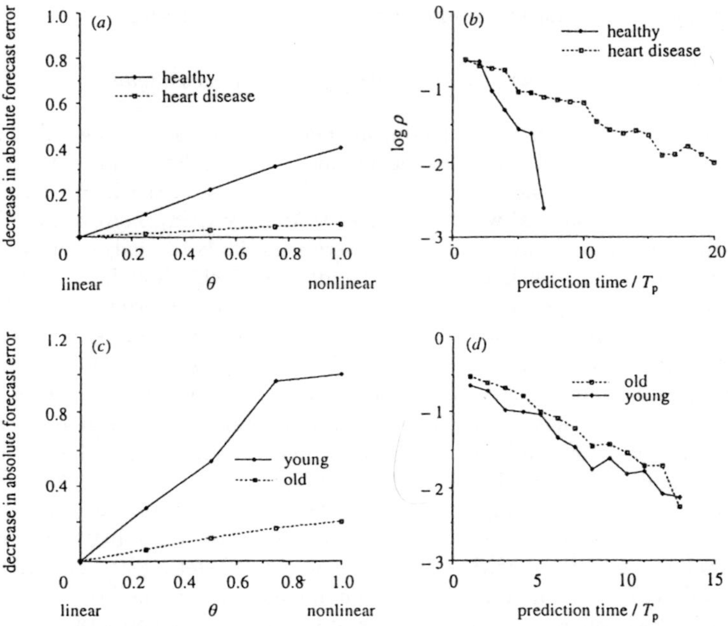


Fig. 6. Illustrative example of the classification of human heart rates by age and health using nonlinearity (decrease in absolute forecast error) and prediction–decay slope. (*a*) Difference plot showing how this healthy patient (patient OH1 in Lefebvre *et al.* 1993) has a greater decrease in absolute forecast error with increasing nonlinearity (hence, is more nonli near) than a patient who has heart disease (patient NB1). (*b*) Prediction–decay curves for the same patients, showing how healthy patients have steeper prediction–decay curves than heart-diseased patients. (*c*) Difference plot showing how a young healthy patient (patient YH1 in Lefebvre *et al.* 1993) has a greater decrease in absolute forecast error with increasing nonlinea rity (hence, is more nonlinear) than an older healthy patient (patient OH3). (*d*) Prediction–decay curves for the same subjects, showing how prediction–decay slope does not distinguish age (see table 1 for ensemble statistics).

group means and the significance levels obtained for these two comparisons. As can be seen, both nonlinearity (as measured by the difference in absolute forecast error) and decay are highly significant discriminators of heart disease, however, only nonlinearity is a good discriminator of age. The more rapid prediction decay in healthy versus diseased patients is consistent with the suggestive result reported by Lefebvre et. al. (1993) using the simplex predictor.

Figure 7 shows a weak positive relationship between nonlinearity and decay. Note that this relationship only becomes significant when the axes are log transformed. Although the evidence is not compelling in this specific example, such ‘difference-decay’ plots may be a useful tool when exploring for chaos in real data. With chaotic dynamics one might expect a tendency toward a positive relationship between nonlinearity and decay.

Table 1: EKG study – summary of results for individual maps: mean values and significance levels
(Mean values and significance levels (t-test) for the difference in absolute forecast error realized with a nonlinear model ($\theta = 1$) against a global linear model ($\theta = 0$), and for prediction–decay slopes.)

| | nonlinearity | | pred.–decay | |
|---|---|---|---|---|
| | diff. (error) | $p$ | slope | $p$ |
| healthy $n = 14$ | 1.065 | | 0.163 | |
| | | $p < 0.005$ | | $p < 0.005$ |
| disease $n = 15$ | 0.314 | | 0.102 | |
| young $n = 6$ | 1.670 | | 0.162 | |
| | | $p < 0.005$ | | $p < 0.4$ (n.s.) |
| old $n = 8$ | 0.609 | | 0.164 | |

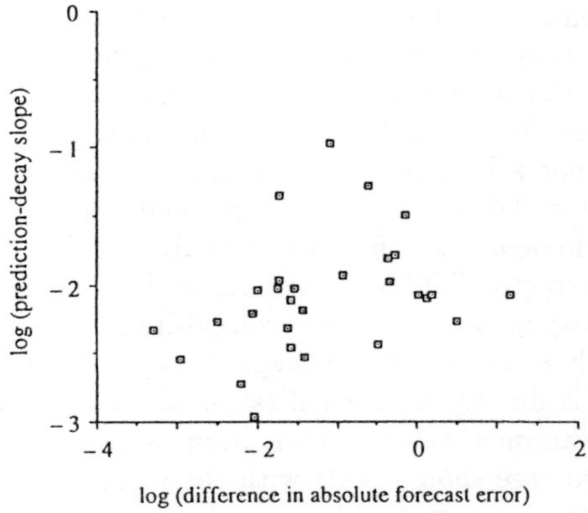


Fig. 7. Difference–decay plot showing a weak relationship ($\rho = 0.333$, $0.25 < p < 0.05$) between prediction–decay slope and nonlinearity for the 29 patients taken as a group. Note that the axes are log-transformed.

4.1.2. Modeling Composite Attractors

When individual time series are short, one can imagine constructing a composite attractor by piecing together segments from individuals (samples) belonging to a particular homogeneous group. Thus one could construct composite maps that are intended to capture the dynamical mean of a group. This might be most useful when individual time series are short or for data-intensive calculations such as measuring the correlation dimension. Indeed, one can use variations of this method to test for homogeneity of groups, by testing whether within-patient forecastability (looking at residuals when the library and prediction segments are taken from a single patient) is different from between-patient forecastability (looking at residuals when the library and prediction segments are each taken from different patients, i.e., building a model from one patient, and using it to forecast another patient).

Here I have created composite attractors for each of the four groups. Heteroscedasticity is minimized by normalizing the means (= 0) and variances (= 1) for each group before concatenating the time series. To avoid the problems created by the discontinuities at the seams between time series, I eliminate $m \times tau + max(T_p)$ time series values before and after the joints from all calculations (libraries and predictions). To speed calculations, and to normalize comparisons, each composite S-map was constructed from 5000 randomly sampled vectors.

Figure 8 shows the prediction-decay curves and the difference plots obtained by comparing changes in absolute forecast error for the young attractor versus the old healthy map ($8c, d$) and for the healthy versus heart-diseased map ($8a, b$). Table 2 summarizes the comparisons. One should take special note that by normalizing the variances of all groups we have effectively factored out variance as a discriminating variable. Thus, any effect that is seen here is over and above the effect of variance in classifying heart rhythms. As with the individual maps, with the composite maps, prediction decay slope is not a significant discriminator of age but is a significant discriminator of health ($p < 0.01$). The healthy attractor had a more precipitous decay (reaching the noise floor at $T_p = 6$ beats, roughly the time for one breathing cycle) than the diseased attractor (hit the noise floor at $T_p = 11$ beats). As with the individual maps, nonlinearity in the normalized composite maps was a good discriminator of age however, unlike the individual maps it was not a good discriminator of health. This difference is due to the normalization of variances in the composite maps. Just as mean and variance scale together, improvements in absolute forecast error seen in the individual case should scale with the magnitude of the excursions (standard deviation) in the time series. That is, where the absolute magnitude of the excursions is large, the absolute magnitude of the improvement in predictability is large. Therefore, in terms of diagnosing heart-disease, nonlinearity per se does not contain significantly more information than is already contained in the variance. However, prediction-decay, does contain more information than is already contained in the variance. As such, prediction-decay should be a good diagnostic tool for heart disease used in conjunction with variance.

As shown in table 2, predictability with the nonlinear model ($\theta = 1$) is much higher in the normalized map for the young subjects than for the old subjects. It is interesting that the heart rates for the young subjects which appeared more variable, are in fact more predictable when the data are normalized. This is likely due to there being more homogeneity in the young group than in the other groups (Sugihara in prep). The strong trends in predictability shown here are not found in the unnormalized data (Lefebvre et. al. 1993), where differences in the variances of the RR-intervals will confound this result.

To summarize, these preliminary results show that over and above variance, prediction-decay is a good discriminator of health; healthy patients have a much steeper decay curve than those with heart disease. And nonlinearity is a good discriminator of age; younger adults have more nonlinearity to their heart rhythms than older ones. In addition, nonlinear predictability appears to be a good classifier of age in the normalized maps.

*4.2. Infant Heart Rhythms:*

As a suggestive element to add to the results above, I shall briefly mention recent work (Sugihara, Allan & Sobel, in prep.) on the development of heart rhythms in infants that shows a very different pattern.

Allan & Sobel (1992) hypothesize that because other autonomic nervous system functions such as pupilary reflex seem to mature at about 30 weeks, that somewhere between 25 and 35 weeks sympathetic and parasympathetic balance evolves and contributes to irregularity in heart rate. Goldberger (1991), moreover, hypothesized that the source of irregularity in human heart rhythms is the sympathetic-parasympathetic balance acting on the sino-atrial node. Thus, one might expect nonlinearity to be absent in early gestational infants, and to be acquired in later development.

Figure 9 shows an illustrative example of this possibility for an infant aged 25 weeks and one aged 40 weeks. Here the S-map was applied to heart rate data taken by chest leads, $n = 1000$. This general pattern was found in 6 additional infants where paired observations could be made at 26 weeks and again at 37 weeks. The early gestational infants had nearly linear heart rhythms, and in all 6 cases, the more mature infants had more nonlinear heart rhythms. It is intriguing to hypothesize that heart rhythms might follow an ontogenetic sequence, starting out simple with high short-term linear autocorrelation, maturing to nonlinearity at birth; then from young adulthood onwards simplifying with age to a signal that is once again more strongly dominated by linear characteristics.

## 5. Summary

In summary, despite concerns over the feasibility of a rigorous test for chaos in real data (from the calculation of meaningful Lyapunov exponents) there are methods available that allow one to gather qualitative evidence for the operation of such dynamics. Moreover, the narrow question of chaos aside, there are useful and important ways that nonlinear methods can be applied in the natural sciences and especially

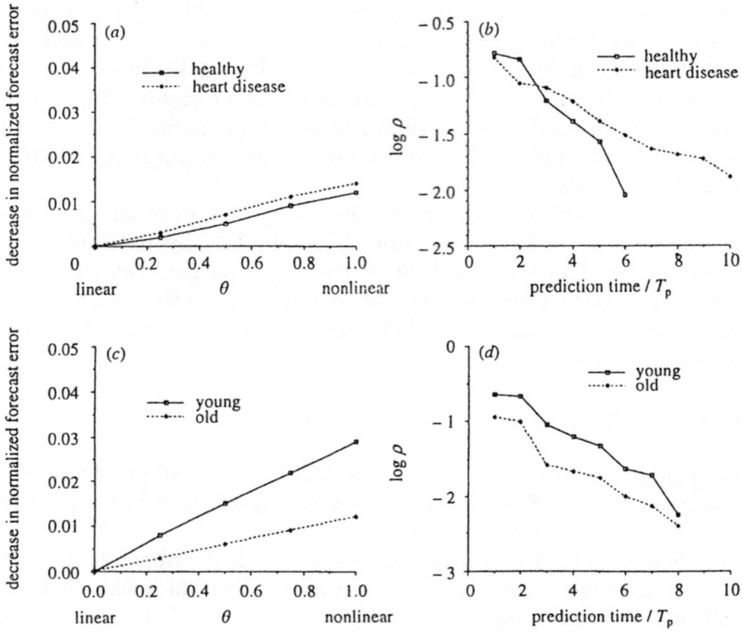


Fig. 8. Use of normalized composite maps for the classification of human heart rates (see table 2 for summary statistics).

Table 2: EKG study – summary of results from normalized composite mapS ($N = 5000$ in all cases)

| | nonlinearity | | predictability | | pred.–decay | |
|---|---|---|---|---|---|---|
| | diff. (error) | $p$ | $\rho$ | $p$ | slope | $p$ |
| healthy | 0.015 | | 0.420 | | 0.309 | |
| | | $p < 0.2$ (n.s.) | | $p < 0.0005$ | | $p < 0.025$ |
| disease | 0.016 | | 0.351 | | 0.111 | |
| young | 0.023 | | 0.522 | | 0.266 | |
| | | $p < 0.0001$ | | $p < 0.0001$ | | $p < 0.5$ (n.s.) |
| old | 0.011 | | 0.378 | | 0.260 | |

in the normalized map for the young subjects than for the old subjects. It is interesting that the heart rates for the young subjects which appeared more variable, are in fact more predictable when the data are normalized. This is likely due to there being more homogeneity in the young group than in the other groups (Sugihara in prep). The strong trends in predictability shown here are not found in the unnormalized data (Lefebvre et. al. 1993), where differences in the variances of the RR-intervals will confound this result.

To summarize, these preliminary results show that over and above variance, prediction-decay is a good discriminator of health; healthy patients have a much steeper decay curve than those with heart disease. And nonlinearity is a good discriminator of age; younger adults have more nonlinearity to their heart rhythms than older ones. In addition, nonlinear predictability appears to be a good classifier of age in the normalized maps.

### *4.2. Infant Heart Rhythms:*

As a suggestive element to add to the results above, I shall briefly mention recent work (Sugihara, Allan & Sobel, in prep.) on the development of heart rhythms in infants that shows a very different pattern.

Allan & Sobel (1992) hypothesize that because other autonomic nervous system functions such as pupilary reflex seem to mature at about 30 weeks, that somewhere between 25 and 35 weeks sympathetic and parasympathetic balance evolves and contributes to irregularity in heart rate. Goldberger (1991), moreover, hypothesized that the source of irregularity in human heart rhythms is the sympathetic-parasympathetic balance acting on the sino-atrial node. Thus, one might expect nonlinearity to be absent in early gestational infants, and to be acquired in later development.

Figure 9 shows an illustrative example of this possibility for an infant aged 25 weeks and one aged 40 weeks. Here the S-map was applied to heart rate data taken by chest leads, $n = 1000$. This general pattern was found in 6 additional infants where paired observations could be made at 26 weeks and again at 37 weeks. The early gestational infants had nearly linear heart rhythms, and in all 6 cases, the more mature infants had more nonlinear heart rhythms. It is intriguing to hypothesize that heart rhythms might follow an ontogenetic sequence, starting out simple with high short-term linear autocorrelation, maturing to nonlinearity at birth; then from young adulthood onwards simplifying with age to a signal that is once again more strongly dominated by linear characteristics.

## 5. Summary

In summary, despite concerns over the feasibility of a rigorous test for chaos in real data (from the calculation of meaningful Lyapunov exponents) there are methods available that allow one to gather qualitative evidence for the operation of such dynamics. Moreover, the narrow question of chaos aside, there are useful and important ways that nonlinear methods can be applied in the natural sciences and especially in

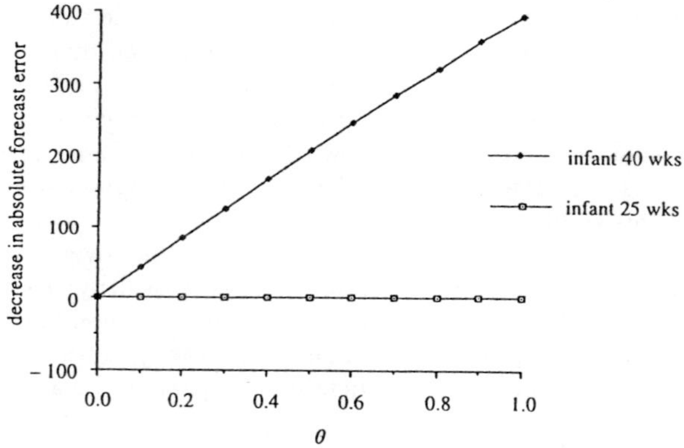


Fig. 9. An anecdotal comparison of heart rate nonlinearity in an infant 25 weeks and an infant 40 weeks. The younger infant did not have t he nonlinearity that was present in the older infant.

pragmatic classification problems and in medical diagnosis.

## Acknowledgements

I would like to thank Martin Casdagli, Jim Crutchfield, Dennis Grace, Cleridy Lenhert, Richard Penner and John Sidorowich for helpful and challenging discussions. I am indebted to Julie Lefebvre, Mark Kamath and D. G. Goodings for their kindness and generosity in providing me with their data on adult heart rates. I also wish to thank Walter Allan and Daniel Sobel for letting me use some of their data on infant heart rates for figure 9. Finally, I would like to thank Howell Tong and Bryan Grenfell for their thorough and insightful comments on the final manuscript. This work was funded in part by ONR:N00014-92-J-4068, and endowment income from the John Dove Issacs Chair in Natural Philosophy.

## References

Abarbanel, H.D.I., R. Brown, J.J. Sidorowich & L.Sh. Tsimrin 1993 The analysis of observed chaotic data in physical systems. *Rev. Mod. Phys.* **64(4)**: 1331-1392.

Allan, W.C., & D.B. Sobel 1992 The Sans of Time *J. Amer. Med. Assoc.,* **268**: 984.

Ascioti, F.A., E. Beltrami, T.O. Caroll & C. Wirick 1993 Is there chaos in plankton dynamics? *Jour. Plank. Res.* **15**(6): 603-617.

Bottou, L., 1993 Local learning for optical character recognition. *Neural Information Processing Systems,* **5** .

Brock, W., 1986 Distinguishing random and deterministic systems: Abridged version. *J. Econ. Theory,* **40,** 168-195.

Brock, W. & S.M. Potter 1993 Nonlinear time series and macroeconometrics. In *Handbook of Statistics, Vol. 11.* G.S. Maddala, C.R. Rao and H.D. Vinod. Elsevier Science Publishers. pp. 195-229.

Brock, W., D. Hsieh, & B. LeBaron 1991 Nonlinear dynamics, chaos, and instability: Statistical theory and economic evidence. MIT Press.

Brock, W., W. Dechert & J. Scheinkman 1987 A test for independence based upon the correlation dimension. Department of Economics, University of Wisconsin, University of Houston, and the University of Chicago.

Casdagli, M., 1989 Nonlinear prediction of chaotic time series. *Physica D,* **35**: 335.

Casdagli, M., 1991 A dynamical systems approach to modeling input-output systems. In *Nonlinear Modeling and Forecasting. Santa Fe Institute Studies in the Sciences of Complexity, Proceedings XII.* M. Casdagli & S. Eubank. New York: Addison-Wesley. pp. 265-281.

Casdagli, M., 1992 Chaos and deterministic versus stochastic modeling. *J. Roy. Stat. Soc. B*: 303-328.

Casdagli, M., S. Eubank, J.D. Farmer & J. Gibson 1991 State space reconstruction in the presence of noise. *Physica D*, **51**: 52-98.

Casdagli, M., D.D. Jardins, S. Eubank, J.D. Farmer, J. Gibson, N. Hunter & J. Theiler 1992 Nonlinear modeling of chaotic time series: theory and applications. *Los Alamos National Laboratory Technical Report LA-UR-91-1637.*

Chan, K.S., & H. Tong 1994 A note on noisy chaos. *J. Roy. Stat. Soc. B*, **56**: 301-311.

Cleveland, W.S., 1979 Robust locally weighted regression and smoothing scatterplots. *J. Amer. Stat. Assoc.,* **74**: 829-836.

Crutchfield, J.P., 1979 Prediction and stability in classical mechanics. Bachelor's Thesis. University of California, Santa Cruz.

Crutchfield, J.P., & B.S. MacNamara 1987 Equations of motion from a data series. *Complex Systems,* **1**: 417-452.

Diebold, F.X., & J.A. Nason 1990 Nonparametric exchange rate prediction? *J. International Economics*, **28**: 312-332.

Ellner, S., 1991 Detecting low-dimensional chaos in population dynamics data: a critical review. In *Chaos and insect ecology* (ed. J.A. Logan & F.P. Hain), pp. 65-92.

Farmer, J.D., & J.J. Sidorowich 1987 Predicting chaotic time series. *Phys. Rev. Lett.,* **59**(8): 845-848.

Gershenfeld, N.A., & A.S. Weigend 1993 The future of time series: learning and understanding. In *Time Series Prediction: Forecasting the Future and Understanding the Past* (ed. A.S. Weigend & N.A. Gershenfeld). Reading, MA: Addison-Wesley. pp. 1-70.

Glass, L., & M.C. Mackey 1988) *From Clocks to Chaos: the Rhythms of Life.* Princeton: Princeton University Press.

Goldberger, A.L., D.R. Rigney & B.J. West 1990 Chaos and fractals in human physiology. *Sci. Am.* **262**: 43-49.

Goldberger, A.L. 1991 Is the normal heartbeat chaotic or homeostatic? *Int. Union Physiol. Sci. / Am. Physiol. Soc.* **6**: 87-91.

Goldberger, A.L., L.J. Findley, M.R. Blackburn & A.J. Mandell 1984 Nonlinear dynamics in heart failure: implications of long-wavelength cardiopulmonary oscillations. *Am. Heart J.* **107**(3): 612-615.

Gough, N.A.J., 1992 Fractals, chaos and fetal heart rate. *Lancet,* **339**: 182-183.

Hardle, W. 1990 *Applied Nonparametric Regression* Cambridge: Cambridge University Press.

Hastings, A., C.L. Hom, S. Ellner, P. Turchin & C. Godfrey 1993 Chaos in ecology: is mother nature a strange attractor? *Ann. Rev. Ecol. and Syst.* **24**: 1-33.

Kaplan, D.T., M.I. Furman, S.M. Pincus, S.M. Ryan, L.A. Lipshitz, & A.L. Goldberger 1991 Aging and the complexity of cardiovascular dynamics, *Biophys. J.,* **59**: 945-949.

Kleiger, R.E., J.P. Miller, J.T. Bigger, A.J. Moss & the Multicenter Post-Infarction Research Group1987 Decreased heart rate variability and its association with increased mortality after acute myocardial infarction. *Am. J. Cardiol.,* **59**: 256-262.

Kravtsov, Y.A., & V.S. Etkin 1984 Dynamic Correlation degree and the problem of determining the dynamic nature of random processes, *Radiotekhnika i electrotechnika,* **12**: 2358-2364.

Lefebvre, J.H., D.A. Goodings, M.V. Kamath & E.L. Fallen 1993 Predictability of normal heart rhythms and deterministic chaos. *Chaos,* **3**: 267-276.

Lipshitz, L.A., & A.L. Goldberger 1992 Loss of 'complexity' and aging: potential applications of fractals and chaos theory to senescence. *J. Amer. Med. Assoc.,* **267**: 1806-1809.

Nychka, D., S. Ellner, A.R. Gallant, D. McCaffrey 1992 Finding chaos in noisy systems *J. Royal Stat. Soc. B,* **54**: 399-426.

Olsen, L.F., G.L. Truty, W.M. Schaffer 1988 Oscillation and chaos in epidemics: a nonlinear dynamic study of six childhood diseases in Copenhagen, Denmark. *Theor. Pop. Biol.* **33**: 344-70.

Osborne, A.R., & A. Provenzale 1989 Finite correlation dimension for stochastic systems with power-law spectra. *Physica D,* **35**: 357-381.

Rand, D.A., & H.B. Wilson 1991 Chaotic stochasticity: a ubiquitous source of unpredictability in epidemics. *Proc. R. Soc. Lond. B* **246**: 179-184

Schaffer, W.M. 1985a Can nonlinear dynamics elucidate mechanisms in ecology and epidemiology? *IMA J. Math. Appl. Med. Biol.,* **2**: 221-252.

Schaffer, W.M. 1985b Order and chaos in ecological systems. *Ecology,* **66**: 93-106.

Schaffer, W.M., L.F. Olsen, G.L. Truty, S.L. Fulmer & D.J. Graser 1988 Periodic and chaotic dynamics in childhood infections. In *From chemical to biological organization.* (ed. M. Markus, S.C. Muller & G. Nicolis New York: Springer-Verlag. pp. 331-347.

Schaffer, W.M., L.F. Olsen, G.L. Truty & S.L. Fulmer 1990 The case for chaos in childhood epidemics. In *The Ubiquity of Chaos.* (ed. S. Krasner Washington: AAAS pp. 138-166.

Sidorowich, J.J. 1992 Repellors Attract Attention. *Nature,* **355**: 584.

Skinner, J.E., A.L. Goldberger, G. Mayer-Kress, & R.E. Ideker 1990 Chaos in the heart: implications for clinical cardiology. *Bio/Technology,* **8**: 1018-1024.

Stokbro, K., & D. Umberger 1992 Forecasting with weighted maps. In *Nonlinear Modeling and Forecasting* (eds M. Casdagli & S. Eubank). Redwood City: Addison-Wesley.

Stone, L. 1992 Coloured noise or low-dimensional chaos? *Proc. R. Soc. Lond. B,*

**250**: 77-81.

Subba Rao, T., & M.M. Gabor 1984 An introduction to bispectral analysis and bilinear time series models. In *Lecture Notes in Statistics* New York: Springer-Verlag.

Subba Rao, T. 1991 Analysis of nonlinear time series (and chaos) by bispectral methods. In *Nonlinear Modeling and Forecasting* (eds M. Casdagli & S. Eubank). Reading, MA: Addison-Wesley.

Sugihara, G. 1994 Nonlinear forecasting for the classification of natural time series. *Phil. Trans. of the Roy. Soc. London Series A,* **348**:(1688) 477-495.

Sugihara, G., & R. May 1990 Nonlinear forecasting as a way of distinguishing chaos from measurement error in a data series. *Nature,* **344**: 734-741.

Sugihara, G., B. Grenfell & R. May 1990 Distinguishing error from chaos in ecological time series. *Phil. Trans. Roy. Soc. Lond. B,* **330**: 235-251.

Theiler, J., B. Galdrikian, A. Longtin, S. Eubank & J.D. Farmer 1991 Using surrogate data to detect nonlinearity in time series. In *Nonlinear Modeling and Forecasting* (eds M. Casdagli & S. Eubank). Reading, MA: Addison-Wesley.

Tong, H. 1990 *Nonlinear Time Series Analysis: A Dynamical Systems Approach.* Oxford University Press.

Tont, S.A. 1981 *J. Mar. Res.,* **39**: 191-201.

Tsonis, A.A., & J.B. Elsnor 1992 Nonlinear prediction as a way of distinguishing chaos from random fractal sequences. *Nature,* **358**: 217-220.

Wales, D.J. 1991 Calculating the rate of loss of information from chaotic time series by forecasting. *Nature, Lond.,* **350**: 485-488.

Yao, Q., & H. Tong 1994 On prediction and chaos in stochastic systems. *Phil. Trans. of the Roy. Soc. London Series A,* **348**:(1688) 357-369.

CHAPTER 14

# MEASURING AND CHARACTERISING SPATIAL PATTERNS, DYNAMICS AND CHAOS IN SPATIALLY-EXTENDED DYNAMICAL SYSTEMS AND ECOLOGIES.

D. A. Rand,
Nonlinear Systems Laboratory,
Mathematics Institute, University of Warwick,
Coventry CV4 7AL, UK.

**Abstract**

In this paper I discuss space-time chaos in both locally mixing continuum systems (reaction-diffusion equations, coupled map lattices and functional maps) and individual-based models (probabilistic cellular automata and artificial ecologies). I particularly emphasise quantification and data-analysis and attempt to address the characterisation of spatial structure and dynamics in such disordered systems. I discuss the relevance of these ideas to ecology, evolution and epidemiology. The artificial ecologies I consider motivate a new definition of space-time chaos for such systems and new data analysis techniques.

## 1. Introduction.

In this paper I want to address a number of basic issues about the dynamics of spatially-extended systems which are disordered in time or space. I am particularly interested in the characterisation of spatial structure and its dynamics and how to quantify and measure this. In particular, I want to discuss the relevance of these ideas to fields such as ecology, evolution and epidemiology. These are areas where virtually all the real systems are spatially-extended but where very few of the models take this into account. There are important practical implications for data collection and analysis.

Ecological models to study spatial effects fall into two broad categories. In the first, one assumes local mixing in space and models this using either a reaction-diffusion equation (e.g. Okubo 1980 and Murray 1990) or a coupled map lattice (e.g. Hassell *et al.* 1991). Those 'patch' models which assume a large number of patches, homogeneous mixing within a patch and dispersal between patches (e.g. Hassell 1978, Reeve 1990, Hastings 1990, Levin *et al.* 1993) also fall into this class.

The other category of models are those which are individual-based. In these one models the stochastic behaviour of individuals which are distributed in space. Dynamics and competition are built into the model through the use of simple transition rules which only depend upon the local configuration. One is interested in the way in which the spatial dynamics are integrated to produce community behaviour. There are now a number of interesting applications of such systems to ecology and epidemiology. These include contact processes and spatial epidemics (e.g. Durrett 1993), host-parasitoid and host-pathogen dynamics (e.g. Comins *et al.* 1993 and Rand *et al.* 1994b), ecological systems (e.g. Durrett & Levin 1994, DeRoos *et al.* 1991 and Rand & Wilson 1994a) and game theory (e.g. Durrett & Levin 1993 and Nowak & May 1992).

I want to discuss both these classes in this paper. I start by very briefly discussing some illustrative examples of continuum systems drawn from (i) mathematics and physics and (ii) ecology, and epidemiology. Then I discuss some basic theory of disordered continuum systems and give a brief introduction to recent results on spatio-temporal chaos. Finally, in this section I consider some question of quantification and data analysis for these systems.

Then I move on to individual-based systems. After discussing some ecological examples and related general theory, I consider some new ideas about the data analysis of such systems. I describe a new method for identifying spatial scales. I outline a method (Rand & Wilson 1994a) for immense data compression in such systems and discuss its application to detecting structural change in ecosystems. I apply this approach to a resource-predator-prey system and a spatial model of succession in a forest community.

These examples motivate a new definition of space-time chaos for such individual-based systems. This is given in section 3.5.

In the discussion I will consider a number of basic questions of general applicability. A basic problem is to understand when such spatially-extended systems admit a good description in terms of low-dimensional dynamics. How can we understand, describe and characterise the spatio-temporal behaviour of systems for which this low-dimensional description does not hold? How should we characterise spatial order and disorder? What is the nature of (a) the continuum limit (see section 2.2.2) and (b) the thermodynamic limit (see section 2.2.3) in such systems? How do we deal with the immense amounts of data in the states of these models and maximise the information gained from this data? Are there general techniques for identifying appropriate spatial and temporal scales?

## 2. Locally mixing continuum systems.

In this half of the paper I will be concerned with systems which are modelled either by reaction-diffusion equations, functional maps or coupled map lattices. Reaction-diffusion equations have the following general form: $\partial_t u = D\nabla^2 u + X(u)$. Here $u = u(x,t) \in \mathbf{R}^m$ and the diffusion matrix $D$ may be complex. An example is the complex Ginzburg-Landau (CGL) equation below.

Coupled map lattices (CMLs) are defined as follows. Let $\Lambda$ be $\mathbf{Z}^\nu$ or a finite rectangle in $\mathbf{Z}^\nu$. The state $u = (u_x)_{x\in\Lambda}$ associates to each point $x$ in $\Lambda$ an element $u_x$ of a vector space $V$. The simplest CMLs are mappings $\Phi$ of the form

$$\Phi : (u_x)_{x\in\Lambda} \mapsto (\sum_{y\in\Lambda} s_\varepsilon(|y-x|)f(u_y))_{x\in\Lambda}.$$

Here $f$ is a mapping from $V$ to $V$ and the interaction $s_\varepsilon$ satisfies (i) $s_\varepsilon(n)$ is exponentially small in $|n|$, (ii) $s_\varepsilon(n) < \varepsilon$ for all $n \neq 0$ and (iii) $s_\varepsilon(0) \in [1-\varepsilon, 1]$. A much studied example is that for nearest-neighbour coupling where $s_\varepsilon(0) = 1-\varepsilon$, $s_\varepsilon(1) = \varepsilon/N$ (where $N$ is the number of immediate neighbours) and $s_\varepsilon(n) = 0$ for all other $n$. A CML $\Phi = \Phi_\varepsilon = A_\varepsilon \circ L$ is a composition of two maps where the *local map* is $L((u_x)) = (f(u_x))$ and $A_\varepsilon$ is the *interaction* or *coupling map* defined by $s_\varepsilon$. For the nearest neighbour example above

$$A_\varepsilon(u)_x = u_x + \varepsilon\nabla^2(u)_x$$

where $\nabla^2(u)_x = \sum_{\text{nbrs } y \text{ of } x}(u_y - u_x)$. Weak coupling means $\varepsilon \approx 0$.

The natural phase space for a CML is usually an open subset $\ell^\infty(E)$ of the $\ell^\infty$-space $\ell^\infty(V)$. Here $\ell^\infty(V)$ is the Banach space of all $u = (u_x)$ such that $u_x \in V$ for all $x$ and $||u|| = \sup_x ||u_x|| < \infty$. And $\ell^\infty(E) = \{u : u_x \in E \text{ for all } x\}$.

Functional maps are similarly defined but in this case $x$ is continuous and ranges over a rectangle in $\mathbf{R}^\nu$. For the interaction $A_\varepsilon$ we take standard smoothing operators such as $\exp(\varepsilon\Delta)$, truncations of this or operators defined by related kernals. For example, the CGL map (see table 1) gives a functional map with closely analogous behaviour to the CGL equation in one and two dimensions (Bohr *et al.* 1989, 1990a,b).

In general we use CMLs to probe the thermodynamic limit and functional maps to probe the continuum limit.

### *2.1. Some illustrative examples.*

For my theoretical discussion I want to refer to the following examples. The mathematical formula for each example is given in table 1.

| | Local (reaction) dynamics $L$ | Coupling $A_g$ |
|---|---|---|
| $x \in [0,1]$ | $x' = 4ax(1-x)$ | $x'_i = x_i + g\nabla^2(\underline{x})_i$ |
| $u = re^{i\phi} \in \mathbf{C}$ | $\partial_t u = \varepsilon u + (1+ic_1)\nabla^2 u - (1-ic_2)\|u\|^2 u$ | |
| $u = re^{i\phi} \in \mathbf{C}$ | $f(re^{i\phi}) = \sqrt{(\mu r^2/(\lambda\mu + (1-\lambda)r^2))}$ $\times \exp(\phi - \tau\alpha r^2)$ | $(1+\tau_0(1+i\beta)\Delta/M)^M$ $\approx \exp(\tau_0(1+i\beta)\Delta)$ |
| prey $v$ | $v' = v(1+b_1)$ $\times \exp(-\alpha p/r - c_1 p/(1+h_1 v))$ | $v'_i = v_i + g_v\nabla^2(\underline{v})_i$ |
| predator $p$ | $p' = p(1-d)\exp(c_2 v)$ | $p'_i = p_i + g_p\nabla^2(\underline{p})_i$ |
| resource $r$ | $r' = r(1+b_3)\exp(-r/k - c_3 p)$ | $r'_i = r_i + g_r\nabla^2(\underline{r})_i$ |
| host $h$ | $h' = \lambda h e^{-p}$ | $h'_i = h_i + g_h\nabla^2(\underline{h})_i$ |
| parasite $p$ | $p' = qh(1-e^{-p})$ | $p'_i = p_i + g_p\nabla^2(\underline{p})_i$ |

Table 1: Some examples of the CMLs, functional maps and pdes used to illustrate the theoretical discussion. They are (i) coupled logistic maps, (ii) the complex Ginzburg-Landau (CGL) equation, (iii) the CGL map, (iv) the resource-predator-prey system and (v) the host-parasitoid system

2.1.1. Mathematical and physical examples.

1. *Coupled logistic maps.* This is perhaps the most studied example, although there is little rigorous theory. Many of the interesting phenomena found in this system are of a very general nature. Much of this was discovered by Kaneko (e.g. Kaneko 1991). It has regimes of fully-developed space-time chaos and also space-time intermittency as described in section 2.3 below. The former is extremely stable and, for typical initial conditions, one sees no periodic windows of the form found in single logistic maps. This is due to the stabilisation of chaotic repellers by spatial extension as described below in section 2.4. Quantities such as Kolmogorov-Sinai entropy, fractal dimension, Lyapunov dimension and the number of positive Lyapunov characteristic exponents are extensive in the sense of section 2.2.4.

2. *Hyperbolic CMLs.* In these each local map has a hyperbolic attractor. For them one can prove rigorous theorems about their geometric, dynamical and statistical behaviour when the coupling is weak (Bunimovich & Sinai 1988, Pesin & Sinai 1991, Gundlach & Rand 1993*a-d*, Campbell & Rand 1994). See section 2.2.3. In particular, in this regime they have fully-developed space-time chaos and possess a natural space-time measure.

3. *The complex Ginzburg-Landau (CGL) equation and map.* The CGL equation describes the modulations of a system that has undergone a bifurcation to an oscillatory state and is therefore very general and ubiquitous. For 1-dimensional physical space, in the various regions of its phase space it exhibits (a) *'amplitude'* or *'defect' turbulence* in which there is a chaotic 'gas' of defects, (b) weaker *'phase' turbulence* in which the amplitude never vanishes and where the patterns are modulated by a signal that is well-described by a phase equation of Kuramota-Sivashinsky type and

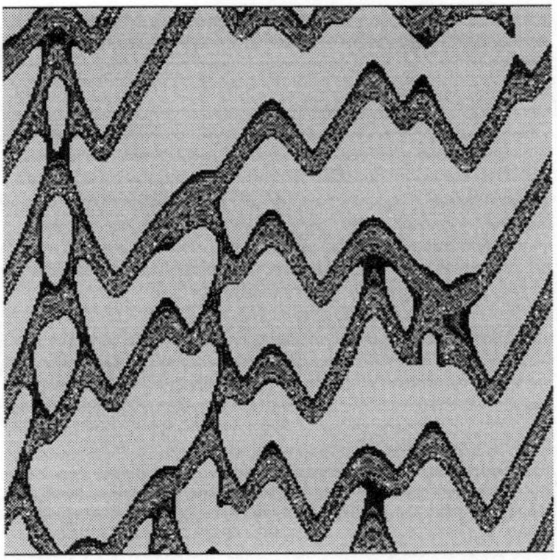

Fig. 1. A typical space-time plot of the prey in the 1-dimensional resource-predator-prey system. It shows how the bursts of prey nucleate and convect across the system.

(c) *space-time intermittency* where hole-like objects randomly nucleate from the background plane-wave absorbing states (e.g. Schraiman *et al.* 1992, Chaté 1994). The CGL map displays similar behaviour. In two dimensions there are very interesting regular and chaotic vortex states (Bohr *et al.* 1989, 1990a, 1990b, Aranson *et al.* 1991, Weber *et al.* 1992).

2.1.2. Biological examples.

4. *A resource-predator-prey system.* The local dynamics for this system are chaotic (e.g. Rand, Wilson & McGlade 1994d). The corresponding diffusively coupled CML displays space-time intermittency. The background state has zero prey and plentiful resource. From this state one gets random nucleation of prey outbreaks followed by predators waves. These structures then propagate with a characteristic velocity across the ecology. This is shown in the space-time plot in Figure 1. We discuss quantifiers of this state below (e.g. in section 2.3).

5. *Host-parasitoid system* (Comins, Hassell & May 1992). The local dynamics for this involve diverging Lotka-Volterra cycles which always lead to extinction. However, in large 2-dimensional systems they can survive for a very long time and there are attractors with infinite lives. In not too small 1-dimensional systems, by adding a very small random number of parasites at the boundaries one can put the system into an attractor which for infinitely long times has a relatively large minimum population size. The states in this model are space-time intermittent rather than fully-developed space-time chaotic. They are similar to those in the resource-predator-prey system above. The background state is that where both parasite and host is absent. From this one gets nucleation of bursts of infection that are then convected across the

system.

*2.2. Mathematical theory.*

2.2.1. Basic length scales: dissipation, excitation and correlation length scales.

Let us suppose that our system is defined over a region of $\nu$-dimensional physical space represented by a rectangle of length $L$. We shall be particularly interested in the behaviour of our systems as the volume $V = L^\nu$ of the system is increased.

Since all the systems we consider are dissipative, there is a length scale $\ell_D$ below which all modes are damped. This is the *dissipation length.* For typical low-dimensional systems $\ell_D \approx L$ and there are only a few excited modes. However, in large systems $\ell_D \ll L$ and this would appear to be the case in many ecological and epidemiological situations. This scale is very important for data analysis. In general one expects that it is not necessary to resolve the spatial structure at scales well below $\ell_D$. I discuss the mathematical meaning of $\ell_D$ in section 2.2.2 below.

Another important length scale is the *excitation length* $\ell_E$. This is the length scale at which energy is injected. It is a particularly important length scale in ecology because it often determines the nature of the instabilities and patterns observed.

Consider the correlation $C_{ij}(\tau, r) = \langle (u_i(x,t) - \bar{u_i})(u_j(x+r, t+\tau) - \bar{u_j})\rangle$ where the averaging $\langle\cdot\rangle$ is over both space $x$ and time $t$. In many spatio-temporal chaotic systems $C_{ij} \sim \exp(-r/\xi - \tau/\gamma)$. This defines the *spatial correlation length* $\ell_C = \xi$ and the *temporal correlation time* $\gamma$.

2.2.2. The continuum limit and the dissipation length.

We consider systems with a finite spatial domain. Suppose that the phase space of our dynamical system is a Hilbert space $\mathbf{H}$. For some systems such as the CGL equation it can be proved (Doering *et al.* 1988) that on the attractor $A$ there is a cone condition of the form $||Q_n u|| \leq ||P_n u||$ where $P_n u$ is the projection of $u \in H$ onto the space $\mathbf{H}_{\leq n}$ spanned by the first $n$ generalised Fourier modes and $Q_n = 1 - P_n$ is the complementary projection onto $\mathbf{H}_{>n}$. It follows from this that if $A' = P_n(A)$ then there is a Lipshitz map $a : A' \subset \mathbf{H}_{\leq n} \to \mathbf{H}_{>n}$ such that every $u \in A$ is of the form $u = u' + a(u')$ for $u' = P_n(u) \in A'$. In other words, the high modes $a(u')$ are *slaved* to the low modes $u'$.

If this holds, we can identify the dissipation length $\ell_D$ as the longest wavelength $\lambda$ of the modes in $\mathbf{H}_{\leq n}$.

When there is slaving on the attractor, it is natural to ask if there is a finite-dimensional dynamical system that mirrors the non-transient behaviour of the full system. Thus the problem becomes one of extending the slaving function $a$ to some sort of open set $U$ in the space of low Fourier modes $\mathbf{H}_{\leq n}$. This can be done if we can extend the cone condition. In this case, standard invariant manifold techniques can be used to prove the existence of an *inertial manifold* which is the graph of a mapping $a : U \subset \mathbf{H}_{\leq n} \to \mathbf{H}_{>n}$.

For a wider class of systems, such as 2-dimensional Navier-Stokes, one can prove that attractors are finite-dimensional. This can be done when one can show that near to the attractor small round balls in phase space are transformed by the dynamics into 'pancakes' which are very thin in all but $m$ dimensions. When this holds one can define $\ell_D$ in a similar way to that described above.

2.2.3. The thermodynamic limit and extensive chaos.

I would like to start this section by proposing a definition of space-time chaos for such continuum systems. (I give the corresponding definition for individual-based systems in section 3.5.) For this definition it is important to keep in mind that we are interested in large systems. Mathematically we are interested in the large volume limit $V \to \infty$. Thus I consider the systems of interest to be parameterised by $V$ or equivalently by the length of the system $L$.

**Definition.** A system is *space-time chaotic* if *(i)* it has a thermodynamic limit at $V = \infty$; and *(ii)* the number of positive Lyapunov characteristic exponents is $\mathcal{O}(V)$.

All the example systems of section 2.1 contain regimes that satisfy this definition. It should be noted that the notion of space-time chaos described here can be further subdivided. For example, we have seen already that in the CGL equation one finds regimes of space-time intermittency and amplitude turbulence, and both of these can satisfy our definition. I discuss these distinctions below.

Let me discuss (ii) first. Of course, the main characteristics of low-dimensional chaos are the existence of an attractor and at least one positive Lyapunov characteristic exponent (LCE) for this attractor. In a spatially-extended system we expect that the number of active degrees of freedom grows roughly linearly in the system's volume $V$. For a system that is both chaotic in space and time we also expect that a roughly constant proportion of these will correspond to unstable modes with a positive exponent. Thus we want the condition (ii).

In many systems a stronger condition holds: there is a limiting *Lyapunov density* $\lambda(\chi) = \lim_{V\to\infty} \lambda_V(\chi)$. Here $\lambda_V(\chi)$ is the ratio of the number of exponents which are $> \chi$ to the number of the positive exponents for the system of volume $V$. An example of such a Lyapunov density is shown in Figure 2 where I draw approximations to that for the host-parasitoid model. All our examples have a Lyapunov density in the appropriate regimes.

We note now that if (ii) holds then it defines a new length scale $\ell_L$ by the relation: number of positive exponents $\sim (L/\ell_L)^{\nu}$. The density of positive characteristic exponents is $\ell_L^{-\nu}$. Since the Lyapunov dimension is determined by the Lyapunov exponents, from the Lyapunov spectrum on can determine a density $\ell_{LD}^{-\nu}$ for the Lyapunov dimension via the relation: Lyapunov dimension of attractor $\sim (L/\ell_{LD})^{\nu}$. For the logistic system with the coupling $g = 0.3$ this is approximately 0.67.

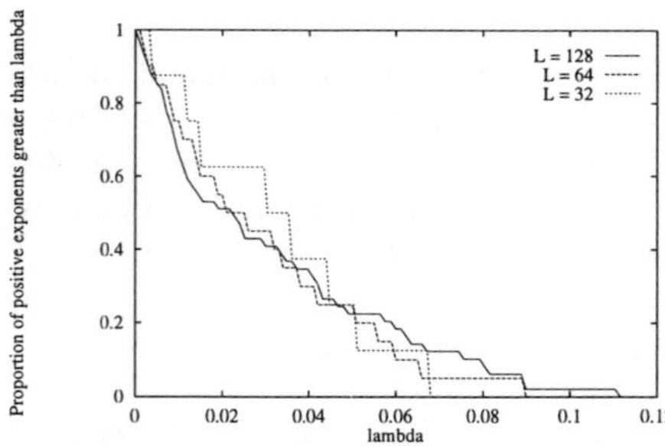


Fig. 2. An approximation of the Lyapunov density for the host-parasitoid system with parameters $\lambda = 2$, $q = 1$ and $g_h = g_p = 0.25$. The graphs are for $L = 32$, $L = 64$ and $L = 128$.

*The thermodynamic limit.*

Now I want to turn to the question of what is the thermodynamic limit for such systems. I will discuss this for coupled map lattices, but it will be clear to the reader how to generalise this to other systems. Consider a CML $\Phi : \ell^\infty(E) \to \ell^\infty(E)$. This is the infinite volume system and has associated with it the finite volume systems $\Phi_V$.

We say that $\Phi$ has a *thermodynamic limit* if it satisfies the following requirements:

(i) $\Phi$ has an attracting invariant set $A$ with open basin $B = B(A)$ in $\ell^\infty(E)$. This restriction on the basin can be relaxed somewhat.

(ii) For typical initial conditions $u(\cdot)$ in $B$ the spatio-temporal statistics are described by a single spatio-temporal ergodic natural measure $\mu$ on $A$.

(iii) If $\mu_V$ is the corresponding natural measure for the system of finite volume $V$, then $\mu_V \to \mu$ as $V \to \infty$ and the number of positive LCE of $\mu_V$ is $\mathcal{O}(V)$.

Here typical has the following meaning: Let $\pi_V : \ell^\infty(E) \to \ell^\infty_V(E)$ be the projection onto the corresponding $V$ sites. Typical means that $u(\cdot) \in B_0$ where $B_0$ is a subset of $B$ such that $\pi_V(B_0)$ has full Lebesgue measure in $\pi_V(B)$ for all $V$. Space-time ergodic means that for all continuous $v : \ell^\infty(E) \to \mathbf{R}$

$$< v(S^x(u)) >_x = < v(\Phi(u(x)) >_t = \int_A v \, d\mu$$

where $S^x$ is the spatial translation: $(S^x u)(y) = u(x + y)$.

It is proved in Bunimovich & Sinai (1988), Pesin & Sinai (1991), Gundlach & Rand (1993*a*-*d*) and Campbell & Rand (1984) that weakly-coupled map lattices with hyperbolic local dynamics have a thermodynamics limit.

2.2.4. Extensive chaos.

Recall the definition of the exponent length $\ell_L$ and the density for Lyapunov dimension. Analogously, we can define the following related length scales:

$\ell_F$: fractal dimension of $A \sim (L/\ell_F)^\nu$

$\ell_{R,q}$: order $q$ Renyi entropy of $A \sim (L/\ell_{R,q})^\nu$

When these quantities and the length scales $\ell_L$ and $\ell_{LD}$ for the Lyapunov exponents and Lyapunov dimension are well defined and finite we say the we have *extensive chaos.* Then, in the language of physics, each of the classical dynamical quantities (fractal dimension, Lyapunov dimension, order $q$ Renyi entropy and number of positive LCEs) is extensive.

It should be noted that these various lengths $\ell$ are independent of the correlation length $\ell_C$. For example, there are examples of one and two-dimensional systems (e.g. Miller and Huse 1993) where $\ell_C \to \infty$ as the coupling is increased to some finite critical value while $\ell_L$ remains finite.

It should not be too difficult to use the results of Campbell & Rand (1984) to show that weakly-coupled map lattices with hyperbolic local dynamics have extensive chaos. All the examples of section 2.1 are extensive in the appropriate regimes.

2.2.5. Spatial velocity of entropy production.

We recall a basic equation from the Grassberger-Procaccia theory for dimension calculations. Consider a time series $u(t)$ arising from a deterministic systems. Let $u_d(t)$ denote the associated $d$-dimensional time-delayed vector $(u(t), u(t+\tau), \ldots, u(t+(d-1)\tau)$. Let $C_d(\varepsilon)$ denote the proportion of pairs of vectors $u_d(s)$, $u_d(t)$ which are within $\varepsilon$ of each other. Then we expect that $C_d(\varepsilon) \sim \varepsilon^{D_2} \exp(-d\tau K_2)$ as $d \to \infty$ and $\varepsilon \to 0$ (Grassberger & Procaccia 1983). Here $K_q$ is the order $q$ Renyi entropy.

A similar approach can be used in the spatial context. The time-series is replaced by spatial vectors $u(x)$ and the vectors $u_d(x)$ formed accordingly. Then if $S_d(\varepsilon)$ denotes the proportion of pairs of vectors $u_d(x)$, $u_d(y)$ which are within $\varepsilon$ of each other we expect that

$$\log S_d(\varepsilon) \sim D_2^s \log \varepsilon - d\tau K_2^s.$$

This defines a spatial dimension $D_2^s$ and a velocity $K_2^s$ of entropy production. For calculation, the way to interpret the limits is to plot $\log S_d(\varepsilon)$ *vs.* $\log \varepsilon$ for various values of $d$ hoping for a region in $\log \varepsilon$ of near linearity which one can use to estimate $D_2^s$. The value of $K_2^s$ is then given by the vertical displacement of these curves.

*2.3. Space-time intermittency.*

Space-time intermittency is characterised by a competition between two states. The first state, the 'laminar' or background state acts as an absorbing state and typically, but not necessarily, is linearly stable. The 'turbulent' active state is usually made up of localised objects. These carry the disorder. They erupt from the

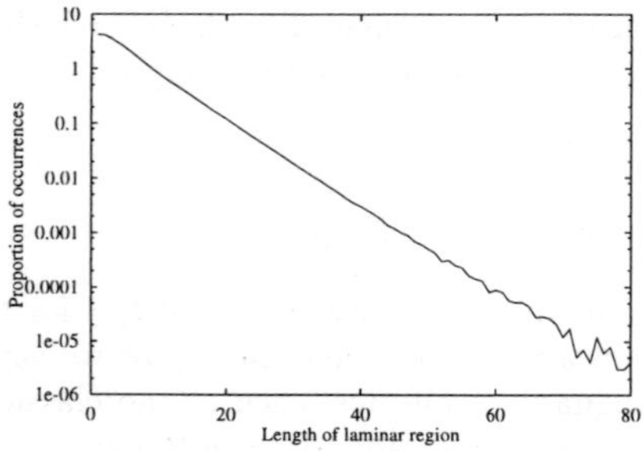


Fig. 3. The distribution $P(\xi)$ of the sizes $\xi$ of the laminar regions separating the localised objects in the host-parasitoid system with parameters $\lambda = 2$, $q = 1$ and $g_h = g_p = 0.25$.

background state in a chaotic fashion. However, note that it is not necessary for the background state to be truly laminar or for the active state to be chaotic.

In the 1-dimensional CGL equations the localised objects are 'holes'. In the resource-predator-prey system they are prey-predator booms which also correspond to holes in the resource. In the host-parasitoid system they are localised lump-like parasite outbreaks. In this system it can be shown (Keeling 1994) that if the host and parasite are both low then the system will become extinct. This state (both populations low) is therefore the stable absorbing state. In the 1-dimensional system most initial conditions will give you just this state and the populations go extinct. However, you can excite the active state by tickling the system with some very small localised random noise. It is easily excited and once active stays present even when the noise is turned off.

In this regime these localised nucleating objects which carry the disorder are the building blocks for a statistical theory. For example, two interesting quantities are the distributions $P(\xi)$ and $P(\tau)$ of the sizes $\xi$ and lifetimes $\tau$ of the laminar regions separating the localised objects (Chaté & Manneville 1988). It is expected that in the intermittent regime these will be exponential i.e. $P(\xi) \sim \exp(-\xi/\ell_1)$ and $P(\tau) \sim \exp(-\tau/\ell_2)$ and this then defines spatial and temporal coherence lengths $\ell_1$ and $\ell_2$. The length distribution for the host-parasitoid system is shown in Figure 3. In this case $\ell_1 \approx 5.4$ and $\ell_2 \approx 2.33$. Note that $\ell_1^{-1}$ corresponds to a spatial entropy since we are using a symbolic coding into laminar and turbulent states. Then, assuming that the localised objects do not vary too much, $\ell_1^{-1}$ and $\ell_2^{-1}$ are related to the spatial and temporal KS-entropies of this system.

### *2.4. Stabilisation of chaotic repellers.*

Consider the coupled logistic map example of section 2.1 and suppose that the parameter of the logistic map is chosen so that the uncoupled logistic maps have a stable 3 cycle. Then the coupled system has an attractor which is the homogeneous state corresponding to this 3-cycle. However, if the system is reasonably large and started from random initial conditions then this attractor is never seen. For the individual maps there is a horseshoe repeller around the 3-cycle and there is a corresponding horseshoe in the coupled system (at least for weak coupling, Campbell & Rand 1994). The states get tangled up in this repeller and the escape time for this grows exponentially or faster with the system size. Thus, for large $V$, the repeller acts effectively as an attractor while for the infinite system it is an attractor.

This is a very common phenomenon in a wide range of systems. Moreover, there are examples, where these escape times become infinite at finite system sizes. An example for the CGL map is given in Bohr *et al.* (1989, 1990a,b)

*2.5. Reconstruction of attractors for spatially-extended systems.*

Recall the various embedding theorem for a time-series coming from a low-dimensional dynamical system (e.g. Takens 1980). Clearly, it is not sensible to attempt reconstruction for large spatially-extended systems using a single scalar observable. However, it has been shown that this does give rise to a practicable method when the observable is a spatially-extended field (Rand & Wilson 1994c). For example in the resource-predator-prey system of section 2.1 for the observable one can take the spatially-extended state of just one of the species and reconstruct the other two from this. If the coupling is strong one can even only monitor the observed state on a sublattice provided that it is well-distributed. This result is of interest for ecology since it offers the possibility of monitoring observable species using remote sensing and then reconstructing the invisible species from the spatially-extended time-series of some observable species.

## 3. Individual-based systems: artificial ecologies.

I now discuss some non-equilibrium individual-based systems. I will concentrate on questions concerned with the data analysis of spatial and temporal patterns in non-equilibrium systems. The main ideas are taken from Rand & Wilson (1994a).

Artificial ecologies (AEs) are a slight generalisation of probabilistic cellular automata (PCAs). Physical space is represented by a 2-dimensional $L \times L$ lattice $\Omega$ of sites. Each site $x$ can be in any one of a number of discrete states $s_1, \dots, s_d$. Therefore, the state of the system is given by a configuration $S = \{S_x\}_{x \in \Omega}$. The state $S(t)$ at time $t$ determines a probability distribution on the potential future states in the following way.

In a PCA each site is updated independently. This is not appropriate for many ecological applications because an event at one site can determine a specific change at another neighbouring site. Thus, in an AE the state of each neighbourhood determines a probability distribution on a finite set of admissible events. This set is called the *event set.* Each of these events is a transformation of the state on the neighbourhood.

There are two possibilities for the updating:

*Synchronous updating* where the state of each neighbourhood determines a probability distribution on the event set. Thus the state $S(t)$ at time $t$ determines a probability distribution of the states at time $t+1$. The whole configuration is then updated synchronously.

*Asynchronous updating* where the state of each neighbourhood determines the rates at which the various events in the event set occur in the neighbourhood. If event $i$ has rate $r_i$ in a neighbourhood then the probability of it occuring in $[t, t+h]$ is $r_i h + o(h)$ where $o(h)/h \to 0$ as $h \to 0$. Once it occurs, the neighbourhood's state may change leading to a change in the rates $r_i$ associated with the neighbourhood.

Note the difference between PCAs and AEs. Wheras in a PCA all events only change the central site, in an AE other sites in the neighbourhood can be changed.

The examples I discuss here are synchronously updated. It appears that for the particular systems I am discussing this does not matter and the corresponding asynchronously updated systems have similar spatial and temporal statistical properties. However, this is not always the case and there are systems where the synchronously and asynchronously updated versions have very different behaviour (e.g. Huberman & Glance, 1993).

### *3.1. Two basic examples.*

In this paper I will consider two examples: a resource-predator-prey system (Rand & Wilson 1994a) and a model for the Middle-European beech forest cycle (Wissell 1991). These examples are chosen for their intermediate-scale non-equilibrium dynamics.

#### 3.1.1. Resource-predator-prey model

Each site can be in one of five states; *resource, predator, predator in a resource, prey or empty.* A predator in a resource is effectively both and its evolution can be described by what happens to each separately. Hence we will ignore this as a separate state. The boundary conditions are periodic.

To define the transition rules, it is simplest to explain our model by focusing on how the individuals within it act. This completely defines the event set and the event probabilities for a given neighbourhood state. In this description by an adjacent site to $x$ we mean one of the sites $x+(1,0)$, $x-(1,0)$, $x+(0,1)$ and $x-(0,1)$.

- A resource site will grow into an adjacent empty site with probability $g$.
- A prey will move into and eat an adjacent resource site. Thus the adjacent resource site will become a prey and the original site will become empty. If there are no resource in adjacent sites a prey will move randomly into one of its adjacent sites.

There is a certain probability of giving birth $p_{V,b}$ into an adjacent empty site. This probability is zero if a prey individual has not eaten for a certain amount of time, $t_{V,b}$. A prey will die with probability one if it has not eaten for a certain amount of time $t_{V,d}$ (where $t_{V,d} > t_{V,b}$).

- An empty site does nothing.
- A predator has a bigger neighbourhood to that of the other species. Its immediate neighbourhood is an 8-site neighbourhood: $\mathcal{N}(x) = \{x + z : |z_1| \leq 1, |z_2| \leq 1, z \in \mathbf{Z}^2, z \neq (0,0)\}$. A predator will move into and eat a neighbouring prey site.

  As for the prey there is a certain probability of giving birth $p_{P,b}$ into a neighbouring empty site. This probability is zero if a predator individual has not eaten for a certain amount of time, $t_{P,b}$. A predator will die with probability one if it has not eaten for a certain amount of time, $t_{P,d}$ ($t_{P,d} > t_{P,b}$), . .

  If there are no neighbouring prey then a predator has a hunting ability in that it can sense prey much further away than it can actually move. A predator can 'see' a prey anywhere in a 48-site neighbourhood, i.e. the neighbourhood is $\mathcal{N}(x) = \{x + z : |z_1| \leq 3, |z_2| \leq 3, z \in \mathbf{Z}^2, z \neq (0,0)\}$. A predator, as the prey, can only move one site in one time step and will thus move to the site in its 8-site neighbourhood which is nearest (in the Euclidean sense) to the located prey.

A typical spatial state for this model is shown in Figure 4. Note the spatial heterogeneity. For these parameter values there is a majority of resource. For different parameter values this can change to be a minority. To obtain this spatial state and in the example used in the following discussion we use the parameter values as given in table 2.

| $g$ | $p_{V,b}$ | $t_{V,b}$ | $t_{V,d}$ | $p_{P,b}$ | $t_{P,b}$ | $t_{P,d}$ |
|---|---|---|---|---|---|---|
| 0.4 | 0.2 | 2 | 8 | 0.1 | 5 | 8 |

Table 2: Parameter values for the resource-predator-prey model. All times are in time steps of the model.

3.1.2. Wissel's Middle European beech forest cycle.

This attempts to model the spatial patterns associated with the natural cycles in the Middle European beech forest (Wissell 1991). The rules for this are given in Figure 5 and a typical spatial state is shown in Figure 6. All my comments on this model are taken from Hendry & McGlade (1994).

*3.2. Dynamics and attractors.*

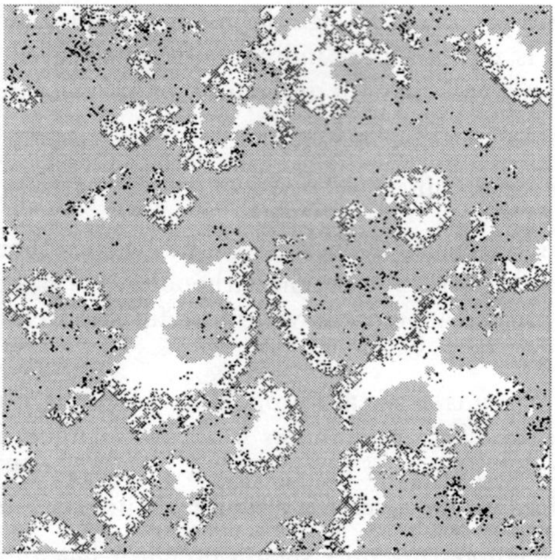

Fig. 4. A typical spatial configuration of the resource-predator-prey AE for the given parameter values. The lattice size shown is 300×300. The lattice size shown is 300×300. The colour-coding is as follows: lightest grey - empty, medium grey - prey, black - predator.

These AEs determine a Markov process on the space of configurations $\underline{S}$. Associated with this is the following transition operator $L$ which acts on probability measures $\rho_n$ on configuration space:

$$\rho_n(d\underline{S}_n) = \int \rho_{n-1}(d\underline{S}_{n-1})P(d\underline{S}_n|\underline{S}_{n-1}) \tag{1}$$

where $P(d\underline{S}_n|\underline{S}_{n-1})$ is the conditional probability of the configuration $\underline{S}_n$ given that the previous one was $\underline{S}_{n-1}$. There is a lot of interest in the limiting behaviour of $\rho_n$ as $n \to \infty$. However, this concerns the large-scale behaviour of the system and, as we will see, we will be considering how to extract information from the intermediate scales as well. In our case the latter information will be more interesting.

Note that for some finite systems the resulting attractor is trivial and consists of an absorbing state $\underline{0}$. However, the expected time for absorbtion is often very large (typically $\mathcal{O}(V)$) and the system has approached a 'statistical stationary state' long before absorbtion destroys it. If this is the case, then one should modify the above approach to be conditioned on the fact that the absorbing state is not reached. In the following discussion we ignore these complications.

The first point to note is that both the above models appear to have attractors onto which they settle once transients have died away. By an *attractor* I mean a subset $A$ of configuration space with the following properties:

*Attractiveness:* With probability one, as $t \to \infty, S_t \to A$. for finite systems this means that $S_t \in A$ for $t$ large.

*Invariance:* Once the system is on the attractor it doesn't leave it, i.e. if $S_t \in A$ then $S_{t'} \in A$, for all $t' > t$.

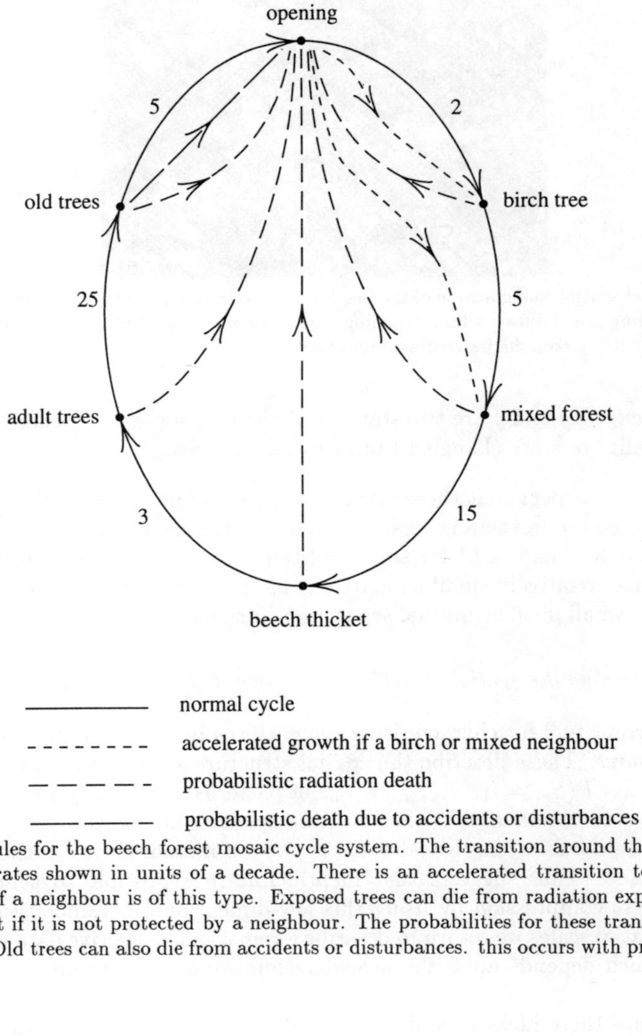


Fig. 5. The rules for the beech forest mosaic cycle system. The transition around the normal cycle occur at the rates shown in units of a decade. There is an accelerated transition to birch tree or mixed forest if a neighbour is of this type. Exposed trees can die from radiation exposure to East, South or West if it is not protected by a neighbour. The probabilities for these transitions are $p_E$, $p_S$ and $p_W$. Old trees can also die from accidents or disturbances. this occurs with probability $p_D$.

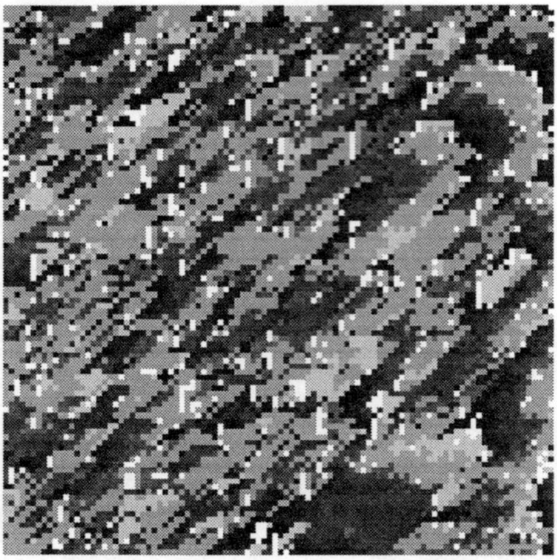

Fig. 6. A typical spatial configuration of the beech forest cycle system. The lattice size is $150 \times 150$. The colour-coding is as follows: white - opening, lightest grey - birch, light grey - mixed forest, dark greys - beech (with darker shades representing older trees).

*Ergodicity:* If $i$ and $j$ are two states in $A$ then for some $n \geq 1$ there is a positive probability of state changing from $i$ to $j$ in $n$ steps.

Moreover, we expect that these attractors are small in the sense that the fraction of states $\rho$ which are in them is small. Analytical arguments and simulation leads us to believe that $\rho \leq \exp -cL^2$ for some constant $c > 0$. This observation is related to the fact that relatively small subpatterns of the state have interactions that are exponentially small in their mutual separation in space.

### *3.3. Characterising the spatial structure associated with the attractor.*

An important tool for characterising the patterns in such systems is given by the *pattern measures.* These describe the spatial structure statistically. They determine the probabilities $P(S_{x_1} = i_1, \ldots, S_{x_m} = i_m)$ of patterns when the state is $\underline{S} = (S_x)$.

A remarkable observation is that for this system and for the states on the attractor, on intermediate length scales, these pattern measures vary within some low-dimensional set. Thus suggests that in principle it is possible to achieve a large amount of data-compression by using this low-dimensional parameterisation. After discussing spatial scales we return to describe a remarkably effective data-compression technique which depends upon the near-determinism of the system at intermediate length scales.

We now use these ideas to address a number of questions. In particular, I want to consider how to (a) characterise spatial structure, (b) quantify the dynamics, (c) deal with the immense amounts of data in the states of these models, (d) identify appropriate spatial and temporal scales, (e) maximise the information gained from this data and (e) use these ideas to detect change in ecosystems. To facilitate this, I will describe a correspondence between the set of pattern measures in the attractor and a set of low-dimensional vectors.

A crucial idea is the idea of our analysis is to look for spatial scales where the ratio of 'deterministic' information to stochastic fluctuations is maximised. Another, slightly different, way of saying this is that we search for a length scale which maximises non-trivial determinism.

*3.4. Window time-series.*

To understand this we consider the time series $x(t) = x_V(t)$ obtained by counting the total number of one of the species (say the resource) inside a fixed window of size $V = \ell \times \ell$. If the window is too small then the time-series is dominated by stochastic effects. If it is too large the interesting effects are averaged out and all we observe is small stochastic fluctuations about a constant mean value.

However, in each of our systems there is a clearly-defined intermediate length scale $\ell_S$ on which to observe the dynamics which gives us the maximum amount of information about the system and its dynamics relative to the amount of data and work necessary to describe it. This scale is an inherent characteristic of the ecosystem and is also the appropriate scale of measurement. Roughly speaking it is characterised by the following properties:

- non-overlapping $\ell_S$-windows are nearly independent, and
- the dynamics of any two points in an $\ell_S$-window are significantly correlated.

This raises the question of whether there is a systematic method for finding this scale. We discuss one which uses the time-series.

Consider an $V$-window $W_V$ where $\ell = \sqrt{V}$ is large compared to any correlation length. Divide this into $m = V/V_S$ $V_S$-windows. Then the time-series for $W_V$ is the sum of the time series for the $\ell_S$-windows. Let $Var(V)$ denote the variance of the time-series in a $\ell$-window. Then (Rand & Wilson 1994a) we expect that if $V_S$ is large enough

$$\frac{Var(V)}{V} = \frac{Var(V_S)}{V_S} + C_0 + C_1 V \tag{2}$$

where $C_0 + VC_1$ is a small error term due to the correlations between windows due to the fact that the time-series are finite and are all acting out dynamics on the same attractor. As the length $T$ of the time series is increased $C_1 \to 0$. We have plotted $Var(V)/V$ for our resource-predator-prey model in Figure 7. The crossover from the small-scale behaviour to the scaling behaviour given by equation (2) is clear and occurs around $V = 10,000$. Therefore, we take $\ell_S = 100$. In the asymptotic regime $C_1 \approx 6.65 \times 10^{-4}$.

*3.5. Determinism and dimension.*

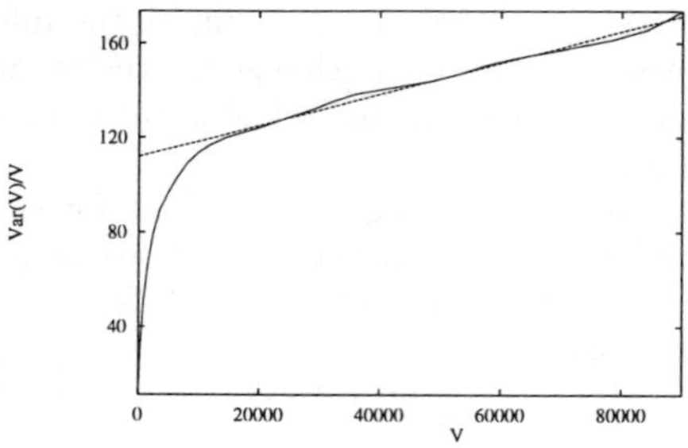


Fig. 7. The relative variance $Var(V)/V$ of the resource time series as viewed through a window of size $V$. The straight line shown is a least-squares fit to the data for $V \geq 12,000$. It has slope 0.000665 and intercept 111.6.

Let $x(t)$ be the time-series from a $V_S$-window. If this is plotted it appears smooth. Therefore, it is natural to ask whether the population dynamics are deterministic. By this I mean that there is some time-independent functional relationship of the form

$$x(t) = f(x(t - d\tau), x(t - (d-1)\tau), \ldots\ldots, x(t - \tau)) + \xi(t - 1). \tag{3}$$

where $\xi(t)$ represents a very small stochastic fluctuation whose magnitude is so small that $\xi(t) \ll x(t)$ and is below the scale at which $f$ is linear. This last condition is very important because when it holds there are practical techniques to detect determinism. In stable equilibrium, periodic or quasiperiodic systems the noise will be stabilised by the local linear contraction and in chaotic systems the Lyapunov exponents can be probed. The techniques of Wilson & Rand (1993) provide a tool for checking for this separation of scales i.e. that the stochastic scale is smaller than the nonlinearity scale.

For our models we verify (3) by using singular value decomposition (SVD) techniques (Bertero & Pike 1982) and direct inspection of the data. For the former, it would be natural to use a version of this that is local in phase space, but it turns out that for this example it suffices to use global techniques.

Firstly, form the time-delayed vectors

$$\underline{x}(t) = (x(t - d\tau), x(t - (d-1)\tau), \ldots, x(t - \tau)).$$

The first piece of evidence comes from the low dimension of the resulting vector time-series. Using SVD techniques we search for an orthogonal basis of $\mathbf{R}^d$ with the following property: Let $e_1, \ldots, e_d$ be any orthogonal basis $\underline{e}$ of $\mathbf{R}^d$ and consider the 'error':

$$e_k(\underline{x}, T) = T^{-1} \int_0^T ||\underline{x}(t) - \underline{y}_k(t)||^2 \, dt$$

of the projection $\underline{y}_k(t) = \sum_{i=1}^k \langle \underline{x}(t), e_i \rangle e_i$ onto the subspace spanned by $e_1, \ldots, e_k$. Define $\sigma_i(\underline{e}) = T^{-1} \int_0^T \langle \underline{x}(t), e_i \rangle^2 \, dt$. Then $e_k(\underline{x}, T) = \sum_{i \geq k} \sigma_i(\underline{e})$. We seek a basis $\underline{e}$

that minimises the error $\sum_{i\geq k} \sigma_i(\underline{e})$ for all $k$. If the dynamics of $\underline{x}$ is ergodic then we can also expect that the basis found in this way will be largely independent of the vector time-series $\underline{x}(t)$. This basis can be found using the SVD techniques described in Rand & Wilson (1994a). The important point is that all but 6% of the variation is in the projection $\underline{x}^{\text{proj}} = \underline{y}_4$ onto $e_1, \ldots, e_4$ (i.e. $\sum_{i=1}^{4} \sigma_i(\underline{e}) / \sum_{i=1}^{\infty} \sigma_i(\underline{e}) \geq .94$). Thus, even globally in phase space, our system is largely 4-dimensional and we can project the time-series onto the first four eigenvectors in order to reconstruct the time-series minus most of the noise. This means that virtually all the dynamics in the original signal are occurring in a four-dimensional embedded space. Such a result is a strong indicator of determinism. It is also surprising considering the high-dimensionality of the original spatial system.

We can now perform an extra direct test of determinism. Choosing some state at time $t$, $\underline{x}(t)$, we find a previous state, $\underline{x}(t-t')$, very close to our original state. We check that, as in a deterministic system, the evolution of these two states will be very similar. We can also perform this test on the projections $\underline{x}^{\text{proj}}(t)$ of these vectors onto $e_1, \ldots e_4$.

In our case the deterministic part of the population dynamics are chaotic. This means that nearby states diverge from each other exponentially fast at a rate determined by the largest Lyapunov exponent. We check that this is the case (Rand & Wilson 1994a).

The fact that the time-series are chaotic is of some general interest because there has long been a controversy about whether ot not cellular automata could be chaotic. Here is an example of an AE which is chaotic on time-series in $V_S$-windows and which we therefore expect has a Lyapunov density in the sense given above. Thus I would like to suggest the following definition of space-time chaos for PCAs and AEs.

**Definition.** A PCA or AE is *space-time chaotic* if it has an attractor satisfying the properties of section 3.2 and an invariant stationary measure $\rho$ (where $\rho_n \to \rho$ as $n \to \infty$ whenever $\rho_n$ is as in equation (1)) and for $\rho$-typical realisations *(i)* there is a volume $V_S$ such that the $V_S$-window time-series are deterministic in the sense of equation (3) with the appropriate separation of stochastic and nonlinear scales and *(ii)* there are positive Lyapunov characteristic exponents associated with the $V_S$-window and the values of these exponents have a dependence upon the realisations of the dynamics in far-away windows which is exponentially weak in their separation distance.

### *3.6. Application 1: Detecting structural change in ecosystems.*

I outline a technique to detect long-term structural change or drift in an ecosystem (Rand & Wilson 1994a). This is an important problem in the management of ecosystems because one wants to detect degradation of the ecosystem before it's too late. The point is to be able to distinguish between structural change, by which we mean change in the parameters governing the dynamics, and the natural dynamical variation of the ecosystem. The detection of change is also important in the analysis of ecological data because nearly all data analysis techniques assume stationarity in the time series being analysed. Our method can detect drift of the order of one part in a thousand per period (i.e. about $10^{-6} - 10^{-5}$ per iteration).

Two different problems are of interest. In the first, we are given a continuous time series from an ecology and wish to detect change occurring within it. In the second we have two different time-series and we wish to identify if there is a significant difference between the two. We only consider the second here. For the first see Rand & Wilson (1994a).

The idea for both problems is to combine our data reduction techniques described above with a technique introduced in Eckmann, Kamphorst and Ruelle (1987) for detecting non-stationarity in data sets. This uses *recurrence plots*. To address the second problem, we count the recurrences within one of the time series and compare this with cross-recurrences that occur between the two. We expect the latter to be fewer and therefore check for a significant difference between the two.

Given the projection $\underline{x}^{\text{proj}}(t) = \underline{y}_4(t)$ introduced above of the time-delayed vectors $\underline{x}(t)$ we look for pairs $t$, $t'$ with such that $\| \underline{x}^{\text{proj}}(t) - \underline{x}^{\text{proj}}(t') \| < \varepsilon$. We calculate the fraction $N_{\varepsilon,\text{within}}(s)$ of such pairs $t$, $t'$ with $|t' - t| = s$ and where $\underline{x}^{\text{proj}}(t)$ and $\underline{x}^{\text{proj}}(t')$ are taken from just within the same time-series. For $\varepsilon$ small these correspond to recurrences in the dynamics. We then compare this with the fraction $N_{\varepsilon,\text{cross}}(s)$ of such pairs but where $\underline{x}^{\text{proj}}(t)$ and $\underline{x}^{\text{proj}}(t')$ are taken from different time-series.

I show two graphs for the resource-predator-prey artificial ecology. Figure 8(a) is the internal recurrence plot plotting $N_v(s)$ against $s = |t - t'|$ for a time series using the parameter values as in table 2. Figure 8(b) is the cross-recurrence plot between the first time series and one where $g = 0.6$. This figure clearly indicates a substantial decrease in the number of recurrences. An inspection of the phase space indicates that the attractor has only moved by a very small amount. Thus it is encouraging that this method still detects such a small change.

### *3.7. Application 2: Reconstructing invisible species.*

In principle this approach to representing patterns can also be used to reconstruct invisible species from the visible ones. I have in mind here a situation where say the resource in our resource-predator-prey model could be monitored but one could not see the predators and prey.

The determinism implies that our 4-dimensional vector $\underline{x}^{\text{proj}}(t)$ determines the pattern measure not only of the visible species but also of the other species. One way to check this is to do the SVD analysis on time-delayed vectors of the form

$$(x(t - d\tau), \ldots, x(t - \tau), y(t - d\tau), \ldots, y(t - \tau))$$

where $x$ is the visible species as discussed above and $y$ is the invisible species. If this has similar $\sigma_k$ as for the single species then this suggests that the second species is slaved to the first. Thus if we know the first then in principle we can predict the second. This slaving appears to be present in the forest model and should be

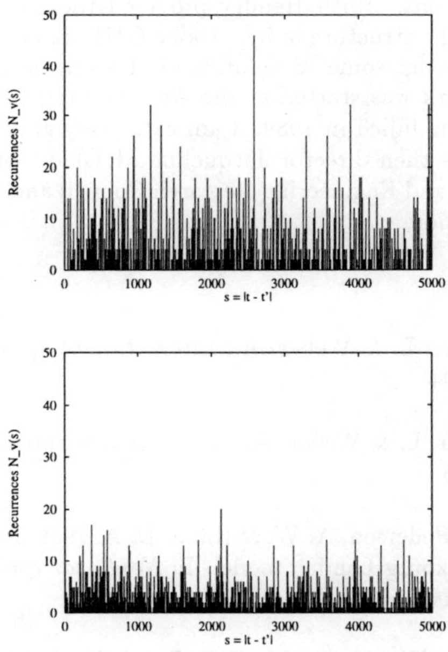


Fig. 8. (a) The recurrence plot for the resource time series using the parameter values from table 2. The lattice size is 150x150, the time series length is $10,000$ points, $d = 4$, $\tau = 5$ and $\varepsilon = 80$. (b) The cross-recurrence plot for the resource time series using $g = 0.4$ and $g = 0.6$ and the rest of the parameter values from table 2 and Figure 8(a).

present in the resource-predator-prey system if the window size is increased so that the prey and predator numbers are not too small. Note however that even if the species are slaved together, it will require a lot of work to approximate the actual slaving function.

## Acknowledgements

I am indebted to the work of my students. I am particularly indebted to Howard Wilson because most of the work described in the second half of this paper is joint with him. Matthew Keeling did the calculations of Lyapunov spectra and laminar state length distributions. Ruth Hendry provided the data from Wissell's model. Much of my work on the structure of hyperbolic CMLs is joint with Kevin Campbell. I also wish to acknowledge some stimulating discussions on these topics with Tomas Bohr. Part of this work was started at the Arbeitsgruppe Theoretische Ökologie of the Forschungszentrum Jülich in 1989. I am extremely grateful for their hospitality and particularly to its then director Jacqueline McGlade for her encouragement. I thank the UK Science and Engineering Research Council and the Wolfson Foundation for their financial support.

## References

Aranson, I. S., Kramer, L. & Weber, A., Interaction of spirals in oscillatory media. *Phys. Rev. Lett* **67** 404

Aranson, I. S., Kramer, L. & Weber, A., Interaction of spirals in nonequilibrium media. Physica D **53** 376.

Bohr, T., Jensen, M., Pedersen, A. W. & Rand, D. A. 1989 Turbulence and linear stability in a discrete Ginzburg-Landau model. In *Nonlinear evolution of spatio-temporal structures.* (eds. F Busse and L Kramer).

Bohr, T., Pederson, A. W., Jensen M. H. & Rand, D. A. 1990a Vortex dynamics in a coupled map lattice. in *New Trends in Nonlinear Dynamics and Pattern Forming Phenomena,* (eds. P. Coullet and P. Huerre) New York:Plenum.

Bohr, T., Pederson, A. W. & Jensen, M. H. 1990b Transition to turbulence in a discrete Ginzburg-Landau model. *Phys. Rev. A* **42** 3626.

Bunimovich, L. A. & Sinai, Ya. G. 1988 Spacetime Chaos in Coupled Map Lattices. *Nonlinearity* **1** (1988) 491-516.

Campbell K. & Rand, D. A. 1994 Natural spatio-temporal measures. Warwick Preprint.

Campbell, K. & Rand, D. A. 1994 A thermodynamic formalism for spatially-extended dynamics. Manuscript.

Caponeri, M. & Ciliberto, S. 1992 Thermodynamic aspects of the transition to spatiotemporal chaos. *Physica* **58D** 365-383.

Chaté, H. 1994 Spatiotemporal intermittency regimes of the one-dimensional Complex-Ginzburg-Landau equation. *Nonlinearity* **7** 185.

Chaté, H. & Manneville, P. 1987 Transition to Turbulence via Spatiotemporal Intermittency. *Phys. Rev. Letts.* **58** 112-115.

Chaté, H. & Manneville, P. 1988 Spatio-temporal intermittency in coupled map lattices. Physica D **32** 409.

Comins, H. N., Hassell M. P. & May R. M. 1992 The spatial dynamics of host-parasitoid systems. *J. Animal Ecol.* **61** 735-748.

Cross, M. C. & Hohenberg, P. C. 1993 Pattern formation outside of equilibrium. *Rev. Mod. Phys.* **65** 851-1112.

DeRoos, A. M. McCauley, E., & Wilson, W. G. 1991 Mobility versus density-limited predator-prey dynamics on different spatial scales. *Proc. R. Soc.* B **246** 117-122.

Durrett, R. 1993 Spatial epidemic models. Preprint.

Durrett, R. and Levin, S. 1993 The importance of being discrete. Tech. Report 93-5 Math. Sci. Inst., Cornell University.

R Durrett and S Levin, 1994 Stochastic Spatial Models: A Users Guide to Ecological Applications. *Phil. Trans. R. Soc. Lond.* **343** 329-350.

Eckmann, J.-P., Kamphorst, S.O. and Ruelle, D. 1987. Recurrence plots of dynamical systems. Preprint: Universite de Geneve.

Grassberger, P. & Procaccia, I. 1983 Measuring the strangeness of strange attractors. *Physica* **9D** 189.

Gundlach, V. M. & Rand, D. A. 1993*a* Spatio-temporal chaos. 1. Hyperbolicity, structural stability, spatio- temporal shadowing and symbolic dynamics. *Nonlinearity* **6** 165-200.

Gundlach, V. M. & Rand, D. A. 1993*b* 2. Unique Gibbs states for higher-dimensional symbolic systems. *Nonlinearity* **6 201-213.**

**Gundlach, V. M. & Rand, D. A. 1993*c* 3. Natural spatio-temporal measures for weakly- coupled map lattices defined by circle maps.** *Nonlinearity* **6 215-230.**

**Gundlach, V. M. & Rand, D. A. 1993*d* 4. Erratum and corrigendum. To be submitted to** *Nonlinearity.*

*Hassell, M. P. 1978 Arthropod Predator-Prey Systems.* Princeton University Press.

Hastings, A. 1990. Spatial heterogeneity and ecological models. *Ecology* **71** (2) 426-428.

Hendry, R. & McGlade, J. 1994 Using Probabilistic Cellular Autcmata to Analyse Mechanisms in a Forest Mosaic Cycle. Manuscript.

Huberman, B. A. & Glance, N. S. 1993 Evolutionary Games and Computer Simulations. *Proc. Nat. Acad. Sci. U. S. A.* **90** 7716-7718.

Kaneko, K. 1990 Simulating physics with coupled map lattices. In *Formation, Dynamics and Statistics of Patterns* (Edited by K. Kawasaki, M. Suzuki and A. Onuki) Vol. **1**. Singapore: World Scientific

Keeling, M. 1994 Personal communication.

Levin, S. Powell, T. M. & Steele, J. (eds.) 1993 *Patch Dynamics.* Lecture Notes in Biomathematics 96. (Springer-Verlag, New York).

Miller, J. & Huse, D. 1993 Hydrodynamics of extended chaotic systems with Ising symmetry. To appear in *Phys. Rev. E.*

Mollinson, D. (ed.) 1994 *epidemic Models: their Structure and Relation to Data.* Cambridge:CUP (to appear).

Nowak, M. A. & May, R. M. 1992 Evolutionary Games and Spatial Chaos. *Nature* **359** 826-829 and to appear in *Int. J. Chaos Bifurc.*

Okubo, A. 1980 *Diffusion and ecological problems: mathematical models.* New York: Springer.

Pesin, Ya.B. & Sinai, Ya.G. 1991 Space-time Chaos in Chains of Weakly Interacting Hyperbolic Systems. *Adv. Soviet Math.* **3** 165-198.

Wilson, H. B. & Rand, D. A. 1994a Using spatio-temporal chaos and intermediate-scale determinism in artificial ecologies to quantify spatially-extended ecosystems. *Proc. Roy. Soc. Lond. B* (to appear).

Rand, D. A., Keeling, M. & Wilson, H. B. 1994b Invasion, Stability and Evolution to Criticality in Spatially Extended Host-Pathogen Systems. *Proc. Roy. Soc. Lond. B* (to appear).

Rand, D. A. & Wilson, H. B. 1994c Reconstructing ecologies from satellite images. In preparation.

Rand, D. A., Wilson, H. B. & McGlade, J. M. 1994d Dynamics and evolution: Evolutionarily stable attractors, invasion exponents and phenotype dynamics. *Phil. Trans. R. Soc. Lond. B.* **343** 261-283.

Reeve, J.D. 1990. Stability, variability, and persistence in host- parasitoid systems. *Ecology* **71** (2) 422-426.

Shraiman, B., Pumir, A., van Saarloos, W., Hohenberg P. C., Chaté H. & Holen M. 1992 Spatio-temporal chaos in the one-dimensional complex Ginzburg-Landau equation. *Physica D* **57** 241.

Takens, F 1981 Detecting strange attractors in turbulence. In *Dynamical Systems and Turbulence, Warwick 1980.*(eds. D. A. Rand and L.-S. Young) pp. 366-381 Berlin:Springer

Weber, A., Kramer, L., Aranson, I. S., & Aranson, I. B. 1992 Stability limits of travelling waves and the transition to spatiotemporal chaos in the complex Ginzburg-Landau equation. *Physica D* **61** 279.

Wilson H. B. & Rand D. A. 1993 Detecting chaos in a noisy time series. *Proc. Roy. Soc B.* **253** 239-244.

Wissell, Ch., 1991 A Model for the Mosaic-Cycle Concept. In *The Mosaic-Cycle Concept of Ecosystems.*(ed. H. Remmert) Berlin:Springer-Verlag.

CHAPTER 15

# NON-LINEAR FORECASTING AND CHAOS IN ECOLOGY AND EPIDEMIOLOGY: MEASLES AS A CASE STUDY

B.T. Grenfell, A. Kleczkowski*, S. Ellner†
and B.M. Bolker‡
Zoology Department
Cambridge University
Downing Street
Cambridge CB2 3EJ, UK

**Abstract**

This paper uses measles incidence in developed countries as the basis of a case study in nonlinear forecasting and chaos. It employs a combination of epidemiological modelling and nonlinear forecasting to explore a range of issues relating to the predictability of measles before and after the advent of mass vaccination. A comparison of the pre-vaccination self-predictability of measles in England and Wales indicates relatively high predictability of these predominantly biennial epidemic series, compared to New York City, which shows mixtures of 1,2 and 3 year epidemics. This analysis also indicates the importance of choosing correct embeddings in order to avoid bias in prediction. Forecasting for English cities indicates significant spatial heterogeneity in predictability before vaccination and an overall drop in predictability during the vaccination era. The interpretation of predictions of observed measles series by epidemiological models is discussed and a simple model for the effects of demographic variations on measles incidence is introduced.

*Department of Plant Sciences, Cambridge University, Downing Street, Cambridge CB2 3EA, UK
†Statistics Department, North Carolina State University, Raleigh NC, USA
‡Department of Ecology and Evolutionary Biology, Princeton University, Princeton NJ 08544-1003, USA

## 1. Introduction

The pattern of measles epidemics in developed countries is one of the best documented population cycles in ecology. The availability of long disease notification series and the comparatively simple natural history of infection have prompted an extensive empirical and theoretical literature on the population biology of the infection. This work has concentrated on two main areas. First, the public health importance of measles (still a major killer in developing countries (Mclean & Anderson, 1988) has generated a large and distinguished body of quantitative epidemiological research, focusing in particular on the origin and perpetuation of epidemics, the persistence of measles as a function of community size and the impact on vaccination strategies of various heterogeneities in transmission (Hamer, 1906; Soper, 1929; Bartlett, 1957; Bartlett, 1960; Dietz & Schenzle, 1985; May, 1986; London & Yorke, 1973; Anderson & May, 1985; Anderson & May, 1991; Dietz, 1976; Anderson & May, 1982). Second, the last decade has considerable interest in the pattern of measles incidence as a potential ecological candidate for deterministic chaos (Schaffer & Kot, 1985; Olsen *et al.* 1988; Olsen & Schaffer, 1990; Kot *et al.* 1988; Schaffer, 1985b; Rand & Wilson, 1991; Sugihara & May, 1990). It is still an open question as to whether measles dynamics are chaotic but, in any case, the debate and in particular the work of W. M. Schaffer and his coworkers (Schaffer & Kot, 1985; Olsen *et al.* 1988; Olsen & Schaffer, 1990; Kot *et al.* 1988; Schaffer, 1985b) , has been highly stimulating to population ecologists and epidemiologists alike.

Recently, Tidd *et al.* (1993), building on the work of Sugihara and May (Sugihara & May, 1990; Sugihara *et al.* 1990), have used nonlinear prediction to examine evidence for chaos in measles. They analyzed observed pre-vaccination time series from a number of cities and towns in developed countries, comparing the self-predictability of these series with forecasts generated by a range of empirical and mechanistic epidemiological models. From this analysis, they proposed evidence for chaos, reflected by a decline in internal predictability of the observed series with the time step of prediction. They also found that the mechanistic epidemiological models out-performed empirical statistical formulations in predicting the observed series.

Tidd et al's paper raises a number of issues about nonlinear forecasting, which we examine in this paper.

*(1) How does the detailed pattern of pre-vaccination epidemics affect the prediction process ?*

The main contrast here is between a predominantly biennial sequence of major epidemics (typified by the pattern in English cities during the 1950s and 60s; Fig. 1 (b-i)) and time series such as that for New York City (Fig. 1(a)), which show occasional epidemics with a period of 3 years (Bolker & Grenfell, 1993; Olsen *et al.* 1988; Grenfell *et al.* 1994). Tidd *et al.* (1993) concentrate on the latter case, focusing mainly on developed country cities, such as New York, which contain 3 year epidemics. Here, we assess the predictive implications of these different patterns, based on a comparative analysis of series from England and Wales and New York City.

*(2) How does mass vaccination affect patterns of predictability ?*

Mass vaccination against measles began in the 1960s and is now applied (to a greater or lesser extent) throughout the world. Assessing the predictability of measles epidemics in the face of vaccination is obviously the most important applied forecasting issue. In this paper, we therefore examine the impact of vaccination on the nonlinear predicability of measles epidemics in England and Wales.

*(3) How do epidemiological models perform in predicting observed measles series ?*

As reviewed in detail below, the comparatively simple natural history of measles infection has led to considerable success in the construction of epidemiological models. These range from the simple homogeneous SEIR model to models incorporating age, spatial structure and other heterogeneities (Bartlett, 1957; Bartlett, 1960; Black, 1966; Hamer, 1906; Soper, 1929; Schenzle, 1984; Anderson & May, 1991). Tidd *et al.* (1993) found that epidemiological models perform better as predictors of observed measles series than empirical linear models. Such comparisons are valuable, however, as discussed below, there are various statistical and epidemiological complexities which affect their interpretation.

*(4) Could long term demographic changes affect patterns of predictability ?*

The birth and subsequent recruitment of children to the susceptible population is the 'engine' which ultimately drives measles epidemics. Recently, Grenfell *et al.* (1994) have shown associations between birth rate variations and average pre-vaccination measles incidence for English cities and Copenhagen. Here, we present a preliminary exploration of how birth rate variations affect infection dynamics and therefore the prospects for forecasting.

The following section defines the data sets and prediction algorithm used. We then briefly review current knowledge about measles dynamics, based on an assessment of how the available models perform in reflecting observed patterns. This serves as a

---

Fig. 1. **See next page.** (a)-(i) Time series of measles notifications in New York City, England and Wales and seven English cities. The monthly New York data were kindly provided by L.F. Olsen and the weekly UK data are from the Weekly Returns of the Office of Population Censuses and Surveys. (j) Comparison of the observed biennial pattern of cases in England and Wales, scaled up by a factor 1.4 to allow for notification efficiency (Bolker and Grenfell, 1992) with the best fit of the RAS model as described in the text (see Bolker and Grenfell, 1992 for full details). The observed mean pattern is shows as a dotted line (bars represent $\pm$ 1 standard deviation) and the model dynamics (a stable biennial limit cycle) as a solid line.

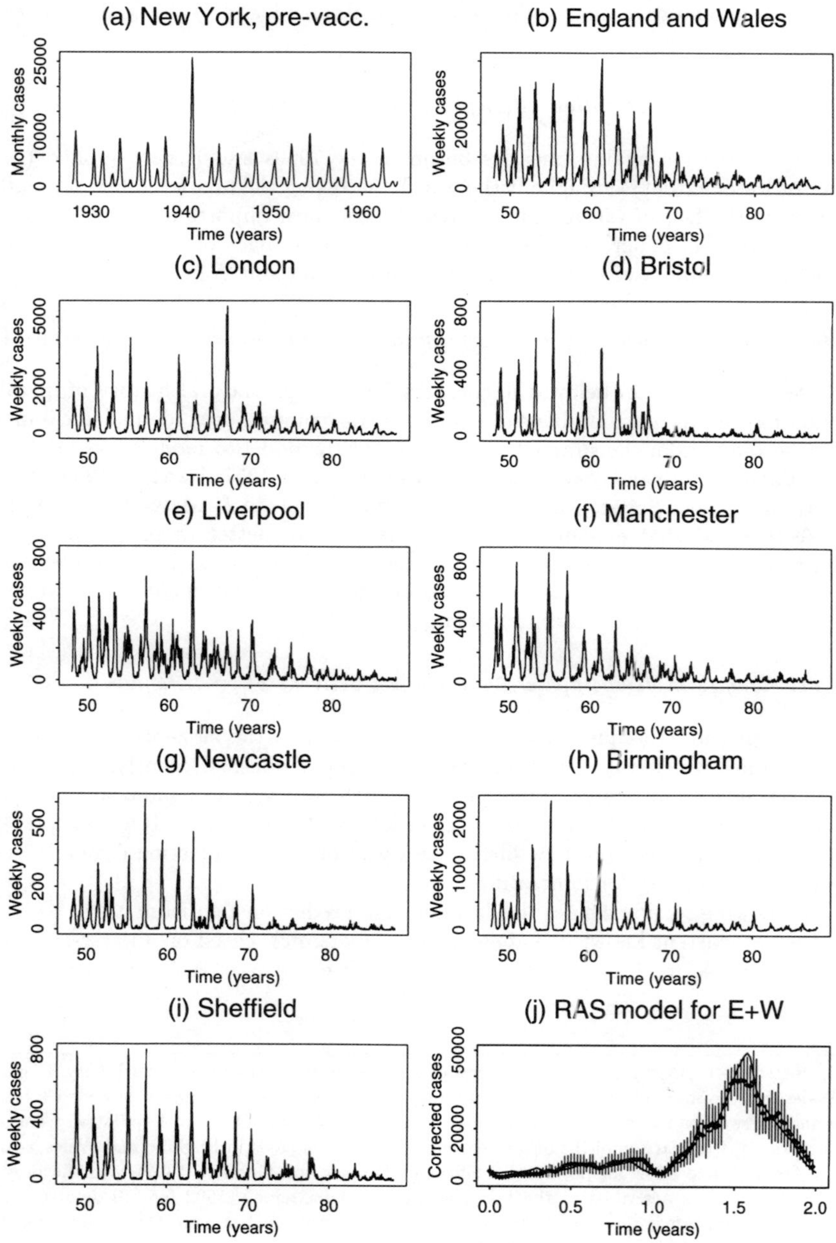
(a) New York, pre-vacc.
Monthly cases
0
10000
25000
1930
1940
1950
1960
Time (years)
(b) England and Wales
Weekly cases
0
20000
50
60
70
80
Time (years)
(c) London
Weekly cases
0
2000
5000
50
60
70
80
Time (years)
(d) Bristol
Weekly cases
0
400
800
50
60
70
80
Time (years)
(e) Liverpool
Weekly cases
0
400
800
50
60
70
80
Time (years)
(f) Manchester
Weekly cases
0
400
800
50
60
70
80
Time (years)
(g) Newcastle
Weekly cases
0
200
500
50
60
70
80
Time (years)
(h) Birmingham
Weekly cases
0
1000
2000
50
60
70
80
Time (years)
(i) Sheffield
Weekly cases
0
400
800
50
60
70
80
Time (years)
(j) RAS model for E+W
Corrected cases
0
20000
50000
0.0
0.5
1.0
1.5
2.0
Time (years)

conceptual basis for subsequent sections, which address the questions set out above.

## 2. Data sources and statistical methods

### *(a) The data*

The analysis centres on notification time series for New York City in the era before vaccination and for seven English cities (and the England and Wales aggregate) prior to and following the advent of vaccination. These data are displayed and documented in Fig. 1. Measles vaccination in England and Wales began in earnest in 1968 (Fine & Clarkson, 1982b) and caused a significant reduction in average disease incidence, with a relative stabilization of average cases from 1972 until the mid 1980s (Fig 1 (b-i)). We therefore take the pre- and post-vaccination series from the periods 1948-67 and 1972-86 respectively. To facilitate comparison with the (monthly) New York data, as well as to avoid possibilities of oversampling (Sugihara *et al.* 1990), the weekly England and Wales series shown in Fig.1 are summed into 4 weekly periods before analysis.

### *(b) Estimating nonlinear predictability*

Extensive reviews of nonlinear forecasting techniques are given by Tong (1990) and Yao and Tong (this volume). For purposes of comparison, we use the same algorithm as Tidd *et al.* (1993). This is a simple zero order $\epsilon$-ball method (Tong, 1990; Farmer & Sidorowich, 1987), with slight modifications, as described below.

(a) To assess the self-predictability of an observed time series, scaled on (0,1), separate the data into two halves; the first half (the atlas) is used to predict the second (the target series). To predict observed series from simulated model series, scale the observed and model series independently on $(0, 1)$ (Tidd, *et al.*, 1993), then take the simulated series as the atlas and the full observed series as the target.
(b) embed the data in a delayed coordinate space (embedding dimension E and delay, d time units)
(c) For each point in the embedded target series, select all atlas points within a radius $\epsilon$ of the observed point. By default, $\epsilon$ is set at 0.1 in scaled coordinates and this common radius for all target points is expanded so that at least one atlas point is within the $\epsilon$-ball. Following Tidd *et al.* (1993), we restrict the maximum number of points in the atlas to 10.
(d) For each lead time step, the target and atlas points are projected forwards one step and the mean of the latter is used as the prediction. If the maximum lead time is T, the last T points in the target (and the initial atlas) are omitted, so that all predictions are based on the same number of atlas points and the same number of predictions are done for each prediction time step. The relative

success of prediction is assessed by an $R^2$ between observed and predicted points; the regression slope between observed and predicted series is calculated to assess bias in the predictions.

Sugihara, Grenfell and May (1990) used differenced data to assess predictability in English cities. Here, we follow Tidd *et al.* (1993), analysing the raw notification data, for comparison with their results.

## 3. Modelling measles dynamics

### The SEIR model

The simplest description of measles infection dynamics is the well known SEIR (Susceptible/Exposed/Infectious/Recovered) model, which has a long history of study in mathematical epidemiology (Hamer, 1906; Soper, 1929; Kermack & McKendrick, 1927; Kermack & McKendrick, 1932; Kermack & McKendrick, 1933; May, 1986; Anderson & May, 1991; Dietz & Schenzle, 1985). It is expressed as a set of three non-linear ordinary differential equations

$$\begin{aligned}\frac{dS}{dt} &= \mu N(1-p) - (\mu + \beta I)\,S \\ \frac{dE}{dt} &= \beta I S - (\mu + \sigma)\,E \\ \frac{dI}{dt} &= \sigma E - (\mu + \gamma)\,I \qquad (1)\end{aligned}$$

$S$, $E$, $I$ and $R$ respectively represent the density of Susceptible, Exposed, Infectious and Recovered individuals, in a constant total population of size $N = S + E + I + R$. Average per capita mortality rate due to all causes is $\mu$ years$^{-1}$; given that infection is assumed not to cause extra mortality, the net birth ($\mu N$) and death ($\mu[S + E + I + R]$) rates are equal. In this simplest version of the model (which ignores maternally-derived immunity in infants), individuals are assumed to be born susceptible, with vaccination effectively at birth moving a proportion $p$ of susceptibles into the recovered class.

During natural measles infection, susceptible individuals are assumed to move through the Exposed and Infectious classes, at rates $\sigma$ and $\gamma$ year$^{-1}$ respectively, and then become immune to reinfection, entering the Recovered class for life. The net infection rate, IS, is controlled by the infection parameter $\beta$ (year$^{-1}$ infective$^{-1}$).

*Seasonality.* There is a significant annual variation in measles incidence, mainly associated with changes in the contact rate of children, caused by the seasonal pattern of school terms (Fine & Clarkson, 1982a; London & Yorke, 1973). A number of workers have modelled the effects of seasonal variations in infection rate (London

& Yorke, 1973; Dietz, 1976; Grossman, 1980; Aron & Schwartz, 1984; Aron, 1990; Schaffer & Kot, 1985; Schenzle, 1984). The simplest approach is to replace the constant infection parameter, $\beta$ , of Eqs. 1 with a time-varying periodic function, such as

$$\beta(t) = b_0 \left[1 + b_1 \cos(2\pi t)\right] \tag{2}$$

Here, $b_0$ is the average infection rate and $b_1$ controls the amplitude of variation around it.

## Modelling the impact of age-structure: the RAS model

The infection parameter, $\beta(t)$, of the seasonally forced SEIR model is assumed to be independent of age, spatial structure or other heterogeneities. Along with spatial structure, heterogeneities in infection rate with age - and specifically the concentration of infection in school children - have been shown to be of considerable importance, both to measles dynamics and the projected impact of vaccination strategies (Dietz & Schenzle, 1985; Fine & Clarkson, 1982a; Anderson & May, 1991). These age-specific heterogeneities in transmission are also intimately bound up with the seasonal variations in infection rates noted above (Fine & Clarkson, 1982a). This interaction has been modelled by Schenzle (1984), who developed a continuous time age-cohort formulation which mimics the average pre-vaccination pattern of measles epidemics in England and Wales very closely (Fig. 1(j)), as well as the qualitative behaviour of the infection after the onset of vaccination. We now briefly review the dynamics of this relatively complex formulation (which is called the RAS - Realistic Age-Structured model - below), compared to those of the standard SEIR model.

### *Deterministic dynamics*

*SEIR model.* For high degrees of seasonal forcing ($b_1 > 0.28$), the SEIR model exhibits large-amplitude chaos (Aron & Schwartz, 1984; Schaffer, 1985a; Schaffer & Kot, 1985; Olsen & Schaffer, 1990; Rand & Wilson, 1991). Recent work also indicates the possibility of complex dynamics at lower degrees of forcing, particularly under the influence of process noise (Rand & Wilson, 1991) and spatial heterogeneity (Engbert & Drepper, 1993). Although a strongly-forced SEIR model can generate patterns of 1, 2 and 3 year cycles which are qualitatively similar to real measles time series (Olsen & Schaffer, 1990), the troughs between major 3 year epidemics are generally much too deep to be realistic (Bolker & Grenfell, 1993; Grenfell, 1992). Although this problem is alleviated somewhat by allowing for immigration of infectives (Engbert & Drepper, 1993), there are still more potential fadeouts of infection (i.e., periods with incidence below one infective) than observed.

*RAS model.* Paradoxically, the deterministic dynamics of the more biologically complex RAS model are dynamically simpler than those of the SEIR (Bolker & Grenfell, 1993). The RAS model characteristically generates limit cycles (Fig. 1(j)), without the unrealistically deep troughs of the SEIR model.

*Stochastic dynamics*

Given the propensity of seasonally-forced measles models for deep troughs in incidence, it is important to explore the impact of demographic stochasticity on their dynamics. Fig. 2 compares the dynamics of stochastic realizations of the seasonally-forced SEIR model and the RAS model, assuming a total population size of N=1 million. These simulations use a standard Monte Carlo framework (first applied to measles in the seminal work of Bartlett, 1957, 1960); full details are given by Bolker and Grenfell (1993). Model parameters are chosen so that the corresponding deterministic dynamics generate chaotic (SEIR) and stable biennial (RAS) patterns of epidemics. Fig. 2 displays the dynamics of the models using both representative time series and phase plots (plotting $S$ *vs.* $I$ at the start of each year, based on 1000 years of simulation after a suitable transient - see Fig. 2 for more details). The $S$ *vs.* $I$ plane for the SEIR model (Fig. 2(a)) shows a relatively simple structure, reflecting mixtures of 1,2 and 3 year cycles (Rand & Wilson, 1991; Engbert & Drepper, 1993), as displayed in the associated time series (Fig. 2(b)). The vertical lines at low values of $I$ in Fig. 2(b) reflect the discrete stochastic nature of the model. The large amplitude 3 year cycles generate frequent extinctions of infection, which are not observed in practice in cities (Bartlett, 1960) or islands (Black, 1966) of population size 1 million.

The RAS model simulation shows much less structure than the SEIR, when projected onto the $S$ *vs.* $I$ plane (Fig. 2(c)). This probably reflects the complexities of age structure. In particular, the dynamics are expressed here in terms of total susceptibles and infectives; more structure night be revealed by allowing for age in the projection. By contrast to the major difference in their deterministic dynamics, the stochastic RAS and SEIR time series for $N = 1$ million (Fig. 2(b,d)) show qualitatively similar patterns (Bolker & Grenfell, 1993). In particular, the RAS model again shows mixtures of 1,2 and 3 year cycles with unrealistically large numbers of fadeouts of infection. For reasons that are not yet clear, significant demographic noise destabilises the limit cycle behaviour of the deterministic RAS model.

In summary, the deterministic RAS model provides a good fit to observed biennial measles epidemics in England and Wales. However, versions of the RAS and SEIR models allowing for demographic stochasticity cannot capture the observed persistence of measles in urban communities (Bolker & Grenfell, 1993). Though including explicit spatial structure into the stochastic RAS model alleviates this problem somewhat (Bolker, in preparation), further refinements to the models are clearly needed, as discussed below.

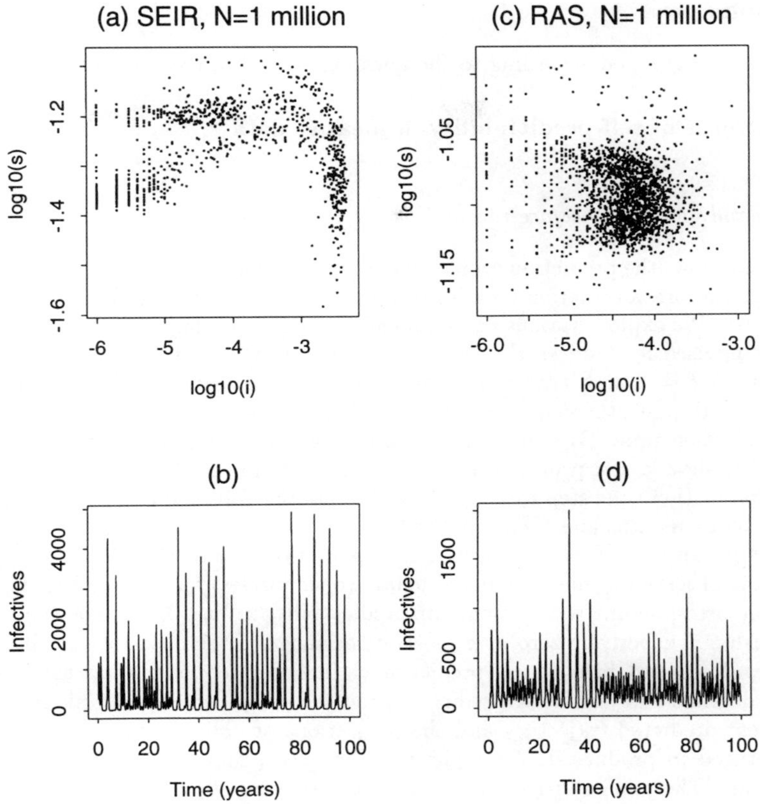


Fig. 2. Monte Carlo simulations of the SEIR and RAS models, assuming a total population size of N=1 million. Simulations, by standard Monte Carlo methods (Grenfell, 1992) were run for 200 years before recording results. The SEIR simulation uses the parameter set: $\mu = 0.02$ yr$^{-1}$, $p = 0$, $b_0 = 0.0010107$ yr$^{-1}$ infective$^{-1}$, $b_1 = 0.28$, $\sigma = 48.67$ yr$^{-1}$, $\gamma = 56.19$ yr$^{-1}$, it and the RAS simulation are described fully by Bolker and Grenfell (1992). Infection is restarted after fadeout by a Poisson 'immigration' of infectives, with mean 21yr$^{-1}$. (a) Phase plot of proportion susceptible (s) vs. proportion infected (i); points were sampled annually from 1000 years of simulation — only non-zero infectives points are shown; (b) a section of the infective time series from the SEIR simulation; (c) and (d) equivalent plots for the RAS model. Although the SEIR and RAS simulations show different mean levels of infection (reflecting slightly different epidemiological assumptions; Bolker and Grenfell, 1993), they exhibit the same qualitative pattern: a mixture of 1,2 and 3 year epidemic cycles.

## 4. Results

These are grouped according to the questions raised in the Introduction

### (a) Patterns of self-predictability in measles data

#### *(i) England and Wales before vaccination*

Fig 3a shows the prediction profile (the pattern of prediction $R^2$ against prediction time) for the pre-vaccination measles incidence time series for the whole of England and Wales. We explore various embedding dimensions and lags, focusing first on the embedding used by Tidd *et al.* (1993) for their New York City analysis: embedding dimensions, $E = 4$ and lag $d = 1$ month. The basic pattern that Tidd *et al.* derive for New York and other cities is a more or less monotonic decline in predictability with prediction time. By contrast, our results for England and Wales (the solid line in Fig 3a) show a sharp decline in predictability, from an $R^2$ of around 0.8 at a 1 month prediction time step to $R^2 \simeq 0.2$ and 6 months, followed by a steady *rise* back to 80% by 24 months and then a decline.

This pattern for $E = 4$, lag $d = 1$ is explored further in Figs 3(b) and (c). Fig 3(b) plots observed against predicted points (open circles) for a prediction time of 6 months (corresponding to the minimum predictability in Fig. 3(a)). The slope of this relationship is effectively zero over most of the range of the observations, indicating a severe downward bias in the predictions. In order to understand this bias, we show a detailed analysis of the prediction process for one target point, along with the equivalent predicted trajectory and the projections of individual atlas points which are averaged to produce it. This picture (Fig. 3(c)) makes the source of the bias very clear. The embedding used is unable to separate adjacent points in the time series, so that the atlas consists of lagged copies of a small number of epidemics. The resulting predicted trajectory shows an annual pattern, which quickly diverges from the observations as the prediction time increases from one month. Though the predicted pattern does somewhat better in representing the observations at 24 months, this is an artifact of the annual periodicity in the data. As reflected in Fig 3 (b), the predictions therefore also consistently underestimate major epidemics.

Returning to Fig 3 (a), this problem is alleviated if either embedding dimension or embedding lag are increased. In particular, if embedding lag is increased to 4 months, the prediction profile is much flatter, and shows only a slow decline over a 24 month prediction time. Fig 3(b) includes a typical plot of observed vs. expected points for $E = 6$, lag $d = 4$ (solid dots), the slope is now not significantly different from unity over most of the range of the observations, indicating that this greater degree of embedding is required to remove the bias. Fig 3(b) also indicates that there is a downward bias in the predictions for very high observed points. This is simply

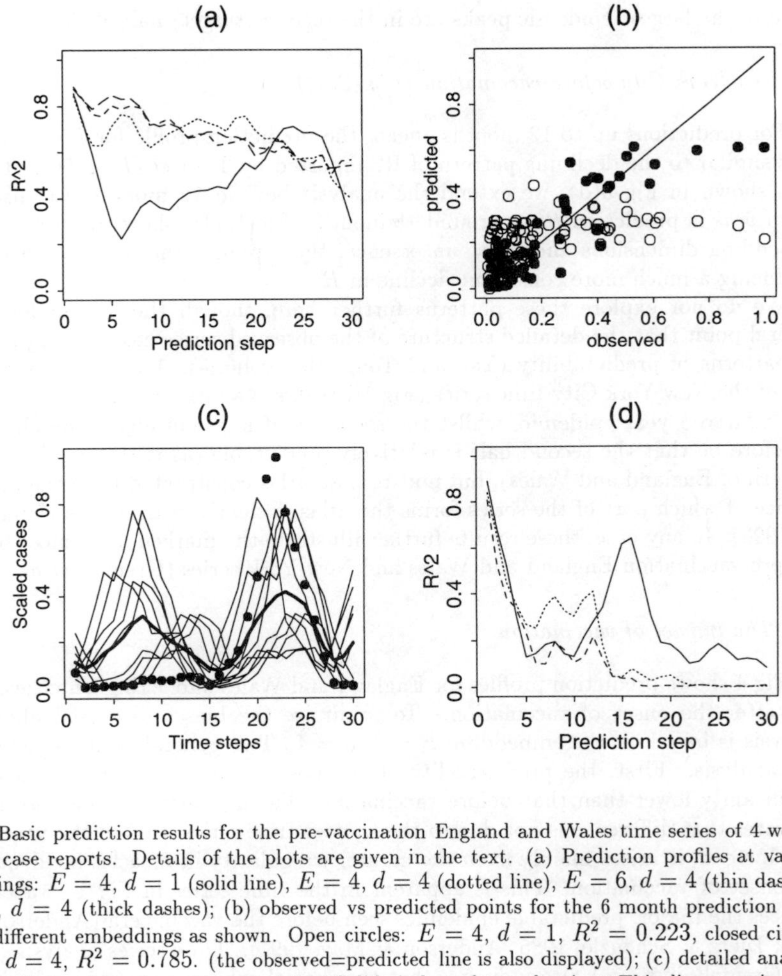


Fig. 3. Basic prediction results for the pre-vaccination England and Wales time series of 4-weekly measles case reports. Details of the plots are given in the text. (a) Prediction profiles at various embeddings: $E = 4$, $d = 1$ (solid line), $E = 4$, $d = 4$ (dotted line), $E = 6$, $d = 4$ (thin dashes), $E = 8$, $d = 4$ (thick dashes); (b) observed vs predicted points for the 6 month prediction step, with 2 different embeddings as shown. Open circles: $E = 4$, $d = 1$, $R^2 = 0.223$, closed circles: $E = 4$, $d = 4$, $R^2 = 0.785$. (the observed=predicted line is also displayed); (c) detailed analysis of the prediction for a typical target point. Solid circles: observed points; Thin lines: projections of atlas points; Thick line: predicted trajectory; (d) prediction profiles for the New York City data for various embeddings. $E = 4$, $d = 1$ (solid line); $E = 6$, $d = 4$ (dotted line); $E = 8$, $d = 4$ (dashed line).

because the largest epidemic peaks are in the second (target) half of the series.

*(ii) New York City before vaccination (Fig. 3(d))*

For predictions up to 12 months ahead, the prediction profile for $E = 4$, $d = 1$ is very similar to the declining pattern of R2 reported by Tidd *et al.* (1990). However, if as shown in Fig 3(d), we extend the analysis beyond 12 months, we discover a sharp *peak* in prediction $R^2$ at around 15 months. Fig 4, also shows results for higher embedding dimensions and lags - in essence, these reduce the peak at 15 months, producing a much more consistent decline in $R^2$.

We do not explore these patterns further here, though they do underline the general point that the detailed structure of the observed series has a strong influence on patterns of predictability (Yao and Tong, this volume). For example, the first half of the New York City time series (Fig 1(a)) shows a mixture of annual epidemics and a large 3 year epidemic, whilst the second half is essentially biennial. It may therefore be that the second half is relatively predictable (as is the regular biennial pattern of England and Wales), but not from an atlas constructed from the first half. Choice of which part of the series forms the atlas is clearly crucial here (Sugihara *et al.* 1990). In any case, these results further illustrate the marked differences between the pre-vaccination England and Wales and New York series (Grenfell *et al.* , 1994).

*(iii) The impact of vaccination*

Fig 4 shows prediction profiles for England and Wales and 7 English cities, before and after the onset of vaccination. To minimize the biases discussed above, the analysis is based on the embedding $E = 6$, $d = 4$. Two general points emerge from this analysis. First, the predictability of the post-vaccination series is, in general significantly lower than that before vaccination. Further analysis (not shown here) indicates that difference is not due to the shorter time series used in the vaccination era. Instead, it arises mainly from the greater irregularity in measles incidence after the onset of vaccination. This is apparent in the time series in Fig 1 - vaccination reduces the major, predictable epidemics seen before the vaccine era (Anderson *et al.* 1984; Dietz & Schenzle, 1985; Anderson & May, 1991). Time series analysis (Bolker and Grenfell, in prep.) also indicates that the spatial coherency of epidemics is lower after the onset of vaccination than in the pre-vaccine era.

Second (and surprisingly), there is no strong trend for average pre-vaccination predictability to decrease with population size. Though a relatively small city, such

---

Fig. 4. **See next page.** Prediction profiles ($E = 6$, $d = 4$) for the pre-vaccination (solid line) and vaccination eras (dotted line) for England and Wales and English cities. Stars indicate significant ($p < 0.05$) differences between pre-vaccination and vaccination correlation coefficients, based on a standard Fisher's Z transformation.

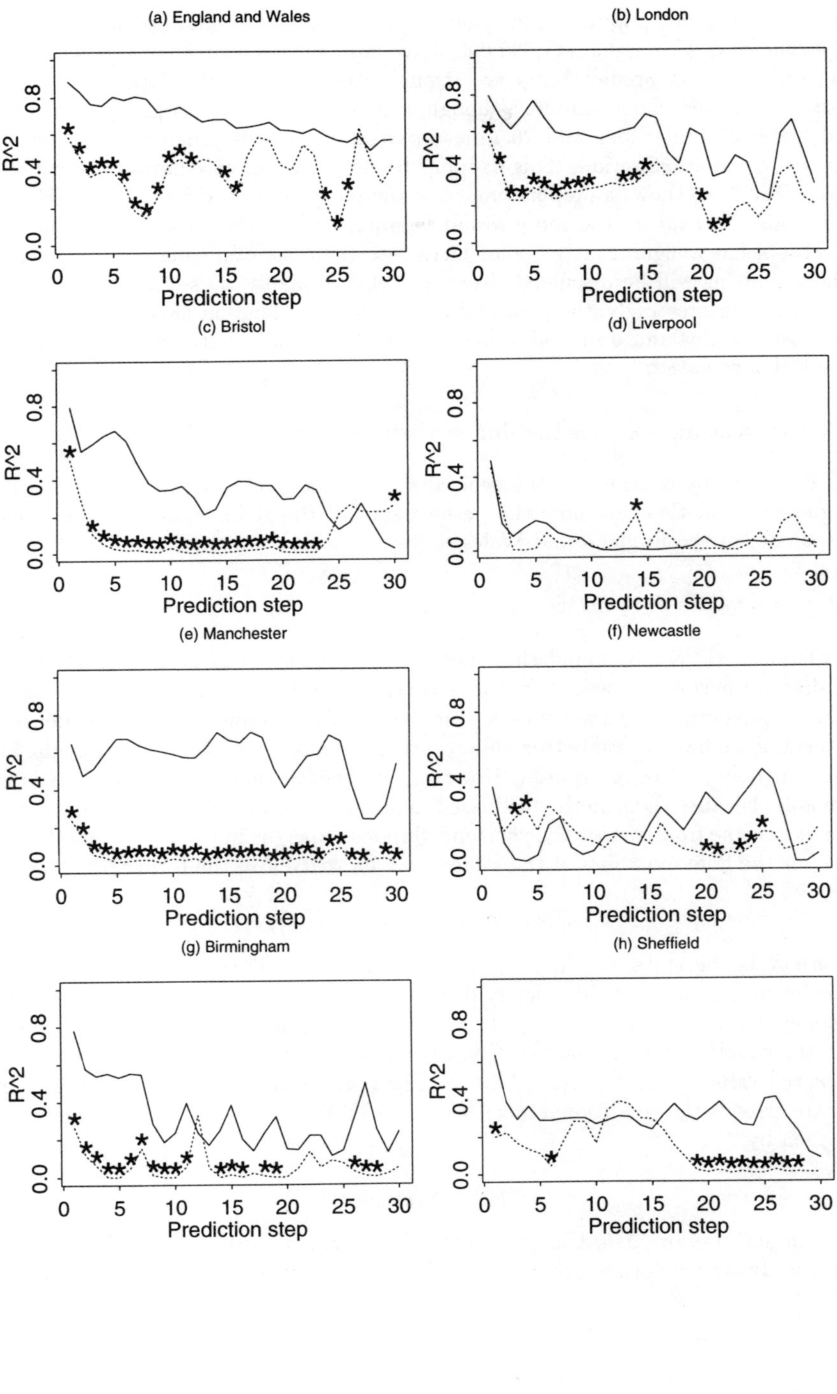
(a) England and Wales
(b) London
(c) Bristol
(d) Liverpool
(e) Manchester
(f) Newcastle
(g) Birmingham
(h) Sheffield
R^2
0.0
0.4
0.8
0 5 10 15 20 25 30
Prediction step

as Newcastle (1960 population 290,000) , is less predictable than London, Manchester (700,000) and Birmingham (1,090,000) have more or less equivalent predictability. In fact differences in predictability are strongly influenced by the detailed structure of individual series. For example, although Manchester (Fig. 4(c)) and Liverpool (Fig. 4(d)) are of similar size and 20 miles apart, Liverpool is much less predictable in the pre-vaccination period. This happens because the pre-vaccination Liverpool time series (Fig 1(e)) shows a preponderance of annual cycles in the first (atlas) half, which is not able to capture the more recent biennial cycles. Over the period considered, Liverpool has a significantly higher birth rate than the other major cities (Grenfell, 1992), and (as will be discussed elsewhere) this probably causes its distinct measles dynamics. Interestingly, the predictability profiles for London (Fig. 4(b)) are similar to those for England and Wales as a whole (Fig. 4(a)), both before and after the onset of vaccination.

## (b) Forecasting measles incidence with epidemiological models

We begin by considering the general strategy of assessing the forecasting ability of models, and then examine the performance of the RAS model in predicting the prevalence of the England and Wales series.

### *(i) Assessing the fit of deterministic models*

Tidd *et al.* (1993) found that mechanistic models, such as the SEIR and RAS models, outperform a variety of stochastic empirical formulations based on surrogate data in predicting observed measles incidence. They adduced this as evidence that deterministic models are better able to predict historic measles epidemics that their phenomenological counterparts. However, we believe that this conclusions is questionable, because their method is biased in favour of deterministic models, even when the data come from a stochastic system. Suppose that (as in the models used by Tidd *et al.* ) the true dynamics of the system can be written in the form

$$X(t+1) = F\left(X(t), \omega e(t)\right) \tag{3}$$

where $X$ is the state vector, $F$ is a nonlinear map, $\omega$ is the noise level and $e(t)$ is a series of independent, identically distributed random variables ('noise'), with $e(t)$ independent of $X(s)$ for $s = t, t-1, \ldots$. (Correlated noise complicates our argument but the conclusion is the same). Consider now how $X(t+1)$ can be predicted from observed values of $X(t)$, $X(t-1)$, .... The optimal predictor in the sense of mean square error is the conditional mean $E(\,X(t+1)|X(t), X(t-1), \ldots)$, which for Eq. 3 is given by

$$E_{e(t)}\left\{X(t), \omega e(t)\right\} \tag{4}$$

(Karlin and Taylor 1975, Chapter 9) (Karlin & Taylor, 1975). Note that expression Eq. 4 is always a deterministic function of past values of $X$. whether Eq. 3 is stochastic

($\omega > 0$) or deterministic ($\omega = 0$): the least-squares optimal predictor is a deterministic function of the information currently available. The same applies for prediction any number of steps into the future, and for continuous time dynamics. Fig. 5 shows a specific example, using the procedures of Tidd *et al.* The true dynamics are a 1-dimensional population model (the Ricker equation) perturbed by noise,

$$x(t+1) = x(t)\exp\left(r\left(1 - x(t)\right)\right)\left(1 + \omega e(t+1)\right) \tag{5}$$

where $e(t)$ is uniformly distributed on the interval [-1, 1]. For the true dynamics, the noise amplitude was $\omega = 0.4$. We varied the value of $\omega$ in the models used to construct the atlas for prediction with the $\epsilon$-ball method of Tidd *et al.* (1993). In line with the arguments above, the optimal prediction atlas is achieved by the model with $\delta = 0$, not by the model with the true value of $\delta$.

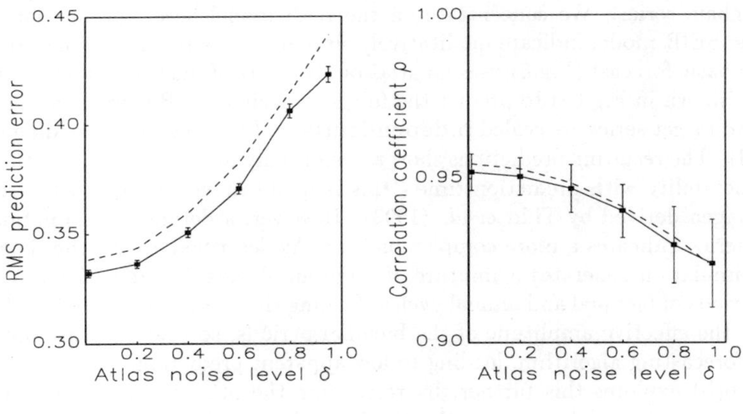


Fig. 5. Prediction error (left panel), and correlation between 1-step ahead predicted and actual values (right panel) in the $\epsilon$-ball method of Tidd *et al.* 1993. The true dynamics are the Ricker model (Eq. 5 in the text) with $r = 3.25$ and noise level $\omega = 0.4$; the atlases were generated by the same model except that the noise level was varied between $\omega = 0$ and $\omega = 0.95$. For each data point, the prediction is based on its 8 nearest neighbours in the atlas. The solid line in each panel shows the mean $\pm$ one standard error over 100 replicate data/atlas pairs, using data series of length 250 and atlas series of length 500; the dashed line shows the expected value in the limit of an infinitely large atlas.

This phenomenon confounds Tidd *et al.*'s method of comparing alternative choices for the prediction model $F$ in Eq. 3. Their procedure is to construct atlases using a variety of candidate models $F_i$. Each candidate $F_i$ is at best an approximation to the true F, so its prediction accuracy (using the correct value of $\omega$ ) would be higher than that using the true dynamics (Eq. 3). Thus, using a smaller- than-true would improve Fi's prediction accuracy, and make it closer to that of the true model. Tidd *et al.* constructed their atlas for epidemiological models using Monte Carlo realizations with relatively low noise levels, whereas for all of the alternative models the atlas was constructed with a higher degree of noise. They then interpreted the results as favouring the deterministic model. However, the comparison is biased by the intrinsic advantage of a deterministic model in their method. To avoid this bias, the comparison could be based instead on a conditional mean (expression Eq. 4), which can be computed by simulation if it is not available in closed form.

*(ii) Model predictions for England and Wales*

Following Tidd *et al.* (1993), we use a Monte Carlo realization of an epidemiological model to forecast observed measles dynamics in a city of size 1 million. Birmingham is chosen as the closest city to this population size in the currently available data set. Fig 6 gives details of various fits of the RAS model to the pre-vaccination Birmingham series. We concentrate on the RAS model here - preliminary studies with the SEIR model indicate qualitatively similar results to those reported below.

The basic forecast (Fig 6) uses an atlas of 100 years of monthly Monte Carlo RAS output (shown in Fig 6a) to predict the full pre-vaccination Birmingham series. The atlas and target series are scaled **independently** on $(0, 1)$ before embedding (Tidd et al, 1993). The resulting predictions show an essentially monotonic exponential decline in predictability with prediction time - this is qualitatively similar to the result for Copenhagen derived by Tidd *et al.* (1993). However, a detailed examination of the model series indicates a more complex picture. As described above, the RAS Monte Carlo simulation generates a mixture of large amplitude 3 year cycles interspersed with periods of biennial and annual cycles. Scaling the whole series on (0.1), therefore reduces the effective amplitude of the biennial periods, so that they are not 'found' by the forecasting algorithm, leading to low apparent predictability.

Fig 6(b) explores this further, by restricting the atlas to an 18 year period of predominantly biennial cycles in the RAS simulation,which visually resembles the observed series (period (II) in the figure). Scaling this period on (0,1) and forecasting from it produces a significant improvement in predictability, compared with predicting from the full atlas (Fig 6b). Finally, choosing an 18 year atlas with triennial cycles (Fig 6b, period (I)) produces a similar decline in predictability to the whole RAS atlas (Fig 6a). Clearly, the choice of scaling has a **major** effect on the performance of the atlas in systems with intermittency.

## 5. Demographic effects on measles dynamics

Any attempt to predict the future from the past is likely to be fraught with danger if we do not take account of long term changes in underlying driving variables of the system. For childhood diseases such as measles, long term changes in birth rate are likely to play a significant role in this context (Grenfell, *et al.* , 1994). Differences

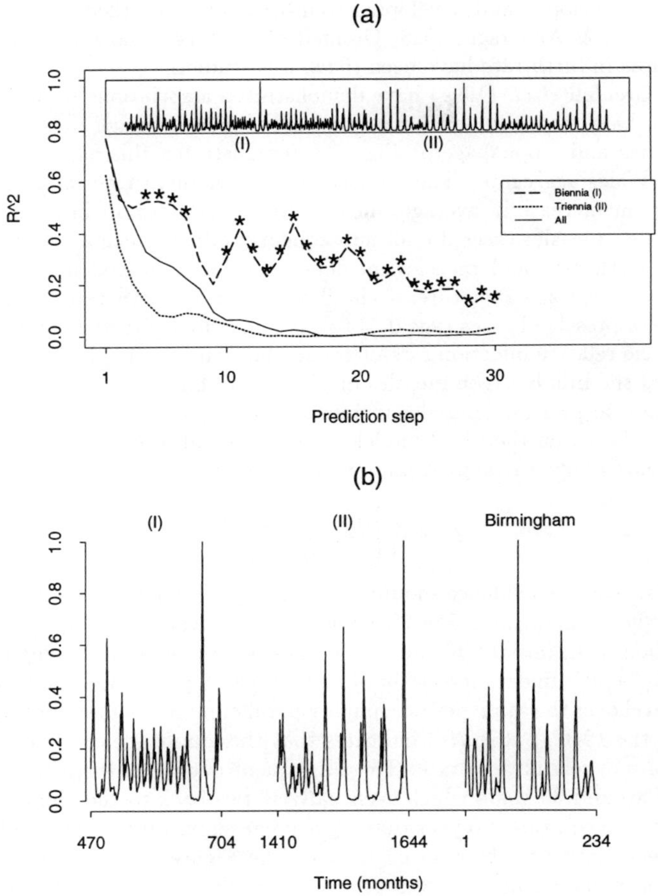


Fig. 6. Predicting the pre-vaccination Birmingham measles time series with the stochastic version of the RAS model. (a) Prediction profiles using the whole 100 year atlas (shown as an inset) and the two subsets (marked as (I) and (II) on the inset) described in the text. (b) Scaled (on $(0, 1)$) versions of the two RAS subsets, compared with the observed series. The stars indicate significant differences ($p < 0.05$) between correlation coefficients for the Birmingham fit based on atlas (II), compared with that for the total atlas.

in birth rate in developed and developing countries have a marked effect on measles dynamics (Mclean & Anderson, 1988; Grenfell *et al.* 1994) , however, the effects of secular **changes** in birth rate have been much less studied.

Recently, Grenfell *et al.* (1994) have demonstrated associations between the post World War 2 'baby boom' in birth rate and subsequent average measles incidence in English cities and Copenhagen. Fig. 7a demonstrates this association for the England and Wales aggregate. The increase in relative birth rate in the late 1940's is followed by an increase in average, measles incidence after an interval of around 5 years, although measles cases do not subsequently fall as much as birth rate. The link between birth rate and measles incidence can be examined more directly by relating the observed age structure of childhood cases to birth rate. Fig 7b shows that the ratio of preschool cases (age 0-5) to cases in primary school children (5-10) - a measure of the relative infection rate of young children - tracks the birth rate. This again supports the link between measles incidence and births.

In order to interpret how measles incidence is influenced by birth rate, we use a simple analysis based on the SEIR model. From the equilibrium of the SEIR model, average incidence obeys the approximate relation (May, 1986)

$$I^\star \simeq \frac{\mu}{\beta}\left[\frac{\beta\gamma N - \sigma\gamma}{\sigma\gamma}\right], \tag{6}$$

in other words, relative incidence should increase in proportion to the 'turnover rate' $\mu$, *ie*, effectively to birth rate. Fig 7(a) shows the equivalent dynamic picture. An SEIR simulation, with initially biennial dynamics (set by seasonal forcing of b1=0.2) is perturbed by a 40% linear increase in birth rate over 8 years (and then a decline to the original level over the same period, roughly paralleling the pattern in England and Wales during the 1940's). Fig 7(c) indicates that the amplitude of cycles tracks the relative increase in recruitment rate. The subsequent decline in births is followed by a transient of around 30 years which eventually settles back to the biennial pattern. Fig 7(d) shows the equivalent 6 year running average superimposed on the birth rate; both are scaled relative to their initial values. The Figure indicates that the above analytical approximation holds - average cases and birth rate scale together for the SEIR model. This result also appears to hold for simulations in the chaotic region of

---

Fig. 7. **See next page.** The impact of birth rate variations on measles dynamics; details are given in the text. (a) time series of observed measles cases (6 year running mean) for England and Wales, compared to Raw births and the recruitment rate, G, derived in the text. All series are scaled relative to the first time point, to provide a crude estimate of relative changes over the period. (b) Raw birth rate for England and Wales, compared to the ratio of measles cases in pre-school (age 0-5) to primary school (age 5-10) children. (c) Deterministic SEIR model simulation, with birth rate variations as shown. Birth rate is increased from year 10 and births and cases are scaled relative to their mean in years 1 to 9. SEIR parameters are as in Fig. 2, except $b_1 = 0.2$; the model was simulated by standard numerical methods. (d) As (c), except the model results are expressed as 6 year running means.

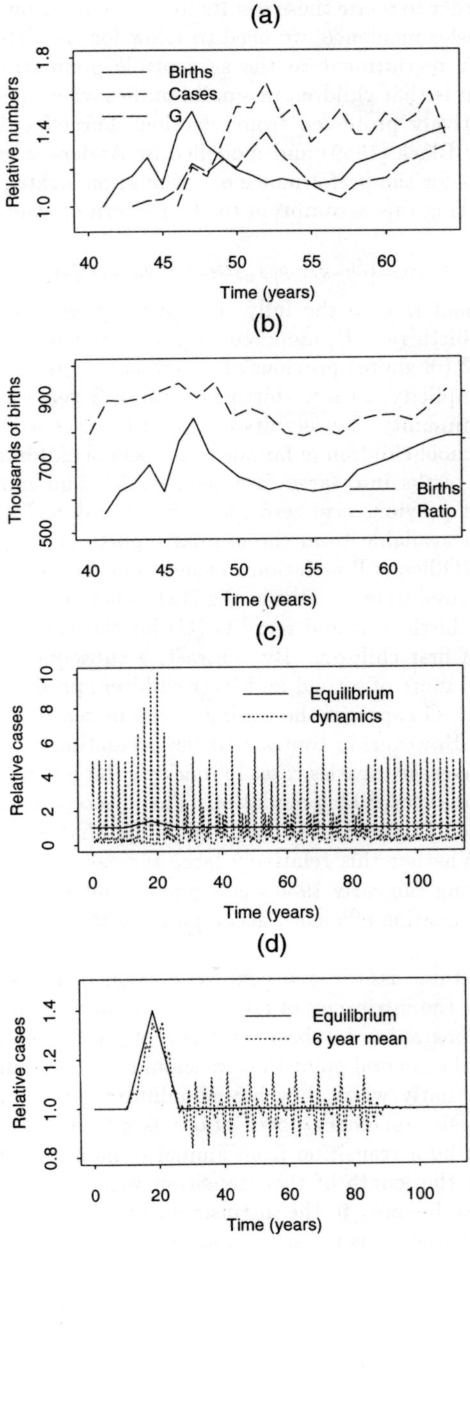
(a)
Births
Cases
G
Relative numbers
Time (years)
(b)
Thousands of births
Preschool Ratio
Births
Ratio
Time (years)
(c)
Equilibrium
dynamics
Relative cases
Time (years)
(d)
Equilibrium
6 year mean
Relative cases
Time (years)

the model. Finally, in order to relate these results to the comparison between observed and birth rate and measles incidence, we need to allow for the delay between births of individuals and their recruitment to the susceptible population. For example, one extreme assumption is that children in small families where no siblings have yet attended school are relatively 'protected' from infection. This effect has been observed in serological studies by Black (1959) and modelled by Anderson and May (1991) in terms of its implications for the performance of vaccination strategies.

A simple model relating this assumption to the pattern of schooling is:

$$G_t = \rho_{0,t-5}B_{t-5} + \rho_{0,t-3}B_{t-3} + \rho_{0,t-1}B_{t-1} \tag{7}$$

where $G_t$ ,the recruitment rate to the fully susceptible population in year $t$ is determined by the crude birth rate $B$, modified by the proportion of children born in families with 0, 1 and 2 (or more) previous children $\rho_0$, $\rho_1$, $\rho_2$ respectively. For example, assuming for simplicity, an inter-birth interval of 2 years and a 1 year delay to allow for maternal immunity, the recruits to the fully susceptible population (i.e. school children or pre-school children in families with school children) are determined by the previous year's births into large families ($\rho_{2,t}B_t$), and more delayed births into smaller families, of previous size zero ($\rho_{0,t-5}B_{t-5}$) and 1 children ($\rho_{1,t-3}B_{t-3}$). Estimates of the $\rho_i$ are available from the annual reports of the Registrar general for England and Wales (Office of Population Censuses and Surveys) and the resulting estimates of G are compared to raw births in Fig 7(a). There is a relatively long delay between the increase in births (B) and recruits (G) for the 1940's baby boom, since this consisted mainly of first children. By contrast, a subsequent peak in births in the late 1950's consisted more of second and later children, producing less of a delay. Fig 7(a) also shows that, G captures the timing of the increase in measles incidence better than raw births. However, in contrast to the predictions of our simple model, the relative increase in average measles cases in markedly larger than the proportional rise in recruitment to the susceptible population. This discrepancy also holds if birth rate changes are included in the RAS model, indicating that age structure alone is not the explanation. Whether this relatively large increase in measles incidence in England and Wales during the early 1950's is a genuine epidemiological effect, or an artifact of increased notification efficiency accompanying the rise in cases, remains to be established.

In order to explore these issues, the next generation of measles models should allow more explicitly for the intricacies of family demography, which are documented in some detail in the demographic database. In terms of prediction, the simple models presented here indicate the general point that we should take account of all important system variables, particularly when they have nonlinear effects. For example, the increase in average measles incidence noted above is reflected in the raw England and Wales data (Fig 1) by a transition from annual dynamics in the late 1950's to biennial epidemics, and the length of this transition varies between cities. Teasing out whether this effect is due only to the intrinsic infection dynamics or is influenced by birth rates or other variables is important, both to the process of prediction and to interpreting any message about nonlinearity that prediction gives us.

## 6. Discussion

This paper has set out to show that the use of nonlinear prediction in epidemiology is a potentially large and intricate area, which we have covered relatively thinly in the space available. We therefore conclude by suggesting areas for future work.

### (a) General forecasting issues

The central point to emerge here is the potential complexity of trying to forecast nonlinear systems which can exhibit more than one sort of behaviour. The situation is especially acute if (as may be the case with measles), the relatively short length of observed series (20 to 40 years) corresponds roughly to the timescale of qualitative changes in behaviour implied by the models. Our preliminary results with a simple prediction scheme depends critically on the detailed structure of the observed series. Not surprisingly, regular biennial patterns (such as in England and Wales) are relatively predictable, compared to more irregular patterns, including 3 year cycles. More work is needed here to refine the comparisons using better algorithms, different choices of atlas to combat non-stationarity (Sugihara *et al.* 1990), data transformation and scaling and comparisons with parametric statistical models. The scaling problem is especially acute when using atlases from models. For example, using atlas and target series on absolute scales might be preferable to scaling on $(0,1)$, but this still depends on the algorithm being sufficiently sensitive to choose the appropriate region of attractor behaviour. Sorting these problems out is essential before we can compare the predictive ability of models such as SEIR and RAS (along the lines of Tidd, *et al.* , 1993).

### (b) Epidemiological issues

#### *(i) Modelling*

Predictions based on models are difficult to interpret, unless we are confident that model behaviour is a correct mirror of real dynamics. For example, no observed measles series to our knowledge shows the periods of regular repeated 3 year cycles generated by the stochastic SEIR and RAS models. This problem points to a need to refine existing measles models. As noted above, there is still much to understand about the spatial dynamics and persistence of measles, though other areas also need to be addressed. In particular, any attempt to predict the future from the past is likely to be fraught with danger if we do not take account of long term changes in underlying driving variables of the system. For childhood diseases such as measles, long term trends in birth rate (the 'engine' which ultimately drives measles epidemics) are likely to play a significant role in this context (Grenfell, *et al.* , 1994). Our preliminary analysis of secular changes in birth rate indicate their potential importance, both for measles dynamics and for the predictability of measles. More explicit models of family demography (Anderson & May, 1991; Becker, 1989) will be required to clarify fully the dynamic effects of demographic changes on measles incidence.

*(ii) Predictability and vaccination*

Finally, on an applied note, the important forecasting issue to clarify is the predictability of measles under the impact of vaccination, rather than in the per-vaccination era (Tidd, et al., 1993). Our preliminary results (based on the short and rather non-stationary series for England and Wales) indicate that predictability falls with vaccination, though this happens less in large centres such as London. More work is needed here, both to clarify the statistical issues and to decide whether achievable levels of forecastability are of any practical use, particularly given changes in vaccine uptake.

## Acknowledgments

We thank Howell Tong for helpful comments and the following for financial support: Isaac Newton Institute for Mathematical Science (BTG and SPE), US National Science Foundation Grant DMS-9217866 (SPE), Royal Society and BBSRC (AK) and Mellon Foundation (BMB).

## References

Anderson, R.M., Grenfell, B.T. & May, R.M. 1984 Oscillatory fluctuations in the incidence of infectious disease and the impact of vaccination: time series analysis. *J. Hyg. (Camb. ),* **93**, 587-608.

Anderson, R.M. & May, R.M. 1982 Directly transmitted infectious diseases: control by vaccination. *Science,* **215**, 1053-1060.

Anderson, R.M. & May, R.M. 1985 Age-related changes in the rate of disease transmission: implications for the design of vaccination programmes. *J. Hyg. (Camb. ),* **94**, 365-436.

Anderson, R.M. & May, R.M. 1991 Infectious diseases of humans: dynamics and control. Oxford University Press.

Aron, J.L. 1990 Multiple attractors in the response to a vaccination program. *Theor. Popul. Biol.,* **38**, 58-67.

Aron, J.L. & Schwartz, I.B. 1984 Seasonality and period-doubling bifurcations in an epidemic model. *J. Theor. Biol.,* **110**, 665-679.

Bartlett, M.S. 1957 Measles periodicity and community size. *J. Roy. Stat. Soc.* **A, 120**, 48-70.

Bartlett, M.S. 1960 The critical community size for measles in the U.S. *J. Roy. Stat. Soc.* **A, 123**, 37-44.

Becker, N.G. 1989 Analysis of Infectious Disease Data. London: Chapman and Hall.

Black, F.L. 1959 Measles antibodies in the population of New Haven, Connecticut. *J. Immunol.,* **82**, 74-83.

Black, F.L. 1966 Measles endemicity in insular populations: critical community size and its evolutionary implication. *J. Theor. Biol.,* **11**, 207-211.

Bolker, B.M. & Grenfell, B.T. 1993 Chaos and biological complexity in measles dynamics. *Proc. R. Soc. Lond. Biol.,* **251**, 75-81.

Dietz, K. 1976 The incidence of infectious diseases under the influence of seasonal fluctuations. *Lect. Notes. Biomath.,* **11**, 1-15.

Dietz, K. & Schenzle, D. 1985 Mathematical models for infectious disease statistics. In A Celebration of Statistics. (ed. A.C. Atkinson & S.E. Feinberg), pp. 167-204. New York: Springer.

Engbert, R. & Drepper, F.R. 1993 Qualitative analysis of unpredictability: a case study from childhood epidemics in "Predictability and nonlinear modelling in natural sciences and economics", April 4-7, 1993, Wageningen, The Netherlands.

Farmer, J.D. & Sidorowich, J.J. 1987 Predicting chaotic time series. *Phys. Rev. Lett.,* **59**, 845-848.

Fine, P.E.M. & Clarkson, J.A. 1982a Measles in England and Wales-I: an analysis of factors underlying seasonal patterns. *Int. J. Epidem.,* **11**, 5-15.

Fine, P.E.M. & Clarkson, J.A. 1982b Measles in England and Wales-II: The impact of the measles vaccination programme on the distribution of immunity in the population. *Int. J. Epidem.,* **11**, 15-25.

Grenfell, B.T. 1992 Chance and chaos in measles dynamics. *J. Roy. Stat. Soc.* **B, 54**, 383-398.

Grenfell, B.T., Bolker, B.M. & Kleczkowski, A. 1994 Demography, seasonality and the dynamics of measles in developed countries. In Epidemic models: their structure and relation to data. (ed. D. Mollison), Cambridge: Cambridge University Press.

Grossman, Z. 1980 Oscillatory phenomena in a model of infectious diseases. *Theor. Popul. Biol.,* **18**, 204-243.

Hamer, W.H. 1906 Epidemic disease in England-the evidence of variability and of persistency of type. *Lancet* **1**, 733-739.

Karlin, S. & Taylor, H.M. 1975 A first course in stochastic processes. New York: Academic Press.

Kermack, W.O. & McKendrick, A.G. 1927 A contribution to the mathematical theory of epidemics. *Proc. R. Soc. Lond.* **A, 115**, 700-721.

Kermack, W.O. & McKendrick, A.G. 1932 A contribution to the mathematical theory of epidemics. Part II. The problem of endemicity. *Proc. R. Soc. Lond.* **A, 138**, 55-83.

Kermack, W.O. & McKendrick, A.G. 1933 A contribution to the mathematical theory of epidemics. Part III. Further studies of the problem of endemicity. *Proc. R. Soc. Lond.* **A, 141**, 92-122.

Kot, M., Graser, D.J., Truty, G.L., Schaffer, W.M. & Olsen, L.F. 1988 Changing criteria for imposing order. *Ecol. Modelling.,* **43**, 75-110.

London, W.P. & Yorke, J.A. 1973 Recurrent outbreaks of measles, chickenpox and mumps. I. Seasonal variation in contact rates. *Am. J. Epidem.,* **98**, 453-468.

May, R.M. 1986 Population biology of microparasitic infections. In Biomathematics vol. 17. (ed. T.G. Hallam & S.A. Levin), pp. 405-442. Berlin: Springer-Verlag.

Mclean, A.R. & Anderson, R.M. 1988 Measles in developing countries. Part I. Epidemiological parameters and patterns. *Epidemiol. Infect.,* **100**, 111-133.

Olsen, L.F., Truty, G.L. & Schaffer, W.M. 1988 Oscillations and chaos in epidemics: a nonlinear dynamic study of six childhood diseases in Copenhagen, Denmark. *Theor. Popul. Biol.,* **33**, 344-370.

Olsen, L.F. & Schaffer, W.M. 1990 Chaos versus noisy periodicity: alternative hypotheses for childhood epidemics. *Science* **249**, 499-504.

Rand, D.A. & Wilson, H. 1991 Chaotic stochasticity: a ubiquitous source of unpredictability in epidemics. *Proc. R. Soc. Lond. Biol.,* **246**, 179-184.

Schaffer, W.M. 1985a Can nonlinear dynamics elucidate mechanisms in ecology and epidemiology? *IMA J. Math. Appl. Med. Biol.,* **2**, 221-252.

Schaffer, W.M. 1985b Order and chaos in ecological systems. *Ecology* **66**, 93-106.

Schaffer, W.M. & Kot, M. 1985 Nearly one dimensional dynamics in an epidemic. *J. Theor. Biol.,* **112**, 403-427.

Schenzle, D. 1984 An age-structured model of pre- and post-vaccination measles transmission. *IMA J. Math. Appl. Med. Biol.,* **1**, 169-191.

Soper, M.A. 1929 The interpretation of periodicity in disease prevalence. *J. Roy. Stat. Soc.* **A, 92**, 34-61.

Sugihara, G., Grenfell, B. & May, R.M. 1990 Distinguishing error from chaos in ecological time series. *Phil. Trans. R. Soc. Lond. Biol.,* **330**, 235-251.

Sugihara, G. & May, R.M. 1990 Nonlinear forecasting as a way of distinguishing chaos from measurement error in time series. *Nature* **344**, 734-741.

Tidd, C.W., Olsen, L.F. & Schaffer, W.M. 1993 The case for chaos in childhood epidemics: II. Predicting historical epidemics from mathematical models. *Proc. R. Soc. Lond. Biol.,* **254**, 257-273.

Tong, H. 1990 Non-linear time series: a dynamical systems approach. Oxford: Oxford University Press.